APTAMERS IN BIOANALYSIS

APTAMERS IN BIOANALYSIS

Edited by

MARCO MASCINI
University of Florence, Italy

WILEY

A JOHN WILEY & SONS, INC., PUBLICATION

1005776699

Library of Congress Cataloging-in-Publication Data:

Mascini, Marco
    Aptamers in bioanalysis /
        p. cm.
    Includes bibliographical references and index.
    ISBN: 978-0-470-14830-3 (cloth)

Printed in the United States of America

10 9 8 7 6 5 4 3 2 1

CONTENTS

III   APPLICATIONS

PREFACE

I am very pleased to act as editor of this book, *Aptamers in Bioanalysis.* Aptamers started as therapeutic agents and in very few years they have become a hot topic in analytical chemistry. In our laboratory we are working on realizing reliable sensors and biosensors, and aptamers appear as optimal components for their assembling. Aptamers appear in this application as a new class of ligands with exceptional binding constants (micromolar to picomolar range). Aptamers are also entering many analytical applications, such as, in the technologies based on separation science (various chromatographic techniques or capillary electrophoresis). Many new exciting analytical problems can be solved with these new compounds.

The increasing presence of contaminants in food, air, or drinking water that are capable of causing intoxication, diseases, or chronic illness has led to the need for analytical systems capable of rapid, and often multianalyte, measurements of complex samples. This need also exists in the medical field where multiparameter diagnostic systems are increasingly required to detect all the well-known and the more recently discovered biomarkers for different diseases. Unfortunately for medical practitioners, disease-related biomarkers are either physiologically present in minute quantities or severely contaminated by nonspecific compounds in a patient's bloodstream or body fluid. Hence, highly sensitive as well as specific recognition elements are required for effective detection of such biomarkers.

When the detection system requires a biomolecular recognition event, antibody-based detection methodologies are still considered the standard assays in environmental, food, and clinical analysis. These assays are well established, and they have been demonstrated to reach the desired sensitivity and selectivity. However, the use of antibodies in multianalyte detection methods and in the analysis of very complex samples encounter some limitations derived mainly from the nature and synthesis of these protein receptors. In order to circumvent some of these drawbacks, other recognition molecules are being explored as alternatives.

The awareness that nucleic acids, RNA in particular, can assume stable secondary structures and that they can be easily synthesized and functionalized, has opened the door for aptamers in several applications.

The main advantage is overcoming the use of animals or cell lines for the production of the molecules. Moreover, antibodies against molecules that are not immunogenic are difficult to generate. Aptamers, on the contrary, are isolated by in vitro methods that are independent of animals: an in vitro combinatorial library can be generated and exploited against any target. In addition, generation of antibodies in vivo means that the animal immune system selects the sites on the target protein to which the antibodies bind. The in vivo parameters restrict the identification of antibodies that can recognize targets only under physiological conditions limiting the extension to which the antibodies can be functionalized and applied.

Moreover, the aptamer selection process can be manipulated to obtain aptamers that bind to a specific region of the target, with specific binding properties, and in different binding conditions. After selection, aptamers are produced by chemical synthesis and purified to a very high degree eliminating the batch-to-batch variation found when using antibodies. By chemical synthesis, modifications in the aptamer can be introduced enhancing the stability, affinity, and specificity of the molecules. Often the kinetic parameters of an aptamer-target complex can be changed for higher affinity or specificity. Another advantage over antibodies can be seen in the higher temperature stability of aptamers; in fact, antibodies are large proteins sensitive to temperature that can undergo irreversible denaturation. On the contrary, aptamers are very stable and they can recover their native active conformation after denaturation.

The selection process itself, with the amplification step, gives some advantages to aptamers with respect to other "nonnatural" receptors, such as oligopeptides, which cannot be amplified during their selection procedure. Therefore, once again the polymerase chain reaction appears as the magic tool to solve the problem of obtaining highly selective ligands. Our colleagues in genetics departments are working to overcome this issue and in the future we will be happy to obtain from libraries other nice ligands of different nature, like polypeptides or polysaccharides, rather than oligonucleotides!

We now have a new class of biosensors, aptasensors, which use aptamers as highly selective recognition elements. As receptor molecules, aptamers allow widespread applicability to a diverse array of target analytes due to their analyte-impartial synthetic generation process. Aptasensors realized on micro- and nanoscale platforms afford many potential advantages, such as miniaturized construction; rapid, sensitive, and specific detection; high throughput; reduced costs; and minimized material consumption. Thus, micro- and nanoaptasensors are highly attractive for a broad range of applications, such as proteomics, metabolomics, environmental monitoring, counterterrorism, and clinical diagnostics and therapeutics.

MARCO MASCINI

CONTRIBUTORS

MORITZ K. BEISSENHIRTZ, Analytical Biochemistry, Institute of Biochemistry and Biology, University of Potsdam, Karl-Liebknecht-Strasse 24–25, Bldg. 25, D 14476 Potsdam-Golm, Germany

JEAN-PIERRE DAGUER, Université Victor Segalen Bordeaux II, INSERM U869, Laboratoire ARNA, Bâtiment 3A 1er étage, 146 Rue Léo Saignat, 33 076 Bordeaux Cédex, France

ERIC DAUSSE, Université Victor Segalen Bordeaux II, INSERM U869, Laboratoire ARNA, Bâtiment 3A 1er étage, 146 Rue Léo Saignat, 33 076 Bordeaux Cédex, France

SHAOJUN DONG, State Key Laboratory of Electroanalytical Chemistry, Changchun Institute of Applied Chemistry, Chinese Academy of Sciences Changchun, Jilin 130022, P.R. China

TIBOR HIANIK, Department of Nuclear Physics and Biophysics, Faculty of Mathematics, Physics and Informatics, Comenius University, Mlynská dolina F1, 842 48 Bratislava, Slovakia

KAZUNORI IKEBUKURO, Department of Biotechnology & Life Science, Faculty of Technology, Tokyo University of Agriculture & Technology, 2-24-16 Naka-cho, Koganei, Tokyo, 184–8588, Japan

KAGAN KERMAN, Department of Chemistry, University of Saskatchewan, 110 Science Place, Saskatoon, S7N 5C9 Saskatchewan, Canada

SERGEY N. KRYLOV, Department of Chemistry, York University, Toronto, M3J 1P3 Ontario, Canada

EIK LEUPOLD, Leibniz Institute of Molecular Pharmacology, Robert-Rössle-strasse 10, D 13125 Berlin, Germany

BINGLING LI, State Key Laboratory of Electroanalytical Chemistry, Changchun Institute of Applied Chemistry, Chinese Academy of Sciences, Changchun, Jilin 130022, P.R. China

FRED LISDAT, University of Applied Sciences Wildau, Biosystems Technology, Bahnhofstrasse 1, D 15745 Wildau, Germany

MARCO MASCINI, Department of Chemistry, University of Florence, Via della Lastruccia 3, 50019 Sesto Fiorentino, Italy

LINDA B. MCGOWN, Department of Chemistry and Chemical Biology, Rensselaer Polytechnic Institute, Troy, NY 12180, USA

MARIA MINUNNI, Department of Chemistry, University of Florence, Via della Lastruccia 3, 50019 Sesto Fiorentino, Italy

OLIVER PÄNKE, Fraunhofer Institute of Biomedical Engineering, Am Mühlenberg 13, D 14476 Potsdam-Golm, Germany

ERIC PEYRIN, DPM UMR 5063 CNRS/Université de Grenoble, Bât E (C) André Rassat, Domaine Universitaire, 301 avenue de la Chimie, BP 53 38 041 Grenoble Cédex 9, France

CORINNE RAVELET, DPM UMR 5063 CNRS/Université de Grenoble, Bât E (C) André Rassat, Domaine Universitaire, 301 avenue de la Chimie, BP 53 38 041 Grenoble Cédex 9, France

FRIEDER W. SCHELLER, Analytical Biochemistry, Institute of Biochemistry and Biology, University of Potsdam, Karl-Liebknecht-Strasse 24–25, Bldg. 25, D 14476 Potsdam-Golm, Germany

KOJI SODE, Department of Biotechnology and Life Science, Faculty of Technology, Tokyo University of Agriculture & Technology, 2-24-16 Naka-cho, Koganei, Tokyo 184–8588, Japan

WALTER STÖCKLEIN, Analytical Biochemistry, Institute of Biochemistry and Biology, University of Potsdam, Karl-Liebknecht-Strasse 24–25, Bldg. 25, D 14476 Potsdam-Golm, Germany

REGINA STOLTENBURG, Centre for Environmental Research Leipzig-Halle GmbH (UFZ), Environmental and Biotechnology Centre (UBZ), Permoserstr. 15, D-04318 Leipzig, Germany

BEATE STREHLITZ, Centre for Environmental Research Leipzig-Halle GmbH (UFZ), Environmental and Biotechnology Centre (UBZ), Permoserstr. 15, D-04318 Leipzig, Germany

EIICHI TAMIYA, Nanobiotechnology and Biodevice Lab, Department of Applied Physics. Graduate School of Engineering, Osaka University, 2-1 Yamadaoka, Suita, Osaka 565–0871, Japan

SARA TOMBELLI, Department of Chemistry, University of Florence, Via della Lastruccia 3, 50019 Sesto Fiorentino, Italy

JEAN-JACQUES TOULMÉ, Université Victor Segalen Bordeaux II, INSERM U869, Laboratoire ARNA, Bâtiment 3A 1er étage, 146 Rue Léo Saignat, 33 076 Bordeaux Cédex, France

HUI WEI, State Key Laboratory of Electroanalytical Chemistry, Changchun Institute of Applied Chemistry, Chinese Academy of Sciences, Changchun, Jilin 130022, P.R. China

ITAMAR WILLNER, Institute of Chemistry, The Hebrew University of Jerusalem, Jerusalem 91904, Israel

ULLA WOLLENBERGER, Analytical Biochemistry, Institute of Biochemistry and Biology, University of Potsdam, Karl-Liebknecht-Strasse 24–25, Bldg. 25, D 14476 Potsdam-Golm, Germany

WATARU YOSHIDA, Department of Biotechnology and Life Science, Faculty of Technology, Tokyo University of Agriculture and Technology, 2 24-16 Nakacho, Koganei, Tokyo 184–8588, Japan

MAYA ZAYATS, Institute of Chemistry, The Hebrew University of Jerusalem, Jerusalem 91904, Israel

PART I

INTRODUCTION

CHAPTER 1

APTAMERS: LIGANDS FOR ALL REASONS

JEAN-JACQUES TOULMÉ, JEAN-PIERRE DAGUER, and ERIC DAUSSE

1.1 INTRODUCTION

Several strategies were described over the last 25 years that make the use of synthetic oligonucleotides in different perspectives of interest for biology and medicine, thanks to the development of very powerful and relatively cheap methods for nucleic acid synthesis on solid support. These strategies generally do not take into account the genetic information borne by the oligonucleotide chain but rather, explore its wide potential of molecular interaction and recognition. The *antisense approach* was the first to be developed (Belikova et al., 1967). It rests on a simple hypothesis: the binding to a target mRNA of a complementary sequence (the antisense sequence), thus leading to the formation of a local double-stranded structure, might interfere with reading the message. This was demonstrated experimentally in the late 1970s by pioneering works of Zamecnik and Stephenson (1978), on the one hand, and Miller et al. (1974), on the other. The progress made in the sequencing of entire genomes offered multiple opportunities for validating this technique, which was used not only against messenger but also against pre-mRNA (Sazani and Kole, 2003) or viral RNA (Toulmé et al., 2001). Very quickly it turned out that regular DNA or RNA oligomers suffered from weaknesses for use in biological media. Numerous chemically modified oligonucleotide analogs were synthesized and evaluated (Wilson and Keefe, 2006). In particular, modifications were introduced for improving resistance to nucleases (Toulmé, 2001). A number of clinical trials have been and are still being carried out (Aboul-Fadl, 2005), but disapointingly, a single antisense

Aptamers in Bioanalysis, Edited by Marco Mascini
Copyright © 2009 John Wiley & Sons, Inc.

oligonucleotide was approved by the U.S. Food and Drug Administration for the treatment of cytomegalovirus–related retinitis (Orr, 2001). More recently, similar modifications were introduced in small interfering RNA (siRNA) that also bind their target RNA through Watson–Crick base pairing (Wilson and Keefe, 2006). siRNAs generally show a higher efficiency than the homologous antisense sequence, due to triggering of the enzymatic destruction of the target transcript by the interference machinery (Scanlon, 2004; Chakraborty, 2007). But both antisense and siRNA suffer from the same types of limitations: target access, specificity of interaction, and cell uptake (Shi and Hoekstra, 2004; Kurreck, 2006).

Antisense oligomers and siRNAs sequences are designed rationally on the basis of Watson–Crick complementarity with the target transcript. During the last 15 years, combinatorial approaches have been developed in both chemistry and biology (Fernandez-Gacio et al., 2003; Frankel et al., 2003; Li and Liu, 2004). In such methods a family of candidates is randomly synthesized. Molecules exhibiting the desired property are then extracted from this pool. There is no prerequisite to the use of such methods: The structure of the target does not need to be known nor does one need to postulate the interactions that will take place between the target and the successful candidate. The interest in a combinatorial approach is related directly to the size of the pool—the molecular diversity—that can be screened. From this point of view, oligonucleotide libraries surpass by several logs the complexity of any other type of library. In vitro selection of oligonucleotides can be undertaken in pools containing up to 10^{15} different candidates (Gold et al., 1995). This is in large part related to the information encoded in the candidate. Consequently, in contrast to any other compound, oligonucleotide candidates from the pool can be amplified and analyzed very easily. Indeed, covalent fusion between an mRNA and the polypeptide it codes for allows the screening of very large libraries (10^{12} to 10^{13} independent fusion candidates) and makes the in vitro selection of proteins by ribosome display a very powerful method (Roberts, 1999).

Pioneering work recognized the interest of either selection or in vitro evolution of nucleic acid mixtures for the identification of protein-binding sites or replication suitability (Mills et al., 1967; Blackwell and Weintraub, 1990). But in 1990, three laboratories independently described a procedure for the identification, within large pools of randomly synthesized molecules, of nucleic acid sequences exhibiting a predetermined property: affinity for a given target (Ellington and Szostak, 1990; Tuerk and Gold, 1990) or enzymatic activity (Robertson and Joyce, 1990). This was achieved through repeated rounds of selection and amplification, thus ensuring directed evolution of the starting pool in response to selection pressure on the population. This method, now known as *SELEX* (systematic evolution of ligands by exponential enrichment) leads to the selection of aptamers [i.e., olig*omers* able (*apt*) to carry out some function]. Since numerous papers described an entire range of applications for aptamers that take advantage of their wide potential due to both the high strength and the specificity of their interaction with their target (Osborne and Ellington, 1997; Famulok, 1999; Jayasena, 1999; Wilson and Szostak, 1999; Brody and Gold, 2000; Toulmé,

2000). Examples are found in many different fields, and excellent reviews were recently published that cover the use of aptamers for the validation of targets (Toulmé et al., 2001; Blank and Blind, 2005; Bunka and Stockley, 2006), for the design of therapeutic agents (Nimjee et al., 2005), for cancer (Ireson and Kelland, 2006), for infectious diseases (Held et al., 2006; James, 2007), for gene therapy (Fichou and Ferec, 2006; Que-Gewirth and Sullenger, 2007), for the development of analytic tools (Tombelli et al., 2005; Mairal et al., 2007), or for the design of probes for imaging (Pestourie et al., 2005). We describe the SELEX process briefly and review a few points dealing with two wide areas of application for aptamers: regulation and sensing.

1.2 THE POWER OF SELECTION AND APTAMER REFINEMENT

Only a few general points are addressed here. The reader is referred to a recent review by Gopinath (2007) for details on procedures for the selection of aptamers. Like any other combinatorial method, the SELEX methodology first requires synthesis of the library. Compared to other libraries, it is easy to prepare an unbiased pool of DNA sequences, as the coupling efficiency of the A, T, G, or C phosphoramidite is very similar. One could compensate for the slightly different incorporation of nucleotides: a mixture of phosphoramidites in a ratio of $1.5 : 1.25 : 1.15 : 1.0$ (A : C : G : U/T) is believed to produce a balanced mixture of sequences (Ho et al., 1996). The diversity of the library is fixed by the length of the random region. One generates 4^n different sequences n nucleotides long. The experimental limit (about 10^{15}) corresponds to the diversity obtained for a random region about 25 nucleotides long. The full theoretical diversity is not covered for pools of candidates with a wider randomized window. For instance, only 1/1000 of the theoretical population will be present in a library of candidates with 30 random nucleotides. However, as the fidelity of the polymerases used in the SELEX process is rather low, each amplification round will generate variants that were not present at the preceding selection step, hence increasing the size of the sequence space explored.

Aptamers are then isolated by an iterative process (typically, 7 to 15 rounds) of binding, partitioning, and amplifying nucleic acid variants (Figure 1.1). The evolution of the population is driven by the selection conditions; the stringency (concentration, incubation time, washes, etc.) is increased progressively from round to round for selecting the candidates exhibiting the highest possible affinity (Gopinath, 2007). Selection is a tedious and time-consuming process when carried out manually. Automated selection was reported about 10 years ago (Cox et al., 1998; Brody and Gold, 2000; Eulberg et al., 2005). Several biotech companies have developed specialized procedures for high-throughput production of aptamers that reduce the isolation time from several months to a few days (Blank and Blind, 2005). An automated microchannel-based platform was recently described (Hybarger et al., 2006). The aptamers generated by automated selection are equivalent to those derived from manual selection, and those isolated against proteins show dissociation constants in the nanomolar range.

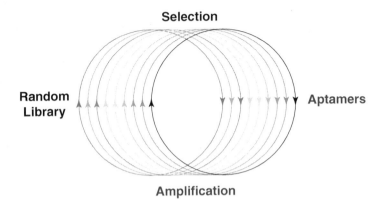

Figure 1.1   Scheme of in vitro selection. (See insert for color representation.)

Intramolecular base pairing defines higher-order structures. Therefore, a nucleic acid library of sequences is actually a library of three-dimensional shapes. Every candidate will display a unique combination of double-stranded helical segments, loops, and bulges. Each nucleotide may contribute hydrogen bonds and electrostatic and van der Waals interactions. The scaffold resulting from the intramolecular folding of oligonucleotides constituting the library offers a three-dimensional potential for interacting with any type of target. The selection process therefore corresponds to the capture of candidates that display a set of interacting groups complementary to that of the target. The association is even optimized through an induced fit mechanism: the aptamer acquires its final shape upon binding to its target (Patel et al., 1997; Hermann and Patel, 2000). This results in both a very strong affinity and a high specificity. For small molecules (e.g., amino acids, nucleosides, dyes, antibiotics), equilibrium dissociation constants in the micromolar range are frequent, whereas K_d values from 10^{-9} to 10^{-12} M^{-1} are typically obtained for proteins (Osborne and Ellington, 1997; Jayasena, 1999).

What makes aptamers ligands of great interest is their exquisite specificity. One of the clearest examples is the aptamer selected against theophylline, a purine derivative used for the treatment of asthma. This aptamer binds with a 10,000-fold-lower affinity to caffeine, another purine analog that differs from theophylline by a single methyl group on the N-7 position (Jenison et al., 1995). This exquisite selectivity was achieved through a careful selection procedure: oligonucleotides bound to the support functionalized with theophylline were first eluted with a solution of caffeine that made if possible to get rid of candidates that did not discriminate between the two purine compounds. A high degree of specificity could be reached even without such a counter-selection step. For example, the pseudoknot aptamer selected by Gold and co-workers against the reverse transcriptase (RT) of the human immunodeficiency virus does not bind to murine or feline RTs (Tuerk et al., 1992). But aptamers have been raised against these enzymes that display similar affinity (K_d = 5 to 20 nM) and do not bind to the HIV enzyme (Chen and Gold, 1994; Chen et al., 1996). Strikingly, these aptamers are specific inhibitors of their cognate enzyme, indicating that they hinder and very likely bind to the catalytic site, as these enzyme are nucleic acid–binding proteins. Despite the identity of the function ensured by these polymerases, specific aptamers have been selected, suggesting that they do not interact with the conserved residues responsible for the catalytic activity. Similar results were obtained for targets that are not natural ligands of nucleic acids: aptamers raised against human immunoglobulin E (IgE) or against human matrix metalloproteinase 9 do not bind to their murine homolog (Mendonsa and Bowser, 2004; Da Rocha-Gomes et al., unpublished results). The same level of specificity can be reached with nucleic acid targets. It has been demonstrated that an aptamer raised against a hairpin structure and interacting with the loop through the formation of six base pairs was far more specific than the antisense sequence generating the same pattern of Watson–Crick pairing (Darfeuille et al., 2006). Compared to the antisense–sense duplex, the three-dimensional structure of the aptamer–hairpin kissing complex provides addional elements of recognition. Interestingly, in vitro selection was used to identify hybridization probes that discriminate strongly between variants of the human papilloma virus (Brukner et al., 2007). Optimizing the probes rather than the hybridization conditions generated oligonucleotides that show a "relaxed" binding potential (i.e., partial complementarity) but that minimize cross-reactivity. This method might be generalizable to the design of genotyping kits.

It has been possible, however, to select aptamers that cross-react with different molecular species. In toggle SELEX, the target is changed during alternating rounds of selection. This strategy yields aptamers that recognize both human and porcine thrombin and display similar properties toward the two proteins: plasma clot formation and platelet activation (White et al., 2001). This approach is useful when one wants to raise ligands recognizing a class of structurally and chemically related targets.

Due to these properties, aptamers are considered to be "chemical antibodies" and actually rival or sometimes surpass antibodies. Indeed, the ease and the reproducibilty of the synthesis as well as several other properties make aptamers interesting alternatives: it is, for instance, possible to raise aptamers against toxic substances.

1.3 THE CHEMISTRY DRIVES THE SHAPE

The initial random library is always synthesized as DNA, but the starting pool used for selection can be made of RNA candidates. In this case the DNA library is transcribed prior to SELEX. Either RNA or DNA aptamers exhibiting similar properties can be selected against a given target. The sequences selected against the HIV-1 reverse transcriptase constitute a good example. DNA and RNA aptamers are competitive inhibitors of each other, indicating that there is a major site for aptamer interaction on the surface of the enzyme that drives the selection (Tuerk et al., 1992; Schneider et al., 1995). Importantly, the anti-HIV-1 RT RNA and DNA aptamers have very different sequences and structures. Whereas the RNA aptamer is a pseudoknot, the DNA aptamer is an imperfect hairpin. The DNA version of the RNA pseudoknot does not bind to the HIV-1 RT. It should be remembered that the conformation of DNA and RNA double helices differs. As aptamers are shapes, not surprisingly, changing the chemistry changes—generally weakens or even abolishes—the binding properties. It is therefore of prime importance to define the chemistry of the library prior to performing the selection. The choice of the aptamer chemistry is partly guided by the intended use of the aptamer. DNA aptamers are far cheaper than RNA aptamers. But if post-SELEX modifications have to be introduced, there are more RNA-mimic oligonucleotide derivatives; in addition, RNA aptamers can be expressed inside cells from DNA expression vectors.

Chemically modified aptamers have been developed due to intrinsic limitations of regular RNA and DNA oligomers (Wilson and Keefe, 2006). In particular, it has long been recognized from studies on antisense sequences that unmodified nucleic acids are short-lived species in biological media. The presence of nucleases in serum leads to the rapid digestion of natural oligonucleotides. Numerous modifications have been described by chemists that render nucleic acid resistant to nucleases (Toulmé, 2001; Wilson and Keefe, 2006). These include substitution at the $2'$ position (e.g., $2'$-O-methyl, $2'$-fluoro) (Kubik et al., 1997; Prakash and Bhat, 2007), phosphate modification (e.g., phosphorothioate, phosphoramidate, morpholino) (Koizumi, 2007), nucleoside modification (e.g., α anomer, bicyclic sugar) (Orum and Wengel, 2001), or even the use of a polyamide backbone (peptide nucleic acid) (Elayadi and Corey, 2001). However, most of these modifications cannot be introduced during the SELEX process, as the modifed nucleotides are not substrates for polymerases and therefore cannot be used by the relevant enzymes either for generating the initial library or for amplifying the oligomers selected.

$2'$-Fluoro- or $2'$-aminopyrimidine derivatives (Figure 1.2a) are widely used for the production of aptamers in which regular purine residues are incorporated (Aurup et al., 1992; Proske et al., 2002; Rhie et al., 2003). The resulting oligonucleotides show improved resistance to nucleases. Alternatively, the four phosphorothiate triphosphates in which a nonbridging oxygen of the internucleoside linker has been substituted by sulfur can be used in polymerase chain reaction (PCR) amplification (Andréola et al., 2000). Similarly, ribonucleoside boranophosphates have been demonstrated to be incorporated by T7 RNA polymerase (Shaw et al., 2003). This enzyme is also able to polymerize transcripts containing $4'$-thiopyrimidines (Figure 1.2a), a modification that increases their stability by about 50-fold relative to unmodified RNA (Kato et al., 2005). It was recently reported that the combination of mutated T7 RNA polymerases, Y639F and Y639F/H784A, allows the efficient incorporation of all four $2'$-O-methyl nucleotides (Chelliserrykattil and Ellington, 2004; Burmeister et al., 2005, 2006).

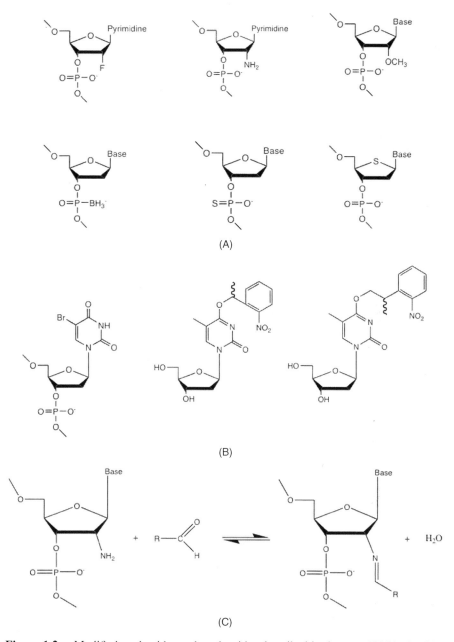

Figure 1.2 Modified nucleotides and nucleosides described in the text. (A) Nucleotides incorporated by polymerases yielding nuclease-resistant oligonucleotides. Top from left to right: 2′-fluoro, 2′-amino, 2′-*O*-methyl. Bottom from left to right: boranophosphate, phosphorothioate, 4′-thio. (B) Photosensitive residues. Left to right: 5-bromo-U, 2-(2-nitrophenyl)ethyl T, 2-(2-nitrophenyl)propyl T. (C) Amino–imino equilibrium used in 2D-SELEX (see the text).

The positions that remain unmodified at the end of the in vitro selection procedure (e.g., the purine residues in a selection carried out with 2′-fluoro-pyrimidine triphosphates) can be modified post-SELEX for further optimization of the aptamers. A systematic study of the 64 variants of the six-membered apical loop of an anti-TAR aptamer led to the identification of locked nucleic acid/2′-O-methyl chimeras fully resistant to nucleases that displayed anti-HIV-1 properties in a cell culture assay (Di Primo et al., 2007). Identification of the few residues that cannot be modified in an RNA aptamer can be carried out by chemical interference, a method used to identify chemical variants of the aptamer originally selected. Such an approach led to the synthesis of a modified anti-HIV-1 reverse transcriptase in which all but two of the positions of the RNA aptamer were substituted by 2′-O-methyl residues (Green et al., 1995). This was also the case for the aptamer used for age-related macula degeneration in human beings (Ruckman et al., 1998).

An original approach developed by Klussmann and co-workers relies on L-enantiomers of aptamers (called *spiegelmers*) (Vater and Klussmann, 2003). L-DNA (or L-RNA) is the mirror image of natural D-DNA (or D-RNA). L-nucleic acids are fully resistant to nucleases, but they cannot be processed by polymerases. Therefore, a natural D-aptamer will first be raised against the mirror image of the target of interest. Once identified, the L version of the sequence selected, the spiegelmer, will be synthesized chemically; it will give rise to a complex with the natural target characterized by properties identical to that formed between the D-aptamer and the mirror image of the target. This strategy is restricted to small molecules for which the enantiomer of the target can be synthesized. It has been applied successfully to amino acids, nucleosides, and peptides (Klussmann et al., 1996; Wlotzka et al., 2002). L-Aptamers targeted to calcitonin gene-related peptide binding and to the monocyte chemoattractant protein CCL2 were shown to be efficient in vivo in an animal model (Denekas et al., 2006; Kulkarni et al., 2007).

A new methodology has been described that aims at increasing the molecular diversity of aptamers by a process that rests on the simultaneous use of SELEX and dynamic combinatorial chemistry (Ganesan, 2002), called *2D-SELEX*. The concept has been validated using oligonucleotides that contain unmodified purine nucleosides and 2′-aminopyrimidine nucleosides (Bugaut et al., 2004). Such oligonucleotides are amenable to standard in vitro selection (i.e., they can be amplified). The 2′-amino group can react reversibly with aldehydes, thus generating imines (Figure 1.2c). A random library of 2′-amino oligonucleotides was prepared as usual for selection. Upon mixing with a small library of aldehydes, this generates a dynamic pool of 2′-amino, 2′-imino oligonucleotides (Figure 1.2c). In the presence of the target, the pool will be enriched in such oligonucleotides. Therefore, both the scaffold (the oligonucleotide sequence) and the pendant groups (the 2′-imino substituents) will be selected at once. Following capture, the iminooligomers selected are hydrolyzed. The regenerated amino oligonucleotides are then PCR-amplified and a new round of 2D-SELEX is carried out. At the end of the process the candidates selected are cloned and sequenced. The oligonucleotides identified are then reacted individually with the aldehydes in the presence of the target, and the mixture is reduced by cyanoborohydride. The 2' pendant groups are then identified by mass spectrometry. 2D-SELEX has been applied successfully to the selection of aptamers to the HIV-1 TAR element, leading to sequences different from that obtained when the SELEX is carried out in the absence of aldehydes (Duconge and Toulmé, 1999; Duconge et al., 2000; Bugaut et al., 2006). The process could be applied to other reversible reactions, such as disulfide bond formation. It could be extended to other types of modifed nucleotides that are incorporated by polymerases such as uracil modified in position 5 (Latham et al., 1994), thus allowing the selection of new fluorescent or electrochemical sensors.

One particular case of modified nucleobase is 5-bromouridin (BrU) (Figure 1.2b), a phoreactive derivative that is used in *photo-SELEX* (Jensen et al., 1995; Golden et al., 2000). The pool of BrU-containing candidates is mixed with the target protein. Following capture of the bound oligomers, the mixture is ultraviolet-irradiated to generate covalent cross-links between the oligonucleotides and the protein. Unbound oligomers are washed away and the photo-cross-linked material is subjected to protease digestion. The free oligonucleotides are then amplified for the next round of selection. Partitioning therefore relies on two features: affinity and accurate positioning of cross-linkable groups, resulting in very high specificity. Photo-SELEX allowed the selection of DNA aptamers against human basic fibroblast growth factor (Jensen et al., 1995).

In contrast to what was stated above about the difficulty of introducing an internal modification in an aptamer, it is fairly easy to modify the 3' or the 5' end of the oligomer during synthesis on the solid support, without altering its binding properties. For example, the optimization of selected sequences may include a terminal 3', 3' cap or disulfide cross-link that renders the oligonucleotide resistant to the abundant 3', 5' exonucleases. Any marker—biotin, fluorescent reporter, or groups that increase the bioavailability, such as polyethylene glycol or cholesterol (Boomer et al., 2005; Dougan et al., 2000)—can also be introduced. This constitutes a major advantage over antibodies and opens the way to the design of probes or sensors as well as to agents for in vivo applications, two highly challenging fields toward therapeutic or diagnostic applications.

1.4 APTAREGULATORS

The high affinity and specificity of recognition displayed by aptamers make them appropriate for the design of regulators of biological function. Indeed, very early on, experiments aiming at the identification of sequences recognized by RNA- or DNA-binding proteins were carried out (Henderson et al., 1994). Frequently and not unexpectedly, sequences that are stronger binders than the natural sequences were identified (Bartel et al., 1991). The biological function requires additional properties, in particular the reversibility of the target–regulator complex over a given physiological concentration range, to turn the process on and off under control. Such aptamers are therefore able to trap the target protein and consequently, to act as efficient decoys. The competition between the aptamer and the natural binding site results in the control of the function ensured by the protein. Interestingly, aptamers selected in the test tube retain their properties inside cells. Therefore, the effect of these decoy aptamers can be investigated in vivo (Famulok et al., 2001). This was demonstrated by aptamers targeted to the Tat (Yamamoto et al., 2000) or Rev proteins of the HIV-1 (Bartel et al., 1991). These proteins are involved in the transcription of the HIV genome and in the nuclear export of incompletely spliced viral mRNA. In situ expression of anti-Rev and anti-TAR RNA aptamers inhibited the retroviral replication by more than 70% in cultured cells (Good et al., 1997).

Similar regulators of protein function can be developed against proteins that are not natural binders of nucleic acids. A very large number of proteins have been used for raising aptamers. One of the most popular examples is the G-tetrad-forming DNA aptamer selected against thrombin, a key regulator in the coagulation cascade (Bock et al., 1992). This aptamer prolonged clotting time in purified fibrinogen and in human plasma and displayed anticoagulant properties in vivo (Griffin et al., 1993).

Regulatory aptamers have also been developed for targeting RNA instead of proteins. Oligonucleotides are poorly adapted to the recognition of RNA structures (Toulmé et al., 2005). Folded RNA regions are not available for intermolecular pairing with the complementary sequence. Consequently, antisense or small interfering RNA targeted to structured regions shows a limited efficiency (Kurreck, 2006). In vitro selection has been carried out to identify aptamers that recognize the folded state of the target RNA region. The numerous interactions identified in tertiary RNA structures, besides Watson–Crick base pairing, suggest that it should be possible to take advantage of nonpaired nucleic acid bases in loops and bulges to engage intermolecular interactions with an aptamer. Additional interactions such as stacking would bring an additional contribution to the binding. Both DNA and RNA aptamers have been identified against the TAR RNA hairpin of HIV-1, an imperfect hairpin involved in the trans-activation of the transcription of the retroviral genome (Boiziau et al., 1999; Duconge and Toulmé, 1999). The binding occurs through the formation of a loop–loop helix between the partially complementary apical loops of the two partners. For the RNA aptamer a six-base-pair helix is formed and a critical noncanonical GA pair closing the aptamer loop has been shown to play a crucial role in the kissing complex formation (Duconge et al., 2000). A high affinity (a K_d value of a few nanomolar) has been obtained. The in situ expression of this aptamer driven by a RNA Pol III promotor was shown specifically to reduce by 60% expression of a reporter gene under the control of the TAR element in cultured HeLa cells (Kolb et al., 2006). Chemically modified aptamers have been designed that show improved biological properties compared to the regular RNA molecule (Darfeuille et al., 2002a,b, 2004). Other RNA structures have been targeted successfully by aptamers, in particular in the internal ribosome entry site of the hepatitis

C virus RNA (Tallet-Lopez et al., 2003; Da Rocha-Gomes et al., 2004; Kikuchi et al., 2005).

It is of great interest to generate reversible regulators that can be activated or deactivated at will, in response to a signal. To this end, regulation can be achieved by aptamers raised against small molecules. When inserted in mRNAs, such aptamers mimic riboswitches that have been identified in prokaryotic mRNAs (Tucker and Breaker, 2005). For these RNAs, conformational changes in response to ligand binding may result in translational regulation, most frequently by switching the ribosome-binding site from a sequestered to a free status. The ligand-binding site on RNA that triggers the conformational rearrangement is functionally equivalent to an aptamer. It was therefore tempting to design artificial regulators of gene expression by inserting an aptamer sequence in the 5′ untranslated region (UTR) of a given gene. The interaction between the aptamer and its target molecule will alter or stabilize the RNA structure, which in turn might prevent the binding of the ribosome or the initiation of the translation. In a pioneering work, Werstuck and Green (1998) validated this approach by inserting an aptamer to Hoechst dye in the 5′ UTR of a reporter gene and demonstrated that they were able to specifically control the translation of the reporter gene by adding the dye either in vitro in wheat germ extract or in cultured CHO cells. Similar results allowing regulation in *cis* have been reported in *Saccharomyces cerevisiae* (Suess et al., 2003). Recently, the replication of a viral genome in which a crucial viral stem–loop was substituted by the theophylline aptamer could be controlled by this purine derivative (Wang and White, 2007).

It is also conceivable to design riboregulators acting in *trans*. For instance, the activity of an antisense RNA was controlled by an aptamer specific for theophylline fused to the antisense sequence (Bayer and Smolke, 2005). In the absence of theophylline, the antisense domain is sequestered in an intramolecular structure. Addition of theophylline induces a conformational change of the riboregulator that allows the antisense to interact with its target mRNA. The expression of a reporter gene (EGFP) in yeast cells expressing such a riboregulator (antiEGFP antisense) was shown to be inhibited in the presence of theophylline. The effect was specific: no inhibition was induced by caffeine that does not bind to the riboregulator.

Analogous strategies can be applied to the reversible control of protein function. The association between an aptamer and its target protein can be prevented by an oligonucleotide that base-pairs with part of the aptamer sequence, thus disrupting the active structure of the aptamer. A 2′-fluoropyrimidine-containing aptamer specific for the human coagulation factor IXa displays anticoagulant properties that can be reverted by the addition of a 2′-*O*-methyl oligonucleotide complementary to the 5′ end of the aptamer (Rusconi et al., 2002, 2004). The antisense oligonucleotide constitutes an antidote to the anticoagulant aptamer. This rational method is of wide interest for designing regulatable aptamers that are suitable for in vivo use, as both the aptamer and the antisense regulator are resistant to nucleases.

An alternative to the development of "aptamer antidotes" is to hide the aptamer's active conformation by reversible chemical modification of some part of the sequence. The reversal of the protection will free the oligonucleotide, which will recover its binding properties as a consequence. The difficulty resides in the use of groups that could be unmasked at will. This was achieved elegantly by introducing caged nucleobases (i.e., bases that are protected by photo-labile groups). Caged analogs of thymidine T^{NPP} containing a photo-labile 2-(2- nitrophenyl) propyl (NPP) group (Figure 1.2b) have been introduced in the well-characterized antithrombine DNA aptamer (Heckel and Mayer, 2005). This G- and T-containing 15-mer folds into a G-tetrad structure that displays anticoagulant properties. The presence of a single residue T^{NPP} prevented the binding of the aptamer analog to thrombin, whereas the K_d value of the photo-reactivated derivative was similar to that of the wild-type aptamer, but a full anticoagulant activity was not restored following irradiation under physiological conditions. Caged residues can be combined with the antidote approach to generating controllable aptamers, as described recently by Mayer and colleagues. The antithrombin DNA aptamer was extended on the 5' side by a short sequence complementary to part of the G-tetrad-forming 15-mer. Initially, the association between the two complementary sequences is prevented by the presence of a C residue caged with an NPE [1-(2-nitrophenyl)ethyl] group (Heckel et al., 2006). The C^{NPE} nucleobase acts as a transient mismatch that can be converted into a match by irradiation at 366 nm, thus allowing intramolecular pairing between the aptamer region and the antidote complementarty sequence. The caged oligomer displays reduced but significant anticoagulant properties, whereas it is totally inactive following uncaging. The use of caged residues therefore allows triggering by light of either activation or deactivation of aptamers.

Formation of an aptamer–protein complex can be controlled by a small molecule that upon binding to the aptamer triggers a conformational change of the aptamer, thus altering its binding properties (Buskirk and Liu, 2005). A conditional RNA aptamer to formamidopyrimidine glycosylase (Fpg), a DNA repair enzyme, was obtained through a clever selection procedure (Vuyisich and Beal, 2002). After a few selection rounds the RNA/Fpg retained on a filter was eluted by the addition of neomycin. Therefore, only Fpg-bound oligoribonucleotides that dissociated in the presence of the aminoglycoside were collected and used for further selection rounds. The aptamer isolated at the end of this process was an inhibitor of Fpg. This inhibition was reverted by the addition of neomycin. The structural analysis of this aptamer demonstrated that it contained two overlapping binding sites for the Fpg and for neomycin, respectively. Interestingly, the neomycin binding site shared similarities with previously identified neomycin aptamer (Wallis and Schroeder, 1997). The rescue of the Fpg activity by neomycin was specific; the addition of kanamycin, an antibiotic structurally related to neomycin that does not bind to the patamer, does not show any effect. Aptamers selected in parallel against the Fpg through a procedure that did not include the elution step by neomycin displayed similar binding and inhibitory properties, except that the inhibition was not reverted by addition of the aminoglycoside (Vuyisich and Beal, 2002). Even though this is of wide potential interest, it remains to be demonstrated whether this approach can be generalized.

1.5 APTASENSORS

Aptamers, which show high affinity of binding and high specificity of target recognition are sometimes described as "chemical antibodies." They actually display a number of advantages over antibodies, as they can in principle be raised against any type of target, including toxic compounds. Moreover, aptamers are smaller than antibodies (molecular mass 5 to 15 kDa) and can get access to regions that are poorly accessible to immunoglobulins. They are easily and reproducibly synthesized and are easy to store. They can be conveniently conjugated to various derivatives. Last, as SELEX is an iterative process (Figure 1.1), it has been possible to adapt in vitro selection to robotic procedures that considerably reduce the time needed to generate aptamers (Cox et al., 1998, 2002). It is therefore not surprising that aptamers became ideal tools for the development of analytical methods. Reports dealing with the purification of molecules, the design of optical, electrochemical, or acoustic sensors, the conception of signaling aptamers (beacons, quantum dots), and the development of aptamer arrays for high-throughput analysis appear on a weekly basis. Multiple applications will be described in detail in the following chapters, and several recent reviews are available (Tombelli et al., 2005; Mairal et al., 2007). We mention here only briefly the different areas in which aptamers have been used and outline a few recent examples that are of high potential interest.

Aptamers have been used for the purification by affinity chromatography of different peptides and proteins: thrombin, thyroid transcription factor 1, L-selectin, and so on (Ravelet et al., 2006). Affinity chromatography of proteins with aptamers is very attractive, as there is no need for a tag that might affect folding and also no need for the tag-cleavage step, thus ensuring quick purification procedures and high yield of recovery, as demonstrated recently for purification of the Taq polymerase (Oktem et al., 2007). Aptamers have been shown to discriminate efficiently between target enantiomers, allowing specific chiral separation. For instance, the D-enantiomer of the oligopeptide arginine–vasopressin was retained on a specific DNA aptamer column while the L-enantiomer was eluted in the void volume (Michaud et al., 2003). The biotinylated aptamer was easily immobilized on a streptavidin polystyrene–divnyl benzene support. Enantiomers of small biomolecules were also resolved efficiently (Michaud et al., 2004). To develop this strategy on a large scale, several problems have to be solved, notably the cost and the problem of the aptamer stability if samples contaminated with nucleases are used. Solutions discussed above (modified oligonucleotides) can be considered.

Aptamers could sense the presence of the target molecule they were selected against if they were combined with a transducer element that converts formation of the aptamer–target complex into a signal that can be measured or at least evaluated. Various formats have been described. Surface plasmon resonance (SPR) is of particular interest, as it is a label-free methodology that is used widely for the quantitative study of selected sequences and analysis of the binding sites. It allowed the characterization of protein–aptamer (Win et al., 2006) and RNA–aptamer complexes (Aldaz-Carroll et al., 2002). It was even used to select RNA aptamers against the human RNaseH1 and the hemaglutinin of the human influenza virus (Pileur et al., 2003; Misono and Kumar, 2005). Depending on the relative size of the two partners, either the aptamer or the target can be immobilized on the SPR chip. Careful analysis of the sensorgrams even allows monitoring the formation of ternary complexes (Di Primo, 2008). However, it is a low-throughput method that requires a sophisticated and rather expensive instrument. Electronic aptamer-based sensors have been developed (Willner and Zayats, 2007), but the most popular sensors are by far optical sensors, based mostly on fluorescence measurements (i.e., intensity, anisotropy, energy transfer, etc.). A very simple assay would make use of a fluorescently labeled analyte (Drolet et al., 1996). Conversely, the aptamer could be associated with a fluorophore whose emission properties will change upon binding to its ligand. The second possibility does not require the synthesis of a target analog that shows binding properties similar to that of the unmodified molecule and is therefore easier to use in real time. Moreover, conjugation of the aptamer to a fluorophore at either the 5′ or 3′ end is trivial. The binding to the analyte can be monitored by changes in the evanescent wave-induced fluorescence anisotropy as described in the pioneering work by Potyrailo et al. (1998). An immobilized FITC-conjugated DNA aptamer allowed the detection of as little as 0.7 amol of thrombin. Differences in the diffusion rates of fluorescently labeled aptamers free or bound to their target can also be measured in solution, as shown for the analysis of IgE (Gokulrangan et al., 2005) and angiogenin (W. Li et al., 2007). Fluorescent analogs of nucleic acid bases could also be incorporated into the aptamer sequence; the binding of the aptamer to its target might eventually induce a modification in the electronic environment of the fluorophore. Generally, multiple variants of fluorescent aptamers should be assayed before a responsive molecule is identified (Jhaveri et al., 2000; Katilius et al., 2006).

A light switch based on a ruthenium complex which shows a strong luminescence emission when it intercalates into double-stranded nucleic acids was used to convert an aptamer into a sensor. The binding of the aptamer to its ligand induces conformational changes that perturb intercalation of the ruthenium complex, leading subsequently to luminescence changes (Jiang et al., 2004). Interestingly, this requires no labeling of either the analyte or the aptamer. Angiogenin and IgE were detected in serum samples using such assays.

Molecular beacons are oligonucleotide probes designed originally for the detection of nucleic acid sequences (Tyagi and Kramer, 1996). They generally have a hairpin shape, which brings into close contact fluorescence reporters bound at the $3'$ and $5'$ ends of the double-stranded stem. These reporters can be either a fluorophore and a quencher or a pair of dyes that constitute donor–acceptor molecules appropriate for fluorescence energy transfer (FRET). This stem–loop structure will unfold upon hybridization to a sequence fully complementary to the loop and stem regions, thus moving the reporters away from each other. This translates into a fluorescence change, either a dequenching of the fluorophore or a decrease in the FRET efficiency. Measurement of the fluorescence therefore allows quantitative detection of the target. Monitoring of fluorescence variation associated with conformational changes has been adapted to aptamers, as these molecules frequently undergo structural variation upon association with their cognate ligand (Hermann and Patel, 2000; Soukup et al., 2000). Beacons were engineered for the detection of thrombin by the G-quartet, forming the DNA aptamer identified previously (Bock et al., 1992). This oligonucleotide is in equilibrium between a random and a folded state; this equilibrium is shifted toward the four-stranded structure by binding to thrombin. This is associated with a significant fluorescent signal when the $3'$ and $5'$ ends of this 15-mer are conjugated to fluorescence reporters (Li et al., 2002). This is unlikely to be generalized, as not every aptamer will undergo a switch from an extended structure in the absence of its ligand to a folded structure in its presence. But a given aptamer might be truncated to give rise to a partly unfolded oligomer that will fold into a closed structure upon binding to the ligand. The three-way junction aptamer for cocaine was converted into a specific sensor by derivatizing a truncated DNA aptamer with fluorescein and dabcyl (Stojanovic et al., 2001). Such beacons signal the presence of their cognate analyte by decreased emission. An off-to-on transition in the presence of the ligand is more desirable, as the detection of the analyte is more sensitive. Many aptamers can be converted into hairpin-shaped beacons by extending one of their extremities by a few nucleotides so as to generate a short sequence complementary to the other end of the oligomer. The length of the resulting double-stranded stem should be adjusted to be opened in the presence of the ligand. Such a design has been used successfully for the antithrombin aptamer (Hamaguchi et al., 2001). The first aptamer beacon that was described was designed as a bipartite molecule from an RNA aptamer targeted to the Tat protein of HIV (Yamamoto et al., 2000). This aptamer was split in two parts, one part of which was reformulated to generate a hairpin-shaped beacon. In the absence of Tat, the two halves are independent. The addition of Tat induces the reannealing of the two RNA halves, which results in fluorescence emission (Yamamoto and Kumar, 2000). This design is appropriate for aptamers that show a rather long double-stranded stretch.

Tripartite beacons can be tailored as described previously for nucleic acid detection. In this case the hairpin structure is not labeled but displays single-stranded $3'$ and $5'$ extensions that are complementary to two oligomers bearing a fluorophore and a quencher, respectively. In the free folded hairpin state, the reporters are in close proximity, but they are moved apart in the extended unfolded state resulting from the binding of the loop region to its complementary sequence, thus generating a fluorescence signal (Nutiu and Li,

2002). This strategy has been adapted to aptamers: The aptamer is labeled with a fluorophore, whereas a short complementary oligonucleotide bears the quenching molecule. Binding the aptamer to its target induces release of the short oligonucleotide and results in fluorescence emission (Nutiu and Li, 2003, 2005a).

Modular biosensors assemble two aptamers into one molecule. The first is a recognition module corresponding to the binding site of the analyte; the second is a signaling domain that accommodates a reporter. The two modules comunicate in such a way that binding of the analyte leads to increased affinity of the signaling domain for the reporter molecule, hence resulting in increased signal. Modular sensors for ATP, FMN, or theophylline were designed in combination with a malachite green sensing domain (Stojanovic and Kolpashchikov, 2004). This is highly reminiscent of small molecule–dependent switches tailored for the control of gene expression (Buskirk et al., 2004).

Very generally, engineered beacons from preidentified aptamers require some knowledge of the secondary or tertiary structure. In addition, design of the signaling aptamer needs critical adjustment of the modified version so that the addition of the ligand shifts the equilibrium efficiently between the two structures corresponding to the free and bound states. It would be much better to select not only on the binding but also on the signaling. A procedure was described for which the selection step involves a capture oligonucleotide. Some of the complexes formed by this oligomer and a complementary region in candidates of the random library are dissociated upon the addition of an analyte. This is assumed to be due to the conformational change of the candidate, which actually constitutes an aptamer for the added analyte. The released oligonucleotides are collected and amplified as usual in the SELEX procedure, thus allowing the identification of aptamers whose binding to their ligand induces a conformational change that in turn leads to disruption of the aptamer–capture oligonucleotide complex. If a quencher is introduced on the capture oligonucleotide and a fluorophore into its complementary region on the aptamer, binding of the analyte to the aptamer will result in increased fluorescence emission. This clever approach has been validated for the direct selection of molecular beacons against oligonucleotides (Rajendran and Ellington, 2003), purine nucleotides (Nutiu and Li, 2005b), and more recently, against the aminoglycoside antibiotic tobramycin (Morse, 2007).

1.6 PROSPECTS

The development of an automated selection process, the intrinsic properties of aptamers combined with their conversion into signaling elements, and the ease of conjugating oligonucleotides on different surfaces make them appropriate for their use in microarray formats. Interest in aptamer chips was outlined long ago (Brody and Gold, 2000). They could be used for the identification and quantification of multiple proteins or biomarkers. There are still few reports on the development of microarray-based aptamer assays. Detection of the analyte can be carried out using fluorescent ligands (Collett et al., 2005), fluorescent aptamers (McCauley et al., 2003), or surface plasmon resonance imaging (Y. Li et al., 2006, 2007b). Photoaptamer-based arrays represent an interesting strategy (Golden et al., 2000). The use of aptamer chips for the simultaneous detection of multiple targets requires further optimization before it can be used for proteomics, but attempts along this line have been undertaken (Cho et al., 2006).

One major challenge is the simultaneous detection of a number of analytes in identical analytical conditions. Aptamers that can be selected at will under predetermined conditions are of particular interest to this end to the extent that they transduce different signals. The association of aptamers to quantum dots (QDs) offers an exciting perspective. QDs are fluorophores that can be excited at the same wavelength and show sharp emission profiles, a large Stokes shift, and a long fluorescence lifetime. They have been used in different beacon formats (Levy et al., 2005; Choi et al., 2006). Nanostructures combining quantum dots, quenching gold nanoparticles, and aptamers allowed the detection of adenosine and cocaine by monitoring the fluorescence emission at 525 and 585 nm, respectively (Liu et al., 2007). Recently, quantum dot–labeled aptamer was reporterd to allow the detection of *Bacillus thurigiensis* spores (Ikanovic et al., 2007).

Increased sensitivity of aptamer-based sensors might be achieved by the inclusion of an amplification step. To this end, modulation of enzymatic activity by the analyte is an interesting perspective. Allosteric ribozymes have been rationally designed by combining a hammerhead or a hairpin ribozyme to an aptamer module. Different formats have been described. Binding of the analyte to the aptamer might regulate the ribozyme activity directly (Tang and Breaker, 1998) or release its association with an inhibitory oligonucleotide (Famulok, 2005). A ribozyme conjugated to two aptamer modules—for the replicase and helicase of the hepatitis C virus—were shown to show high selectivity (Cho et al., 2005). Ribozyme–aptamer conjugates have been designed for screening small-molecule libraries in order to identify inhibitors of viral enzymes. The RNA pseudoknot aptamer inhibitor of the HIV-1 reverse transcriptase (Tuerk et al., 1992) prevented the activity of the conjugated ribozyme when bound to the retroviral polymerase. If a small molecule displaces an aptamer from its target, the ribozyme becomes active, which is visualized by a fluorescent signal, thus allowing identification of potential reverse transcriptase ligands (Yamazaki et al., 2007). The potential of an aptamer-based amplified signal was recently evidenced: a DNA sequence bearing the cocaine aptamer at its 3' end was engineered to constitute a template for DNA polymerase in the presence—but not in the absence—of cocaine. The replication fragment was used to titrate out a fluorescent molecular beacon (Shlyahovsky et al., 2007). This opens the way to the design of highly sensitive sensors.

Whereas oligonucleotide libraries were traditionally partitioned by filtration or capture on beads or on affinity columns, capillary electrophoresis has been validated as a promising alternative. Aptamers displaying a high affinity were obtained in a very limited number of selection rounds. Nonequilibrium capillary electrophoresis of equilibrium mixtures led to the identification of DNA aptamers with a K_d value of about 1 nM against protein farnesyltransferase (Berezovski et al., 2005). Interestingly, this methodology allows the selection to be carried out according to kinetics criteria. Non-SELEX capillary electrophoresis (i.e., selection without amplification) has been described for the selection of DNA aptamers against the h-Ras protein (Berezovski et al., 2006a). This procedure is of particular interest for the identification of chemically modified oligonucleotides that cannot be synthesized by polymerases (Berezovski et al., 2006b). As the selection occurs in solution, there is no background of sequences targeted to the support (i.e., filter, beads, etc.). But this technique is restricted to targets of size larger than about 20 kDa, as complexes between oligonucleotide candidates and smaller molecules have a mobility close to that of oligomers.

Over the last 15 years, aptamers have emerged as compounds of wide potential for both therapeutic and diagnostic purposes. Many applications of interest have not been described here: for example, in the field of cell biology, molecular imaging, and delivery (Chu et al., 2006; Hicke et al., 2006). Last but not least, a targeted antivascular epithelium growth factor aptamer (pegaptanib) has been approved by the U.S. Food and Drug Administration for the treatment of neovascular age-related macular degeneration (Ng et al., 2006). This constitutes a milestone in the development of aptamers. It is anticipated that numerous aptamers will develop into drugs and diagnostic tools in the near future.

Acknowledgments

We thank Sonia Da Rocha-Gomes (INSERM U869) for sharing unpublished data. J.P.D. is the recipient of a fellowship of the European Union. The INSERM Unit 869 "RNA: Régulation Naturelle et Artificielle" is supported by the Conseil Régional d'Aquitaine.

REFERENCES

Aboul-Fadl, T. (2005). Antisense oligonucleotides: the state of the art. *Curr Med Chem* 12, 2193–2214.

Aldaz-Carroll, L., Tallet, B., Dausse, E., Yurchenko, L., Toulmé, J-J. (2002). Apical loop-internal loop interactions: a new RNA–RNA recognition motif identified through in vitro selection against RNA hairpins of the hepatitis C virus mRNA. *Biochemistry* 41, 5883–5893.

Andréola, M. L., Calmels, C., Michel, J., Toulmé, J-J., Litvak, S. (2000). Towards the selection of phosphorothioate aptamers Optimizing in vitro selection steps with phosphorothioate nucleotides. *Eur J Biochem* 267, 5032–5040.

Aurup, H., Williams, D. M., Eckstein, F. (1992). 2'-Fluoro and 2'-amino-2'-deoxynucleoside 5'-triphosphates as sustrates for T7 RNA polymerase. *Biochemistry* 31, 9636–9641.

Bartel, D. P., Zapp, M. L., Green, M. R., Szostak, J. W. (1991). HIV-1 Rev regulation involves recognition of non-Watson–Crick base pairs in viral RNA. *Cell* 67, 529–536.

Bayer, T. S., Smolke, C. D. (2005). Programmable ligand-controlled riboregulators of eukaryotic gene expression. *Nat Biotechnol* 23, 337–343.

Belikova, A. M., Zarytova, V. F., Grineva, N. I. (1967). Synthesis of ribonucleosides and diribonucleoside phosphates containing 2-chloroethylamine and nitrogen mustard residues. *Tetrahedron Lett* 37, 3557–3562.

Berezovski, M., Drabovich, A., Krylova, S. M., Musheev, M., Okhonin, V., Petrov, A., Krylov, S. N. (2005). Nonequilibrium capillary electrophoresis of equilibrium mixtures: a universal tool for development of aptamers. *J Am Chem Soc* 127, 3165–3171.

Berezovski, M., Musheev, M., Drabovich, A., Krylov, S. N. (2006a). Non-SELEX selection of aptamers. *J Am Chem Soc* 128, 1410–1411.

Berezovski, M. V., Musheev, M. U., Drabovich, A. P., Jitkova, J. V., Krylov, S. N. (2006b). Non-SELEX: selection of aptamers without intermediate amplification of candidate oligonucleotides. *Nat Protoc* 1, 1359–1369.

Blackwell, T. K., Weintraub, H. (1990). Differences and similarities in DNA-binding preferences of MyoD and E2A protein complexes revealed by binding site selection. *Science* 250, 1104–1110.

Blank, M., Blind, M. (2005). Aptamers as tools for target validation. *Curr Opin Chem Biol* 9, 336–342.

Bock, L. C., Griffin, L. C., Latham, J. A., Vermass, E. H., Toole, J. J. (1992). Selection of single-stranded DNA molecules that bind and inhibit human thrombin. *Nature* 355, 564–566.

Boiziau, C., Dausse, E., Yurchenko, L., Toulmé, J-J. (1999). DNA aptamers selected against the HIV-1 TAR RNA element form RNA/DNA kissing complexes. *J Biol Chem* 274, 12730–12737.

Boomer, R. M., Lewis, S. D., Healy, J. M., Kurz, M., Wilson, C., McCauley, T. G. (2005). Conjugation to polyethylene glycol polymer promotes aptamer biodistribution to healthy and inflamed tissues. *Oligonucleotides* 15, 183–195.

Brody, E. N., Gold, L. (2000). Aptamers as therapeutic and diagnostic agents. *Rev Mol Biotech* 74, 5–13.

Brukner, I., El-Ramahi, R., Gorska-Flipot, I., Krajinovic, M., Labuda, D. (2007). An in vitro selection scheme for oligonucleotide probes to discriminate between closely related DNA sequences. *Nucleic Acids Res* 35, e66.

Bugaut, A., Toulmé, J-J., Rayner, B. (2004). Use of dynamic combinatorial chemistry for the identification of covalently appended residues that stabilize oligonucleotide complexes. *Angew Chem Int Ed Engl* 43, 3144–3147.

Bugaut, A., Toulmé, J-J., Rayner, B. (2006). SELEX and dynamic combinatorial chemistry interplay for the selection of conjugated RNA aptamers. *Org Biomol Chem* 4, 4082–4088.

Bunka, D. H., Stockley, P. G. (2006). Aptamers come of age—at last. *Nat Rev Microbiol* 4, 588–596.

Burmeister, P. E., Lewis, S. D., Silva, R. F., Preiss, J. R., Horwitz, L. R., Pendergrast, P. S., McCauley, T. G., Kurz, J. C., Epstein, D. M., Wilson, C., Keefe, A. D. (2005). Direct in vitro selection of a 2′-O-methyl aptamer to VEGF. *Chem Biol* 12, 25–33.

Burmeister, P. E., Wang, C., Killough, J. R., Lewis, S. D., Horwitz, L. R., Ferguson, A., Thompson, K. M., Pendergrast, P. S., McCauley, T. G., Kurz, M., Diener, J., Wilson, C., Keefe, A. D. (2006). 2′-Deoxy purine, 2′-O-methyl pyrimidine (dRmY) aptamers as candidate therapeutics. *Oligonucleotides* 16, 337–351.

Buskirk, A. R., Liu, D. R. (2005). Creating small-molecule-dependent switches to modulate biological functions. *Chem Biol* 12, 151–161.

Buskirk, A. R., Landrigan, A., Liu, D. R. (2004). Engineering a ligand-dependent RNA transcriptional activator. *Chem Biol* 11, 1157–1163.

Chakraborty, C. (2007). Potentiality of small interfering RNAs (siRNA) as recent therapeutic targets for gene-silencing. *Curr Drug Targets* 8, 469–482.

Chelliserrykattil, J., Ellington, A. D. (2004). Evolution of a T7 RNA polymerase variant that transcribes 2'-O-methyl RNA. *Nat Biotechnol* 22, 1155–1160.

Chen, H., Gold, L. (1994). Selection of high-affinity RNA ligands to reverse transcriptase: inhibition of cDNA synthesis and RNase H activity. *Biochemistry* 33, 8746–8756.

Chen, H., Mcbroom, D. G., Zhu, Y. Q., Gold, L., North, T. W. (1996). Inhibitory RNA ligand to reverse transcriptase from feline immunodeficiency virus. *Biochemistry* 35, 6923–6930.

Cho, E. J., Collett, J. R., Szafranska, A. E., Ellington, A. D. (2006). Optimization of aptamer microarray technology for multiple protein targets. *Anal Chim Acta* 564, 82–90.

Cho, S., Kim, J. E., Lee, B. R., Kim, J. H., Kim, B. G. (2005). Bis-aptazyme sensors for hepatitis C virus replicase and helicase without blank signal. *Nucleic Acids Res* 33, e177.

Choi, J. H., Chen, K. H., Strano, M. S. (2006). Aptamer-capped nanocrystal quantum dots: a new method for label-free protein detection. *J Am Chem Soc* 128, 15584–15585.

Chu, T. C., Twu, K. Y., Ellington, A. D., Levy, M. (2006). Aptamer mediated siRNA delivery. *Nucleic Acids Res* 34, e73.

Collett, J. R., Cho, E. J., Ellington, A. D. (2005). Production and processing of aptamer microarrays. *Methods* 37, 4–15.

Cox, J. C., Rudolph, P., Ellington, A. D. (1998). Automated RNA selection. *Biotechnol Prog* 14, 845–850.

Cox, J. C., Hayhurst, A., Hesselberth, J., Bayer, T. S., Georgiou, G., Ellington, A. D. (2002). Automated selection of aptamers against protein targets translated in vitro: from gene to aptamer. *Nucleic Acids Res* 30, e108.

Da Rocha-Gomes, S., Dausse, E., Toulmé, J-J. (2004). Determinants of apical loop-internal loop RNA–RNA interactions involving the HCV IRES. *Biochem Biophys Res Commun* 322, 820–826.

Darfeuille, F., Arzumanov, A., Gait, M. J., Di Primo, C., Toulmé, J-J. (2002a). 2'-O-methyl-RNA hairpins generate loop–loop complexes and selectively inhibit HIV-1 Tat-mediated transcription. *Biochemistry* 41, 12186–12192.

Darfeuille, F., Arzumanov, A., Gryaznov, S., Gait, M. J., Di Primo, C., Toulmé, J-J. (2002b). Loop–loop interaction of HIV-1 TAR RNA with N3'→P5' deoxyphosphoramidate aptamers inhibits in vitro Tat-mediated transcription. *Proc Natl Acad Sci U S A* 99, 9709–9714.

Darfeuille, F., Hansen, J. B., Orum, H., Di Primo, C., Toulmé, J-J. (2004). LNA/DNA chimeric oligomers mimic RNA aptamers targeted to the TAR RNA element of HIV-1. *Nucleic Acids Res* 32, 3101–3107.

Darfeuille, F., Reigadas, S., Hansen, J. B., Orum, H., Di Primo, C., Toulmé, J-J. (2006). Aptamers targeted to an RNA hairpin show improved specificity compared to that of complementary oligonucleotides. *Biochemistry* 45, 12076–12082.

Denekas, T., Troltzsch, M., Vater, A., Klussmann, S., Messlinger, K. (2006). Inhibition of stimulated meningeal blood flow by a calcitonin gene-related peptide binding mirror-image RNA oligonucleotide. *Br J Pharmacol* 148, 536–543.

Di Primo, C., Rudloff, I., Reigadas, S., Arzumanov, A. A., Gait, M. J., Toulmé, J-J. (2007). Systematic screening of LNA/2′-O-methyl chimeric derivatives of a TAR RNA aptamer. *FEBS Lett* 581, 771–774.

Di Primo, C., (2008). Real time analysis of the RNAI-RNAII-Rop complex by surface plasmon sequence: from a decaying surface to a standard kinetic analysis. *J. Mol. Recognition* 21, 37–45.

Dougan, H., Lyster, D. M., Vo, C. V., Stafford, A., Weitz, J. I., Hobbs, J. B. (2000). Extending the lifetime of anticoagulant oligodeoxynucleotide aptamers in blood. *Nucl Med Biol* 27, 289–297.

Drolet, D. W., Moon-McDermott, L., Romig, T. S. (1996). An enzyme-linked oligonucleotide assay. *Nat Biotechnol* 14, 1021–1025.

Ducongé, F., Toulmé, J-J. (1999). In vitro selection identifies key determinants for loop-loop interactions: RNA aptamers selective for the TAR RNA element of HIV-1. *RNA* 5, 1605–1614.

Ducongé, F., Di Primo, C., Toulmé, J-J. (2000). Is a closing "GA pair" a rule for stable loop-loop RNA complexes? *J Biol Chem* 275, 21287–21294.

Elayadi, A. N., Corey, D. R. (2001). Application of PNA and LNA oligomers to chemotherapy. *Curr Opin Invest Drugs* 2, 558–561.

Ellington, A. D., Szostak, J. W. (1990). In vitro selection of RNA molecules that bind specific ligands. *Nature* 346, 818–822.

Eulberg, D., Buchner, K., Maasch, C., Klussmann, S. (2005). Development of an automated in vitro selection protocol to obtain RNA-based aptamers: identification of a biostable substance P antagonist. *Nucleic Acids Res* 33, e45.

Famulok, M. (1999). Oligonucleotide aptamers that recognize small molecules. *Curr Opin Struct Biol* 9, 324–329.

Famulok, M. (2005). Allosteric aptamers and aptazymes as probes for screening approaches. *Curr Opin Mol Ther* 7, 137–143.

Famulok, M., Blind, M., Mayer, G. (2001). Intramers as promising new tools in functional proteomics. *Chem Biol* 8, 931–939.

Fernandez-Gacio, A., Uguen, M., Fastrez, J. (2003). Phage display as a tool for the directed evolution of enzymes. *Trends Biotechnol* 21, 408–414.

Fichou, Y., Ferec, C. (2006). The potential of oligonucleotides for therapeutic applications. *Trends Biotechnol* 24, 563–570.

Frankel, A., Li, S., Starck, S. R., Roberts, R. W. (2003). Unnatural RNA display libraries. *Curr Opin Struct Biol* 13, 506–512.

Ganesan, A. (2002). Recent developments in combinatorial organic synthesis. *Drug Discov Today* 7, 47–55.

Gokulrangan, G., Unruh, J. R., Holub, D. F., Ingram, B., Johnson, C. K., Wilson, G. S. (2005). DNA aptamer-based bioanalysis of IgE by fluorescence anisotropy. *Anal Chem* 77, 1963–1970.

Gold, L., Polisky, B., Uhlenbeck, O., Yarus, M. (1995). Diversity of oligonucleotide functions. *Annu Rev Biochem* 64, 763–797.

Golden, M. C., Collins, B. D., Willis, M. C., Koch, T. H. (2000). Diagnostic potential of PhotoSELEX-evolved ssDNA aptamers. *J Biotechnol* 81, 167–178.

Good, P. D., Krikos, A. J., Li, S. X. L., Bertrand, E., Lee, N. S., Giver, L., Ellington, A., Zaia, J. A., Rossi, J. J., Engelke, D. R. (1997). Expression of small, therapeutic RNAs in human cell nuclei. *Gene Ther* 4, 45–54.

Gopinath, S. C. (2007). Methods developed for SELEX. *Anal Bioanal Chem* 387, 171–182.

Green, L., Waugh, S., Binkley, J. P., Hostomska, Z., Hostomsky, Z., Tuerk, C. (1995). Comprehensive chemical modification interference and nucleotide substitution analysis of an RNA pseudoknot inhibitor to HIV-1 reverse transcriptase. *J Mol Biol* 247, 60–68.

Griffin, L. C., Tidmarsh, G. F., Bock, L. C., Toole, J. J., Keung, L. L. K. (1993). In vivo anticoagulant properties of a novel nucleoside-based thrombin inhibitor and demonstration of regional anticoagulation in extracorporeal circuits. *Blood* 81, 3271–3276.

Hamaguchi, N., Ellington, A., Stanton, M. (2001). Aptamer beacons for the direct detection of proteins. *Anal Biochem* 294, 126–131.

Heckel, A., Mayer, G. (2005). Light regulation of aptamer activity: an anti-thrombin aptamer with caged thymidine nucleobases. *J Am Chem Soc* 127, 822–823.

Heckel, A., Buff, M. C., Raddatz, M. S., Muller, J., Potzsch, B., Mayer, G. (2006). An anticoagulant with light-triggered antidote activity. *Angew Chem Int Ed Engl* 45, 6748–6750.

Held, D. M., Kissel, J. D., Patterson, J. T., Nickens, D. G., Burke, D. H. (2006). HIV-1 inactivation by nucleic acid aptamers. *Front Biosci* 11, 89–112.

Henderson, B. R., Menotti, E., Bonnard, C., Kuhn, L. C. (1994). Optimal sequence and structure of iron-responsive elements: selection of RNA stem-loops with high affinity for iron regulatory factor. *J Biol Chem* 269, 17481–17489.

Hermann, T., Patel, D. J. (2000). Adaptive recognition by nucleic acid aptamers. *Science* 287, 820–825.

Hicke, B. J., Stephens, A. W., Gould, T., Chang, Y. F., Lynott, C. K., Heil, J., Borkowski, S., Hilger, C. S., Cook, G., Warren, S., Schmidt, P. G. (2006). Tumor targeting by an aptamer. *J Nucl Med* 47, 668–678.

Ho, S. P., Britton, D. H., Stone, B. A., Behrens, D. L., Leffet, L. M., Hobbs, F. W., Miller, J. A., Trainor, G. L. (1996). Potent antisense oligonucleotides to the human multidrug resistance-1 mRNA are rationally selected by mapping RNA-accessible sites with oligonucleotide libraries. *Nucleic Acids Res* 24, 1901–1907.

Hybarger, G., Bynum, J., Williams, R. F., Valdes, J. J., Chambers, J. P. (2006). A microfluidic SELEX prototype. *Anal Bioanal Chem* 384, 191–198.

Ikanovic, M., Rudzinski, W. E., Bruno, J. G., Allman, A., Carrillo, M. P., Dwarakanath, S., Bhahdigadi, S., Rao, P., Kiel, J. L., Andrews, C. J. (2007). Fluorescence assay based on aptamer-quantum dot binding to *Bacillus thuringiensis* spores. *J Fluoresc* 17, 193–199.

Ireson, C. R., Kelland, L. R. (2006). Discovery and development of anticancer aptamers. *Mol Cancer Ther* 5, 2957–2962.

James, W. (2007). Aptamers in the virologists' toolkit. *J Gen Virol* 88, 351–364.

Jayasena, S. D. (1999). Aptamers: an emerging class of molecules that rival antibodies in diagnostics. *Clin Chem* 45, 1628–1650.

Jenison, R., Gill, S., Polisky, B. (1995). Oligonucleotide ligands that discriminate between theophylline and caffeine. In *PCR Strategies*, M. A. Innis, D. H. Gelfand, J. J. Sninsky, eds., Academic Press, San Diego, CA, pp. 289–299.

Jensen, K. B., Atkison, M. C., Willis, M., Koch, T. H., Gold, L. (1995). Using in vitro selection to direct the covalent attachment of human immunodeficiency virus type 1 Rev protein to high-affinity RNA ligands. *Proc Natl Acad Sci U S A* 92, 12220–12224.

Jhaveri, S., Rajendran, M., Ellington, A. D. (2000). In vitro selection of signaling aptamers. *Nat Biotechnol* 18, 1293–1297.

Jiang, Y., Fang, X., Bai, C. (2004). Signaling aptamer/protein binding by a molecular light switch complex. *Anal Chem* 76, 5230–5235.

Katilius, E., Katiliene, Z., Woodbury, N. W. (2006). Signaling aptamers created using fluorescent nucleotide analogues. *Anal Chem* 78, 6484–6489.

Kato, Y., Minakawa, N., Komatsu, Y., Kamiya, H., Ogawa, N., Harashima, H., Matsuda, A. (2005). New NTP analogs: the synthesis of 4'-thioUTP and 4'-thioCTP and their utility for SELEX. *Nucleic Acids Res* 33, 2942–2951.

Kikuchi, K., Umehara, T., Fukuda, K., Kuno, A., Hasegawa, T., Nishikawa, S. (2005). A hepatitis C virus (HCV) internal ribosome entry site (IRES) domain III–IV-targeted aptamer inhibits translation by binding to an apical loop of domain IIId. *Nucleic Acids Res* 33, 683–692.

Klussmann, S., Nolte, A., Bald, R., Erdmann, V. A., Furste, J. P. (1996). Mirror-image RNA that binds *d*-adenosine. *Nat Biotechnol* 14, 1112–1115.

Koizumi, M. (2007). True antisense oligonucleotides with modified nucleotides restricted in the N-conformation. *Curr Top Med Chem* 7, 661–665.

Kolb, G., Reigadas, S., Castanotto, D., Faure, A., Ventura, M., Rossi, J. J., Toulmé, J-J. (2006). Endogenous expression of an anti-TAR aptamer reduces HIV-1 replication. *RNA Biol* 3, 150–156.

Kubik, M. F., Bell, C., Fitzwater, T., Watson, S. R., Tasset, D. M. (1997). Isolation and characterization of 2'-fluoro-, 2'-amino-, and 2'-fluoro-/amino-modified RNA ligands to human IFN-gamma that inhibit receptor binding. *J Immunol* 159, 259–267.

Kulkarni, O., Pawar, R. D., Purschke, W., Eulberg, D., Selve, N., Buchner, K., Ninichuk, V., Segerer, S., Vielhauer, V., Klussmann, S., Anders, H. J. (2007). Spiegelmer inhibition of CCL2/MCP-1 ameliorates lupus nephritis in MRL-(Fas)lpr mice. *J Am Soc Nephrol* 18, 2350–2358.

Kurreck, J. (2006). siRNA Efficiency: structure or sequence—that is the question. *J Biomed Biotechnol* doi:10.1155/JBB/2006/83757.

Latham, J. A., Johnson, R., Toole, J. J. (1994). The application of a modified nucleotide in aptamer selection: novel thrombin aptamers containing 5-(1-pentynyl)-2'-deoyuridine. *Nucl Acids Res* 22, 2817–2822.

Levy, M., Cater, S. F., Ellington, A. D. (2005). Quantum-dot aptamer beacons for the detection of proteins. *ChemBioChem* 6, 2163–2166.

Li, J. J., Fang, X., Tan, W. (2002). Molecular aptamer beacons for real-time protein recognition. *Biochem Biophys Res Commun* 292, 31–40.

Li, W., Wang, K., Tan, W., Ma, C., Yang, X. (2007). Aptamer-based analysis of angiogenin by fluorescence anisotropy. *Analyst* 132, 107–113.

Li, X., Liu, D. R. (2004). DNA-templated organic synthesis: nature's strategy for controlling chemical reactivity applied to synthetic molecules. *Angew Chem Int Ed Engl* 43, 4848–4870.

Li, Y., Lee, H. J., Corn, R. M. (2006). Fabrication and characterization of RNA aptamer microarrays for the study of protein–aptamer interactions with SPR imaging. *Nucleic Acids Res* 34, 6416–6424.

Li, Y., Lee, H. J., Corn, R. M. (2007). Detection of protein biomarkers using RNA aptamer microarrays and enzymatically amplified surface plasmon resonance imaging. *Anal Chem* 79, 1082–1088.

Liu, J., Lee, J. H., Lu, Y. (2007). Quantum dot encoding of aptamer-linked nanostructures for one-pot simultaneous detection of multiple analytes. *Anal Chem* 79, 4120–4125.

Mairal, T., Cengiz Ozalp, V., Lozano Sanchez, P., Mir, M., Katakis, I., O'Sullivan, C. K. (2007). Aptamers: molecular tools for analytical applications. *Anal Bioanal Chem.* 390, 989–1007.

McCauley, T. G., Hamaguchi, N., Stanton, M. (2003). Aptamer-based biosensor arrays for detection and quantification of biological macromolecules. *Anal Biochem* 319, 244–250.

Mendonsa, S. D., Bowser, M. T. (2004). In vitro selection of high-affinity DNA ligands for human IgE using capillary electrophoresis. *Anal Chem* 76, 5387–5392.

Michaud, M., Jourdan, E., Villet, A., Ravel, A., Grosset, C., Peyrin, E. (2003). A DNA aptamer as a new target-specific chiral selector for HPLC. *J Am Chem Soc* 125, 8672–8679.

Michaud, M., Jourdan, E., Ravelet, C., Villet, A., Ravel, A., Grosset, C., Peyrin, E. (2004). Immobilized DNA aptamers as target-specific chiral stationary phases for resolution of nucleoside and amino acid derivative enantiomers. *Anal Chem* 76, 1015–1020.

Miller, P. S., Barrett, J. C., Ts'o, P. O. P. (1974). Synthesis of oligodeoxyribonucleotide ethyl phosphotriesters and their specific complex formation with transfer ribonucleic acid. *Biochemistry* 13, 4887–4896.

Mills, D. R., Peterson, R. L., Spiegelman, S. (1967). An extracellular Darwinian experiment with a self-duplicating nucleic acid molecule. *Proc Natl Acad Sci U S A* 58, 217–224.

Misono, T. S., Kumar, P. K. (2005). Selection of RNA aptamers against human influenza virus hemagglutinin using surface plasmon resonance. *Anal Biochem* 342, 312–317.

Morse, D. P. (2007). Direct selection of RNA beacon aptamers. *Biochem Biophys Res Commun* 359, 94–101.

Ng, E. W., Shima, D. T., Calias, P., Cunningham, E. T., Jr., Guyer, D. R., Adamis, A. P. (2006). Pegaptanib, a targeted anti-VEGF aptamer for ocular vascular disease. *Nat Rev Drug Discov* 5, 123–132.

Nimjee, S. M., Rusconi, C. P., Sullenger, B. A. (2005). Aptamers: an emerging class of therapeutics. *Annu Rev Med* 56, 555–583.

Nutiu, R., Li, Y. (2002). Tripartite molecular beacons. *Nucleic Acids Res* 30, e94.

Nutiu, R., Li, Y. (2003). Structure-switching signaling aptamers. *J Am Chem Soc* 125, 4771–4778.

Nutiu, R., Li, Y. (2005a). Aptamers with fluorescence-signaling properties. *Methods* 37, 16–25.

Nutiu, R., Li, Y. (2005b). In vitro selection of structure-switching signaling aptamers. *Angew Chem Int Ed Engl* 44, 1061–1065.

Oktem, H. A., Bayramoglu, G., Ozalp, V. C., Arica, M. Y. (2007). Single-step purification of recombinant *Thermus aquaticus* DNA polymerase using DNA-aptamer immobilized novel affinity magnetic beads. *Biotechnol Prog* 23, 146–154.

Orr, R. M. (2001). Technology evaluation: fomivirsen, Isis Pharmaceuticals Inc/CIBA vision. *Curr Opin Mol Ther* 3, 288–294.

Orum, H., Wengel, J. (2001). Locked nucleic acids: a promising molecular family for gene-function analysis and antisense drug development. *Curr Opin Mol Ther* 3, 239–243.

Osborne, S. E., Ellington, A. E. (1997). Nucleic acid selection and the challenge of combinatorial chemistry. *Chem Rev* 97, 349–370.

Patel, D. J., Suri, A. K., Jiang, F., Jiang, L. C., Fan, P., Kumar, R. A., Nonin, S. (1997). Structure, recognition and adaptive binding in RNA aptamer complexes. *J Mol Biol* 272, 645–664.

Pestourie, C., Tavitian, B., Duconge, F. (2005). Aptamers against extracellular targets for in vivo applications. *Biochimie* 87, 921–930.

Pileur, F., Andréola, M. L., Dausse, E., Michel, J., Moreau, S., Yamada, H., Gaidamakov, S. A., Crouch, R. J., Toulmé, J-J., Cazenave, C. (2003). Selective inhibitory DNA aptamers of the human RNase H1. *Nucleic Acids Res* 31, 5776–5788.

Potyrailo, R. A., Conrad, R. C., Ellington, A. D., Hieftje, G. M. (1998). Adapting selected nucleic acid ligands (aptamers) to biosensors. *Anal Chem* 70, 3419–3425.

Prakash, T. P., Bhat, B. (2007). 2′-Modified oligonucleotides for antisense therapeutics. *Curr Top Med Chem* 7, 641–649.

Proske, D., Gilch, S., Wopfner, F., Schatzl, H. M., Winnacker, E. L., Famulok, M. (2002). Prion-protein-specific aptamer reduces PrPSc formation. *ChemBioChem* 3, 717–725.

Que-Gewirth, N. S., Sullenger, B. A. (2007). Gene therapy progress and prospects: RNA aptamers. *Gene Ther* 14, 283–291.

Rajendran, M., Ellington, A. D. (2003). In vitro selection of molecular beacons. *Nucleic Acids Res* 31, 5700–5713.

Ravelet, C., Grosset, C., Peyrin, E. (2006). Liquid chromatography, electrochromatography and capillary electrophoresis applications of DNA and RNA aptamers. *J Chromatogr A* 1117, 1–10.

Rhie, A., Kirby, L., Sayer, N., Wellesley, R., Disterer, P., Sylvester, I., Gill, A., Hope, J., James, W., Tahiri-Alaoui, A. (2003). Characterization of 2′-fluoro-RNA aptamers that bind preferentially to disease-associated conformations of prion protein and inhibit conversion. *J Biol Chem* 278, 39697–39705.

Roberts, R. W. (1999). Totally in vitro protein selection using mRNA-protein fusions and ribosome display. *Curr Opin Chem Biol* 3, 268–273.

Robertson, D. L., Joyce, G. F. (1990). Selection in vitro of an RNA enzyme that specifically cleaves single-stranded DNA. *Nature* 344, 467–468.

Ruckman, J., Green, L. S., Beeson, J., Waugh, S., Gillette, W. L., Henninger, D. D., Claesson-Welsh, L., Janjic, N. (1998). 2′-Fluoropyrimidine RNA-based aptamers to the 165-amino acid form of vascular endothelial growth factor (VEGF165). Inhibition of receptor binding and VEGF-induced vascular permeability through interactions requiring the exon 7-encoded domain. *J Biol Chem* 273, 20556–20567.

Rusconi, C. P., Scardino, E., Layzer, J., Pitoc, G. A., Ortel, T. L., Monroe, D., Sullenger, B. A. (2002). RNA aptamers as reversible antagonists of coagulation factor IXa. *Nature* 419, 90–94.

Rusconi, C. P., Roberts, J. D., Pitoc, G. A., Nimjee, S. M., White, R. R., Quick, G., Jr., Scardino, E., Fay, W. P., Sullenger, B. A. (2004). Antidote-mediated control of an anticoagulant aptamer in vivo. *Nat Biotechnol* 22, 1423–1428.

Sazani, P., Kole, R. (2003). Therapeutic potential of antisense oligonucleotides as modulators of alternative splicing. *J Clin Invest* 112, 481–486.

Scanlon, K. J. (2004). Anti-genes: siRNA, ribozymes and antisense. *Curr Pharm Biotechnol* 5, 415–420.

Schneider, D. J., Feigon, J., Hostomsky, Z., Gold, L. (1995). High-affinity ssDNA inhibitors of the reverse transcriptase of type 1 human immunodeficiency virus. *Biochemistry* 34, 9599–9610.

Shaw, B. R., Dobrikov, M., Wang, X., Wan, J., He, K., Lin, J. L., Li, P., Rait, V., Sergueeva, Z. A., Sergueev, D. (2003). Reading, writing, and modulating genetic information with boranophosphate mimics of nucleotides, DNA, and RNA. *Ann N Y Acad Sci* 1002, 12–29.

Shi, F., Hoekstra, D. (2004). Effective intracellular delivery of oligonucleotides in order to make sense of antisense. *J Control Release* 97, 189–209.

Shlyahovsky, B., Li, D., Weizmann, Y., Nowarski, R., Kotler, M., Willner, I. (2007). Spotlighting of cocaine by an autonomous aptamer-based machine. *J Am Chem Soc* 129, 3814–3815.

Soukup, G. A., Emilsson, G. A. M., Breaker, R. R. (2000). Altering molecular recognition of RNA aptamers by allosteric selection. *J Mol Biol* 298, 623–632.

Stojanovic, M. N., Kolpashchikov, D. M. (2004). Modular aptameric sensors. *J Am Chem Soc* 126, 9266–9270.

Stojanovic, M. N., de Prada, P., Landry, D. W. (2001). Aptamer-based folding fluorescent sensor for cocaine. *J Am Chem Soc* 123, 4928–4931.

Suess, B., Hanson, S., Berens, C., Fink, B., Schroeder, R., Hillen, W. (2003). Conditional gene expression by controlling translation with tetracycline-binding aptamers. *Nucleic Acids Res* 31, 1853–1858.

Tallet-Lopez, B., Aldaz-Carroll, L., Chabas, S., Dausse, E., Staedel, C., Toulmé, J-J. (2003). Antisense oligonucleotides targeted to the domain IIId of the hepatitis C virus IRES compete with 40S ribosomal subunit binding and prevent in vitro translation. *Nucleic Acids Res* 31, 734–742.

Tang, J., Breaker, R. R. (1998). Mechanism for allosteric inhibition of an ATP-sensitive ribozyme. *Nucleic Acids Res* 26, 4214–4221.

Tombelli, S., Minunni, M., Mascini, M. (2005). Analytical applications of aptamers. *Biosens Bioelectron* 20, 2424–2434.

Toulmé, J-J. (2000). Aptamers: selected oligonucleotides for therapy. *Curr Opin Mol Ther* 2, 318–324.

Toulmé, J-J. (2001). New candidates for true antisense. *Nat Biotechnol* 19, 17–18.

Toulmé, J-J., Di Primo, C., Moreau, S. (2001). Modulation of RNA function by oligonucleotides recognizing RNA structure. *Prog Nucleic Acid Res Mol Biol* 69, 1–46.

Toulmé, J-J., Darfeuille, F., Di Primo, C., Dausse, E. (2005). Aptamers to nucleic acid structures. In *The Aptamer Handbook*, S. Klussmann, ed., Wiley, Hoboken, NJ, pp. 185–221.

Tucker, B. J., Breaker, R. R. (2005). Riboswitches as versatile gene control elements. *Curr Opin Struct Biol* 15, 342–348.

Tuerk, C., Gold, L. (1990). Systematic evolution of ligands by exponential enrichment: RNA ligands to bacteriophage T4 DNA polymerase. *Science* 249, 505–510.

Tuerk, C., Macdougal, S., Gold, L. (1992). RNA pseudoknots that inhibit human immunodeficiency virus type-1 reverse transcriptase. *Proc Natl Acad Sci U S A* 89, 6988–6992.

Tyagi, S., Kramer, F. R. (1996). Molecular beacons: probes that fluoresce upon hybridization. *Nat Biotechnol* 14, 303–308.

Vater, A., Klussmann, S. (2003). Toward third-generation aptamers: spiegelmers and their therapeutic prospects. *Curr Opin Drug Discov Dev* 6, 253–261.

Vuyisich, M., Beal, P. A. (2002). Controlling protein activity with ligand-regulated RNA aptamers. *Chem Biol* 9, 907–913.

Wallis, M. G., Schroeder, R. (1997). The binding of antibiotics to RNA. *Prog Biophys Mol Biol* 67, 141–154.

Wang, S., White, K. A. (2007). Riboswitching on RNA virus replication. *Proc Natl Acad Sci U S A* 104, 10406–10411.

Werstuck, G., Green, M. R. (1998). Controlling gene expression in living cells through small molecule–RNA interactions. *Science* 282, 296–298.

White, R., Rusconi, C., Scardino, E., Wolberg, A., Lawson, J., Hoffman, M., Sullenger, B. (2001). Generation of species cross-reactive aptamers using "toggle" SELEX. *Mol Ther* 4, 567–573.

Willner, I., Zayats, M. (2007). Electronic aptamer-based sensors. *Angew Chem Int Ed Engl* 46(34), 6408–6418.

Wilson, C., Keefe, A. D. (2006). Building oligonucleotide therapeutics using non-natural chemistries. *Curr Opin Chem Biol* 10, 607–614.

Wilson, D. S., Szostak, J. W. (1999). In vitro selection of functional nucleic acids. *Annu Rev Biochem* 68, 611–647.

Win, M. N., Klein, J. S., Smolke, C. D. (2006). Codeine-binding RNA aptamers and rapid determination of their binding constants using a direct coupling surface plasmon resonance assay. *Nucleic Acids Res* 34, 5670–5682.

Wlotzka, B., Leva, S., Eschgfaller, B., Burmeister, J., Kleinjung, F., Kaduk, C., Muhn, P., Hess-Stumpp, H., Klussmann, S. (2002). In vivo properties of an anti-GnRH spiegelmer: an example of an oligonucleotide-based therapeutic substance class. *Proc Natl Acad Sci U S A* 99, 8898–8902.

Yamamoto, R., Kumar, P. K. R. (2000). Molecular beacon aptamer fluoresces in the presence of Tat protein of HIV-1. *Genes Cells* 5, 389–396.

Yamamoto, R., Katahira, M., Nishikawa, S., Baba, T., Taira, K., Kumar, P. K. R. (2000). A novel RNA motif that binds efficiently and specifically to the Tat protein of HIV and inhibits the trans-activation by Tat of transcription in vitro and in vivo. *Genes Cells* 5, 371–388.

Yamazaki, S., Tan, L., Mayer, G., Hartig, J. S., Song, J. N., Reuter, S., Restle, T., Laufer, S. D., Grohmann, D., Kräusslich, H. G., Bajorath, J., Famulok, M. (2007). Aptamer displacement identifies alternative small-molecule target sites that escape viral resistance. *Chem Biol* 14, 804–812.

Zamecnik, P. C., Stephenson, M. L. (1978). Inhibition of Rous sarcoma virus replication and cell transformation by a specific oligodeoxynucleotide. *Proc Natl Acad Sci U S A* 75, 280–284.

CHAPTER 2

SELEX AND ITS RECENT OPTIMIZATIONS

BEATE STREHLITZ and REGINA STOLTENBURG

2.1 INTRODUCTION

Nucleic acids are very attractive compounds for combinatorial chemistry. They are able to fold into defined secondary and tertiary structures, and they can easily be amplified by polymerase chain reaction (PCR) or in vitro transcription. Very complex libraries of random sequence oligonucleotides with about 10^{15} different molecules can be produced by chemical synthesis and screened in parallel for a particular functionality, such as high-affinity ligand binding (aptamers) or catalytic activity (ribozymes, DNAzymes). In 1990, Tuerk and Gold described a new in vitro selection and amplification method, named *SELEX* (systematic evolution of ligands by exponential enrichment). They used a combinatorial nucleic acid library to select RNA oligonucleotides that bind very tightly and selectively to a certain non–nucleic acid target (T4 DNA polymerase gp43). In the same year, Ellington and Szostak (1990) used a similar selection procedure to isolate RNA molecules from a random-sequence RNA library which can recognize and bind small organic dyes (Cibacron Blue, Reactive Blue 4). They named these selected individual RNA sequences *aptamers*, derived from the Latin word *aptus*, meaning "to fit." Two years later, the successful selection of single-stranded DNA (ssDNA) aptamers from a chemically synthesized pool of random-sequence DNA molecules could be shown (Ellington and Szostak, 1992).

The functionality of aptamers is based on their stable three-dimensional structure, which is dependent on the primary sequence, the length of the nucleic acid

Aptamers in Bioanalysis, Edited by Marco Mascini
Copyright © 2009 John Wiley & Sons, Inc.

molecule (smaller than 100 nt), and the environmental conditions. Typical structural motives are stems, internal loops, bulges, hairpin structures, tetra loops, pseudoknots, triplicates, kissing complexes, and G-quadruplex structures. In the presence of the target, the aptamers undergo adaptive conformational changes, and their three-dimensional folding creates a specific binding site for the target. The intermolecular interactions between aptamer and target are characterized by a combination of complementarity in shape, stacking interactions between aromatic compounds, and the nucleobases of the aptamers; electrostatic interactions between charged groups; and hydrogen bondings (Patel, 1997; Patel et al., 1997; Hermann and Patel, 2000).

The SELEX technology has been applied to various classes of targets. Inorganic and small organic molecules, peptides, proteins, carbohydrates, and antibiotics, as well as complex targets such as target mixtures or whole cells and organisms, were used for aptamer selection (Famulok, 1999; Wilson and Szostak, 1999; Göringer et al., 2003; Klussmann, 2006; Stoltenburg et al., 2007b). Aptamers can be developed for molecules connected with nucleic acids (nucleotides, cofactors) and for nucleic acid–binding proteins such as enzymes (polymerases) or regulatory proteins, but also for molecules naturally not associated with nucleic acids, such as growth factors (Jellinek et al., 1994; Green et al., 1996) or for organic dyes (Ellington and Szostak, 1990, 1992).

SELEX technology quickly became an important and widely used tool in molecular biological, pharmaceutical, and medical research, and was often modified to make the selection process more efficient and less time consuming, and to select aptamers with particular binding features (affinity, specificity) for different target molecules and for different applications.

2.2 APTAMERS AND THEIR SELECTION BY SELEX

The SELEX process is characterized by iterative cycles of in vitro selection and enzymatic amplification, which mimic a Darwinian type of process driving the selection toward relatively few, but optimized structural motifs as a solution to a given problem (e.g., ligand binding) (Göringer et al., 2003). The basic steps in a SELEX process are shown in Figure 2.1.

The starting point for a typical SELEX process is a chemically synthesized random DNA oligonucleotide library consisting of about 10^{13} to 10^{15} different sequence motifs (James, 2000). Each DNA strand in this library contains a central random region flanked by different specific sequences that function as primer binding sites in the PCR. The central random regions of such libraries typically consist of 20 to 80 nt and primer binding sites of 18 to 21 nt each. For the selection of DNA aptamers, this library can be used directly in the first round of a SELEX process. Sense and antisense primers derived from the specific sequences at the 5′ and 3′ ends enable amplification of the oligonucleotides selected in the SELEX rounds. For the selection of RNA aptamers, however, the DNA library has to be converted into an RNA library prior to starting the

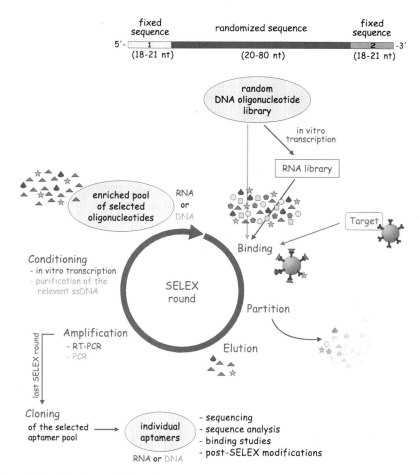

Figure 2.1   In vitro selection of target-specific aptamers using SELEX technology. The starting point of each SELEX process is a synthetic random DNA oligonucleotide library, which is used directly for the selection of DNA aptamers. For the selection of RNA aptamers, this library has first to be transferred into an RNA library. The SELEX procedure is characterized by the repetition of successive steps, consisting of selection (binding, partition, and elution), amplification, and conditioning. In the first SELEX round the library and target molecules are incubated for binding. Unbound oligonucleotides are removed by several stringent washing steps of the binding complexes. The target-bound oligonucleotides are eluted and subsequently amplified by PCR or RT-PCR. A new enriched pool of selected oligonucleotides is generated by preparation of the relevant ssDNA from the PCR products (DNA SELEX) or by in vitro transcription (RNA SELEX). This selected oligonucleotide pool is then used for the next selection round. In general, 6 to 20 SELEX rounds are needed for the selection of highly affine, target-specific aptamers. The last SELEX round is finished after the amplification step. The enriched aptamer pool is cloned and several individual aptamers have to be characterized. (See insert for color representation.)

RNA SELEX process. A special sense primer with an extension at the 5' end containing the T7 promoter sequence and an antisense primer are necessary to convert the ssDNA library into a double-stranded (dsDNA) library by PCR. The dsDNA is then transcribed in vitro by the T7 RNA polymerase, resulting in a randomized RNA library, which is used to start an RNA SELEX. In the first selection round the complex RNA or DNA pool is incubated with the target. Subsequently, the resulting binding complexes are partitioned from unbound and weakly bound oligonucleotides by stringent washing steps. The efficient removal of nonbinding oligonucleotides is one of the most crucial aspects of an aptamer selection process and strongly affects the binding features of the aptamers to be selected. There are a variety of methods (e.g., affinity chromatography, magnetic separation, filtration, or centrifugation), which are based on target immobilization to generate affinity matrices, selective capturing of the binding complexes, or size separation. Target-bound oligonucleotides are eluted and amplified by PCR (DNA SELEX) or RT-PCR (RNA SELEX). The resulting dsDNA has to be transformed into a new oligonucleotide pool by purifying the relevant ssDNA or by in vitro transcription and purifying the synthesized RNA. This new enriched pool of selected oligonucleotides is used again for a binding reaction with the target in the next SELEX round. By iterative cycles of selection and amplification, the initial random oligonucleotide pool is reduced to relatively few sequence motifs with the highest affinity and specificity for the target. In general, 6 to 20 SELEX rounds are required for aptamer selection. The number of rounds that are necessary depends on a variety of parameters, such as target features and concentration, design of the starting random DNA oligonucleotide library, selection conditions, ratio of target to oligonucleotides, or the efficiency of the partitioning method. Additional steps can be introduced into each round of the SELEX process, particularly with regard to the specificity of the oligonucleotides. Negative selection steps or subtraction steps are strongly recommended to minimize an enrichment of non-specifically binding oligonucleotides or to direct the selection to a specific epitope of the target. The affinity of the oligonucleotides to their target can be influenced by the stringency of the selection conditions. Typically, the stringency is increased progressively in the course of a SELEX process. This can be achieved by reducing the target concentration in later SELEX rounds or by changing the binding and washing conditions (i.e., buffer composition, volume, time) (Marshall and Ellington, 2000). The SELEX process is finished when an enrichment of target-specific oligonucleotides is detectable. The last SELEX round is stopped after the amplification, and the PCR products are cloned to get individual aptamer clones from the pool selected. Commonly, about 50 or more aptamer clones are analyzed by sequencing and sequence analysis. Sequence alignments are very useful to assess the complexity of the aptamer pool selected and to identify aptamers with homologous sequences. Based on the alignment data, the aptamer clones with mostly identical sequences, different in only a few single nucleotide positions, can be grouped. In some cases, special sequence patterns or highly conserved regions can be identified among the aptamer groups. These regions are often involved in the specific target binding of the aptamers. Secondary structure analyses of the aptamer sequences also provides information about binding relevant structures (Zuker, 2003). Representative aptamer clones are chosen and used in differently designed binding assays to characterize their binding features in more detail, including the affinities and specificities. Mutation and truncation experiments can be performed to narrow the minimal binding region within the aptamer sequence. The core binding domains of aptamers typically range in length from 15 to 40 nt. Finally, most of the aptamers selected are subjected to some post-SELEX modifications: for example, with a view to enhancing the stability of the aptamers (i.e., incorporation of modified nucleotides, capping), to use the aptamers in analytical detection assays, or for target purification (i.e., attachment of reporter groups, functional groups, or linker molecules).

2.3 MODIFICATIONS OF SELEX TECHNOLOGY

The SELEX process for the selection of target-specific aptamers is a universal process characterized by the repetition of five main steps: binding, partition, elution, amplification, and conditioning; however, for each aptamer selection, the protocol has to be adapted to the current requirements. The SELEX design and the specific selection conditions depend, for example, on the target, the oligonucleotide library, or the desired features and application of the aptamers to be selected. Therefore, a multitude of SELEX variants have been developed since the first description of this technology in 1990 (Table 2.1).

Important aspects for designing the oligonucleotide library are the size of the randomized region, the type of randomization, and chemical modifications of the DNA or RNA. Short libraries are more manageable, more cost-effective in chemical synthesis, and the selection of rather short aptamers is preferred for many applications. However, longer randomized regions give the libraries greater structural complexity. Besides natural nucleic acids, chemically modified oligonucleotide libraries were used in some SELEX experiments, with the goal of increasing the complexity of a library, to introduce new features such as functional groups, which could provide new possibilities for interaction with target molecules, to enhance the stability of oligonucleotide conformations, or to increase the resistance to nucleases, which is important for many applications of aptamers (Jayasena, 1999; Kopylov and Spiridonova, 2000; Kusser, 2000; Klussmann, 2006). Typical modifications concern the $2'$ position of the sugar ($2'$-NH$_2$, $2'$-fluro, $2'$-O-methyl) in RNA libraries. New functionalities can be incorporated in nucleic acids by modifications of the nucleobases commonly at the C-5 position of pyrimidines or at the C-8 position of purines (e.g., covalent SELEX) (Kopylov and Spiridonova, 2000). For example, 5-bromouracil and 5-iodouracil were used to generate photo-cross-linkable aptamers, which can be activated by ultraviolet irradiation to form a covalent linkage with the bound target (Photo-SELEX) (Golden et al., 2000). A common modification of the phosphate backbone of nucleic acids is replacement of the nonbinding oxygen

TABLE 2.1 Examples of Aptamer Selection Procedure Modifications

Designation	Description	Refs.
Negative SELEX	Minimizes the co-selection of unwanted nucleic acid ligands (e.g., for immobilization matrix) Preselection with molecules, which should not be recognized; removing of unwanted oligonucleotide structures from the pool	Geiger et al., 1996; Haller and Sarnow, 1997; Blank et al., 2001; Vater et al., 2003
Counter SELEX or Subtractive SELEX	Generates aptamers that are able to discriminate between closely related structures Introducing a selection step to the related target for elimination of aptamers from the oligonucleotide pool, which are not able to distinguish between related structures	Jenison et al., 1994; Geiger et al., 1996; Haller and Sarnow, 1997; Wang et al., 2003; White et al., 2003; Lee and Lee, 2006
Blended SELEX	To give additional properties to aptamers beyond the binding capability Enlarging nucleic acid molecules by special non–nucleic acid components	Smith et al., 1995; Radrizzani et al., 1999
Expressions cassette SELEX or SELEX-SAGE	Special forms of blended SELEX (with transcription factors) Optimizes aptamer activity for gene therapy applications	Martell et al., 2002; Roulet et al., 2002
Chimeric SELEX	Using two or more different libraries for production of chimerical aptamers with more than one desired feature or function Each parent library is selected for a distinct feature, then the aptamers selected are fused	Burke and Willis, 1998

Multistage SELEX	Special form of chimeric SELEX After fusion of preselected aptamer components, reselection to the entirety of the targets	Wu and Curran, 1999
Deconvolution SELEX	To generate aptamers for complex targets Discrimination between relevant aptamers (binding to distinct target structures within the complex mixture) and irrelevant oligonucleotides	Morris et al., 1998; Blank et al., 2001
Covalent SELEX or Cross-linking SELEX	Aptamers containing reactive groups that are capable of covalent linking to a target protein	Jensen et al., 1995; Kopylov and Spiridonova, 2000
Photo-SELEX	Aptamers bearing photoreactive groups Bind and photo-cross-link to a target and/or photo-activate a target molecule	Jensen et al., 1995; Brody et al., 1999; Johnson and Gershon, 1999; Golden et al., 2000
Spiegelmer technology	Selection with ordinary D-nucleic acids for the mirrored (enantiomer) target Synthesis of the resulting aptamers as L-isomers which now bind to the originally unmirrored target	Klussmann et al., 1996; Eulberg and Klussmann, 2003; Faulhammer et al., 2004
Tailored SELEX	Integrated method to identify aptamers with only 10 fixed nucleotides through ligation and removal of primer binding sites within the SELEX process Useful for selection of short aptamers and spiegelmers	Vater et al., 2003
Signaling aptamers or molecular beacons	Aptamers that report target binding by switching their structures and show a signal (e.g., fluorescent)	Jhaveri et al., 2000; Rajendran and Ellington, 2003

(continued)

TABLE 2.1 *(Continued)*

Designation	Description	Refs.
Genomic SELEX or cDNA-SELEX	Construction of a SELEX library of an organism's genome (e.g., cDNA fragments) Target proteins or metabolites from the same organism are used to identify meaningful interactions Allows the identification of protein targets directly from mRNA pools	Dobbelstein and Shenk, 1995; Gold et al., (1997a, 1997b); Singer et al., 1997; Zolotukhin et al., 2001; Wen and Gray, 2004; Shimada et al., 2005
Toggle SELEX	Switching ("toggling") between targets during alternating rounds of selection	Bianchini et al., 2001; White et al., 2001
Indirect selection	Selection target (e.g., metal ions) is not the actual binding partner of the aptamer, but binding is target dependent (occurs only in presence of the target)	Kawakami et al., 2000
In vivo selection	One possibility for selecting RNA-processing signals Uses transient transfection in an iterative procedure in cultured vertebrate cells	Coulter et al., 1997
Cell SELEX	Cell-based selection with counterselection strategy to collect DNA sequences that interact only with the target sells	Shangguan et al., 2006
Tissue SELEX	Method of generating aptamers capable of binding to complex tissue targets such as collections of cells in diseased tissues	Morris et al., 1998; Daniels et al., 2003
TECS-SELEX	Cell surface displayed recombinant or natural protein is used directly as the selection target	Ohuchi et al., 2006

Method	Description	Reference
FluMag-SELEX	DNA oligonucleotides with a fluorescein modification Target immobilization on magnetic beads	Stoltenburg et al., 2005
CE-SELEX	Use of capillary electrophoresis for separation	Mendonsa and Bowser, 2004; Mosing et al., 2005; Drabovich et al., 2006; Tang et al., 2006b
Non-SELEX	Process that involves repetitive steps of partitioning with no amplification between them Nonequilibrium capillary electrophoresis of equilibrium mixtures (NECEEM) used for partitioning	Berezovski et al., 2006
MonoLEX	One-step selection by affinity chromatography, physical segmentation of the affinity resin, single final PCR amplification step of bound aptamers	Nitsche et al., 2007
EMSA-SELEX	Use of electrophoretic mobility shift assay (EMSA) for partitioning in every round	Tsai and Reed, 1998
Use of nano-manipulator–atomic force microscope (nM-AFM)	Only one round of selection necessary Picking up and visualizing of singled aptamer–target complexes by nM-AFM Affinity measurement of single aptamer–target complexes possible by dynamic force spectroscopy Amplification by single-molecule PCR	Guthold et al., 2002
On-chip selection	Selection (in combination with a method for point mutations) and analyzing of DNA aptamers on chips	Asai et al., 2004
Yeast genetic selection	In vivo optimization of in vitro–preselected aptamers Library with degenerated aptamers Use of a yeast three (one)-hybrid system	Cassiday and Maher, 2003

Source: Stoltenburg et al. (2007a).

in the phosphodiester linkage by sulfur (Andreola et al., 2000). This produces a phosphorothioate linkage and thus increases the resistance of these oligonucleotides against nuclease digestion (phosphorothioate SELEX) (Jhaveri et al., 1998; King et al., 2002). Other modifications of oligonucleotide libraries serve for the quantification of selected oligonucleotides during the SELEX process. In general, radioactive-labeled nucleotides are incorporated or fluorescent molecules are attached to the 5' end of the oligonucleotides.

Modifications often concern specific steps in the SELEX process. Additional selection steps can be introduced into each SELEX round to minimize the co-selection of nonspecifically binding oligonucleotides, to discriminate between closely related target structures, to facilitate the selection of aptamers for a particular epitope of a target molecule, or to remove oligonucleotides, which bind to known but unwanted target structures (e.g., Negative SELEX, Counter SELEX, Deconvolution SELEX, Subtractive SELEX) (Geiger et al., 1996; Haller and Sarnow, 1997; Morris et al., 1998; Blank et al., 2001; Wang et al., 2003).

A crucial step in a SELEX process with outstanding relevance for the selection of aptamers with high affinity and specificity is efficient partitioning between target-binding and target-nonbinding oligonucleotides in each SELEX round. Commonly, affinity chromatography or magnetic separation technology (FluMag-SELEX) (Stoltenburg et al., 2005) is used to remove all nonbinding oligonucleotides efficiently during the SELEX process. For this purpose, the target molecules are immobilized on a matrix via functional groups, special tags, or linker molecules. Affinity chromatography uses column material such as Sepharose or agarose (Liu and Stormo, 2005; Tombelli et al., 2005), whereas magnetic separation technology makes use of magnetic beads. These are available, commercially providing different functionalized surfaces and thus allowing the immobilization of various target molecules. They enable easy handling and offer a potential for parallel manual and semiautomated processing of multiple targets (Lupold et al., 2002; Murphy et al., 2003; Kikuchi et al., 2003; Stoltenburg et al., 2005; Wochner et al., 2007). Other partition methods without target immobilization are nitrocellulose membrane filtration (Tuerk and Gold, 1990; Schneider et al., 1993; Bianchini et al., 2001) and centrifugation (Homann and Göringer, 1999; Rhie et al., 2003), which depend on the size of the target or binding complex. In recent years, special strategies for the identification and separation of aptamer–target complexes during the SELEX process have been described (Gopinath, 2007; Yang et al., 2007) [e.g., flow cytometry (Davis et al., 1996; Yang et al., 2003), capillary electrophoresis (CE-SELEX) (Mendonsa and Bowser, 2004; Mosing et al., 2005; Drabovich et al., 2006; Tang et al., 2006a), gel electrophoresis (EMSA-SELEX) (Tsai and Reed, 1998), or surface plasmon resonance (Misono and Kumar, 2005)]. Capillary electrophoresis as a homogeneous technique can facilitate efficient separation in free solution between target-bound and target-unbound oligonucleotides during aptamer selection. Berezovski et al. (2006) describe a new SELEX protocol that involves the typical repetitive selection rounds but with no amplification step for the target-binding oligonucleotides in each round. They used nonequilibrium capillary electrophoresis of equilibrium mixtures (NECEEM) for partitioning and named this approach Non-SELEX. Only a few steps in NECEEM-based partitioning in the Non-SELEX method were sufficient to improve the affinity of an oligonucleotide pool to a target protein, and thus the aptamer selection time could be reduced remarkably. The MonoLEX method, described by Nitsche et al., comprises a one-step selection by affinity chromatography, followed by subsequent physical segmentation of the affinity resin and a single final PCR amplification step of bound aptamers (Nitsche et al., 2007).

In recent years, a growing number of SELEX experiments have been described using complex target structures or heterogeneous target mixtures for aptamer selections (Cell SELEX or Whole-cell SELEX), (Daniels et al., 2003; Cerchia et al., 2005; Shangguan et al., 2006; Tang et al., 2007b). Cell-based SELEX processes are very useful approaches to discovering specific molecular markers associated with human diseases such as cancer and simultaneously to identify ligands (aptamers) specific for these markers (Lee and Lee, 2006). In these cases, proper knowledge of the nature of markers is not necessary, which is a huge advantage. In general, two or more cell types are used for negative, subtractive, and positive selection steps to receive aptamers able to distinguish between normal cells and pathological cells, due to alterations in the cell surface structures (e.g., of the protein patterns). These aptamers could be used in diagnosis, imaging, and therapy of cancer. Additionally, cell-based SELEX processes can be applied to directly target, for example, transmembrane proteins in their natural environment (Cerchia et al., 2005) or recombinant-expressed proteins displayed at cell surfaces (TECS-SELEX), (Ohuchi et al., 2006).

2.4 ADVANTAGES AND LIMITATIONS OF APTAMERS AND THEIR SELECTION TECHNOLOGY

Since the first presentation of the SELEX technology, the selection of aptamers and their applications, have been described in hundreds of publications, which reflect the great interest in this research field and its enormous potential for pharmacy and medicine as well as environmental analytics. One of the advantages of aptamers is their ability to bind very tightly and specifically to a wide variety of targets. The affinities are often comparable to those observed for antibodies. Most of the K_d values calculated are in the low-nanomolar-to-picomolar range, depending on the measuring principle. Additionally, aptamers are selected by an in vitro process independent of animals or cell lines. Thus, the SELEX process is also applicable under nonphysiological conditions. Selection of aptamers for toxic target molecules or for molecules with no or only low immunogenicity is possible. As mentioned above, various modifications can be introduced into the basic SELEX process to direct the selection to desired aptamer features or to intended applications of the aptamers. Variations of the selection conditions (e.g., buffer composition, temperature, binding time), the design of the randomized oligonucleotide library, or additional selection steps can strongly affect the specific binding features (affinity, selectivity) of aptamers. However, often-observed problems of aptamer selections are the enrichment of nonspecifically binding oligonucleotides or the repeated selection of already known aptamer sequences in different SELEX processes depending on target composition and SELEX conditions. Therefore, the selection conditions have to be adapted accurately to current requirements, and negative selection or subtractive steps are strongly recommended.

The SELEX technology is applicable to a wide variety of targets. Besides defined single targets, complex target structures or mixtures without proper knowledge of their composition are suitable for successful aptamer selection. In reality, however, not any target molecule is suitable for an aptamer selection. Target features that facilitate successful aptamer selection are, for example, positively charged groups (e.g., primary amino groups), the presence of hydrogen-bond donors and acceptors, and planarity (aromatic compounds) (Wilson and Szostak, 1999; Rimmele, 2003). Defined single target molecules should be available in sufficient amount and with high purity for the SELEX performance. Aptamer selection is more difficult for targets with largely hydrophobic character and for negatively charged molecules (e.g., containing phosphate groups). These target requirements are based on the fundamental principles of intermolecular interactions in an aptamer–target complex.

Once selected, aptamers can be produced by chemical synthesis in large quantity and with high accuracy and reproducibility. Denatured aptamers can be regenerated easily within minutes, which is important for many applications. Different postselection modifications of aptamers are possible. Unmodified nucleic acids, especially RNA, are relatively unstable in biological fluids; thus, the use of aptamers as drugs is limited. Different strategies were developed to overcome this problem (e.g., chemical modifications at the nucleotide level or terminal capping can significantly improve the stability of aptamers) (Jayasena, 1999; Andreola et al., 2000; Dougan et al., 2000; Kopylov and Spiridonova, 2000; Kusser, 2000; Marro et al., 2005; Klussmann, 2006). A successful method to select nuclease-resistant aptamers is the spiegelmer technology (Klussmann et al., 1996; Williams et al., 1997; Wlotzka et al., 2002; Vater and Klussmann, 2003). As a result of this method, mirror-image oligonucleotides (RNA or DNA) can be identified that bind a natural target (amino acids or peptides). Other modifications (e.g., attached functional groups, reporter molecules, polyethylene glycol, cholesterol, or lipid tags) serve for the quantification and immobilization of aptamers, or are useful in improving the bioavailability of potential therapeutic aptamers (e.g., by increasing the pharmacokinetic half-life). Further optimizations of the selected aptamers are possible with regard to their affinities and specificities (e.g., truncation, reselection) (Pan et al., 1995; Bittker et al., 2002; Hwang and Lee, 2002; Held et al., 2003).

Aptamers can be delivered into cells or expressed within cells (intramers) by transcription of expression cassettes. Such strategies permit investigation of the function and interplay of proteins in the context of living cells or organisms (Famulok et al., 2001). Aptamers could be used for intracellular detection of target molecules or could function as intracellular inhibitors (Burgstaller et al., 2002). Their small size relative to antibodies enables aptamers to access protein epitopes that might otherwise be blocked or hidden (Lee et al., 2006). Low-molecular-mass aptamers are important tools for imaging analysis or cytometry, because they can reach targets inside cells (Ulrich et al., 2004). An important aspect of aptamers with regard to medical applications is their low to nonexistent immunogenicity (Nimjee et al., 2005b). Using aptamers as therapeutic drugs, matched antidotes can easily be produced, which are able to control drug activity selectively (Rimmele, 2003; Rusconi et al., 2004).

Despite an ever-increasing number of publications about aptamers and the optimization of their selection, SELEX is still a relatively slow and complicated process. No standardized SELEX protocol is applicable for any type of target. The selection conditions have to be adapted to the current circumstances (target, desired features of the aptamers, applications). The entire process to get high-affine and specific aptamers (iterative SELEX rounds, cloning of the aptamer pool, characterization) is very time consuming. The majority of the published aptamers were selected manually. Depending on the SELEX conditions, a limited manual parallelization of the process is possible. Furthermore, automated SELEX protocols were described and established as a robotic workstation (Cox and Ellington, 2001; Cox et al., 2002; Eulberg et al., 2005; Nitsche et al., 2007) or microfabricated chip-based SELEX platform (Hybarger et al., 2006) to make the aptamer selection process more accessible and more rapid.

2.5 APPLICATIONS OF APTAMERS BEING DEVELOPED FOR THE MARKET

Fields of applications of aptamers include mainly clinical diagnostics and therapy, the investigation of protein functions, and use as molecular recognition elements in analytical systems or as specific binders in enrichment and separation systems. Comprehensive overviews about aptamers with therapeutic applications have been given by Ulrich and Nimjee (Nimjee et al., 2005b; Ulrich et al., 2006). The first approved aptamer with therapeutic function is the anti-human VEGF (vascular endothelial growth factor) aptamer (Tucker et al., 1999). The PEGylated form of this aptamer was called pegaptanib and used as the medicinal active component of the newly developed drug for the treatment of wet age-related macular degeneration. The pharmaceutical product Macugen (pegaptanib sodium injection) from Pfizer Inc./OSI Pharmaceuticals was approved in December 2004 in the United States and in January 2006 in Europe (Maberley, 2005; Chapman and Beckey, 2006). Pegaptanib also shows medicinal activity for the therapy of patients with diabetic macular edema. For this application it is now being tested in phase 2 studies (Ng and Adamis, 2006).

The first in-human experience with an RNA aptamer and its complementary oligonucleotide antidote used as an anticoagulant system have been described (Dyke et al., 2006). This anticoagulation system (REG1; Regado Biosciences) comprises a protein-binding oligonucleotide to factor IXa (RB006) and its complementary oligonucleotide antidote RB007. It was evaluated successfully in healthy volunteers and hence should be a novel platform for selective and actively reversible pharmacotherapeutics in patients with thrombotic disorders (Dyke et al., 2006).

Beyond this, the development of aptamers for therapeutics and biomedicine is pushed on a broad front (Yan et al., 2005). Some companies are very active in this field. For example, Archemix Inc. (http://www.archemix.com) and SomaLogic Inc. (http://www.somalogic.com/) in the United States, Noxxon Pharma AG (http://www.noxxon.net) in Germany, and Isis Innovation Ltd. (http://www.isis-innovation.com) in the UK have aptamer-based therapeutics under product development or in clinical studies. In addition, research on aptamers for applications in medicine is far advanced for tumor imaging and therapy (White et al., 2003; Ferreira et al., 2006; Herr et al., 2006; Hicke et al., 2006), influenza virus detection and inhibition (Jeon et al., 2004; Gopinath et al., 2006), and clinical diagnostics in general (Guthrie et al., 2006).

The intracellular penetration of nucleic acid–based drugs in some cases needs to be enhanced. This is the reason why adequate nonviral delivery systems were designed either to ensure cellular penetration, protection against degradation, or to allow long-term delivery (Bloquel et al., 2006; Fattal and Bochot, 2006; Cheng et al., 2007). Using nonviral strategies, oligonucleotides can be delivered naked (less efficient) or entrapped in cationic lipids, polymers, or peptides, forming slow-release delivery systems, which can be adapted according to the organ targeted and the purposes of the therapy (Bloquel et al., 2006).

Aptamers are applied to an increasing degree as biomolecular recognition elements in test kits, assays, or biosensor formats well known from bioanalytical applications. As an example, a disposable electrochemical strip sensor for allergy detection uses an immunoglobulin E (IgE) aptamer, based on the former published aptamer D17.4 (Wiegand et al., 1996; Papamichael et al., 2007) in its optimized version (D17.4ext) (Liss et al., 2002). Biotinylized D17.4ext is used in a competition assay by a sensor with electrochemical detection and by ELONA with optical detection. The ELONA based on this aptamer has better stability and a smaller nonspecific binding effect than those of a comparable ELISA based on a mAb test (Papamichael et al., 2007). The same aptamer (D17.4ext) is used in an aptamer-modified carbon nanotube field-effect transistor (CNT-FET) (Maehashi et al., 2007). Comparison with a mAb-modified CNT-FET for IgE detection provides better results for the aptamer-based assay under similar conditions (Maehashi et al., 2007).

A further electrochemical sensor platform development uses methylene blue connected to one end of the aptamer, which is fixed on the electrode surface by the other end as a redox marker for the detection of aptamer-binding reaction (Baker et al., 2006). The measuring principle is comparable to that of aptamer beacons with optical detection: In the absence of target, the aptamer is thought to remain partially unfolded with a greater distance between the redox marker and the electrode surface; in the presence of target, the aptamer presumably folds into the binding junction with a decrease in the distance between the marker and the electrode surface, altering the electron transfer and increasing the reduction peak observed for methylene blue. Using this principle, aptamer sensors for the detection of cocaine (Baker et al., 2006), thrombin (Xiao et al., 2005a,b), and platelet-derived growth factor (PDGF) (Lai et al., 2007) have been described.

A further example is a simple *dipstick assay*, based on a test-paper-like platform with lateral-flow technology (Liu et al., 2006). This assay format is used mostly with antibody receptors (e.g., as in a well-known commercially available pregnancy test kit). The new aptamer dipstick was developed using the adenosine and cocaine aptamers described earlier (Huizenga and Szostak, 1995; Stojanovic et al., 2000). The cocaine test was also applied to human blood serum samples (Liu et al., 2006).

Efficient immobilization of aptamers on surfaces is necessary for the construction of tough, stable sensors and assay systems as one necessary step to overcome limitations for practical applications (Bini et al., 2007). Bini et al. (2007) have compared thrombin aptamers immobilized on a gold surface by chemisorption (thiolated aptamer) and by biotin–streptavidin interaction (biotinylated aptamer carrying a linker) on a gold surface modified by a thiol–dextran–streptavidin layer. The linker-modified aptamer immobilized via streptavidin–biotin showed better reproducibility and sensitivity results for the quartz crystal sensor. Aptamers can be used for the functionalization of titanium-alloy surfaces (e.g., implant material, scaffolds) to enhance cell adhesion. The aptamers directed to osteoblasts are fixed electrochemically on the surface of the alloy and promote cell adhesion (Guo et al., 2005, 2007).

The use of aptamers for the separation, purification, and quantification of analytes in chromatography, electrochromatography, and capillary electrophoresis techniques in general is described in detail and with examples by Ravelet et al. (2006). DNA and RNA aptamers have been used for the separation and purification of proteins and the separation of small molecules and enantiomers. In capillary electrophoresis, aptamers are used primarily for the separation of species and the characterization of affinity interactions. Ravelet et al. (2006) state the huge potential of these molecular tools in the separation science field. However, for big separation units, a large amount of aptamer is necessary, making it more expensive than other separation materials. Therefore, the use of aptamers for separation units is limited primarily to miniaturized systems.

A very sophisticated application of aptamers aims at the development of logic gates (AND/OR) (Yoshida and Yokobayashi, 2007) and related smart materials, where aptamer targets are used as the chemical stimuli for controllable cooperativeness of aptamer-coated nanomaterials (Liu and Lu, 2006). Some examples of aptamer applications are compiled in Table 2.2.

2.6   FUTURE PERSPECTIVES

The most widespread application of aptamers is in the medical sector. Macugen (Pfizer Inc./OSI Pharmaceuticals), used in the therapy of wet age-related macula

TABLE 2.2 Examples of the Use of Aptamers

Target	Aptamer/Assay	Field of Application	Research or Product State	Refs.
VEGF	RNA, chemically modified	Therapy, wet age-related macula degeneration	Product: Macugen (Pfizer Inc.; OSI Pharmaceuticals)	Maberley, 2005; Chapman and Beckey, 2006; Que-Gewirth and Sullenger, 2007
Thrombin	DNA, thrombin inhibitor ARC-138	Therapy, anticoagulant	Phase 1 studies in August 2004, not in further development, Archemix Corp.	Nimjee et al., 2005a; Que-Gewirth and Sullenger, 2007
	DNA, thrombin inhibitor ARC-2172	Therapy, anticoagulant	Preclinical	Que-Gewirth and Sullenger, 2007
Anti-FIXa aptamer–antidote pair	RNA, factor IXa inhibitor and its antidote, REG1	Therapy, anticoagulant	Product development, phase 1 studies completed in 2006, Regado Biosciences Inc.; Archemix Corp.	http://www.archemix.com; Nimjee et al., 2005a; Dyke et al., 2006; Que-Gewirth and Sullenger, 2007
Nucleolin	AS1411	Therapy, cancer	Phase 2	http://www.archemix.com; Ireson and Kelland, 2006; Que-Gewirth and Sullenger, 2007
PDGF-B	E-10030	Therapy, treatment of neovascular age-related macular degeneration	Phase 1 in 2007	http://www.archemix.com

Target	Aptamer/type	Application	Current status	Reference
Tat protein of HIV	RNA^Tat	Medicine, diagnostic tool	Current status not known	Yamamoto et al., 2000
HIV-1 surface glycoprotein (gp120)	Not specified	Identification of anti-HIV drugs with anti-infectivity characteristics	Product	http://www.isis-innovation.com
Vasopressin	L-RNA, NOX-F37, spiegelmer	Therapy, treatment of diseases that are associated with body fluid overload, congestive heart failure	Product development	http://www.noxxon.net; Purschke et al., 2006
Ghrelin	NOX-B11, spiegelmer	Therapy, treatment of obesity	Product development, animal model, licensed to Pfizer	http://www.noxxon.net
Amylin	NOX-A42, spiegelmer	Therapy, type 2 diabetes	Product development	http://www.noxxon.net
Angiogenin	AL6 (DNA)/fluorescence anisotropy	Medicine, diagnosis and prognosis of diseases, detect. limit: 1 nM	Research	Li et al., 2007
Transcription factor HMGA1b	NOX-A50, spiegelmer	Therapy, inhibits tumor growth	Product development, tumor cell line, animal model	http://www.noxxon.net
Influenza virus hemagglutinin globular region (HA-(91–261) peptide)	DNA	Medicine, inhibition of viral infection by blocking the receptor-binding region of the surface glycoprotein HA	Research (animal model)	Jeon et al., 2004

(continued)

TABLE 2.2 (*Continued*)

Target	Aptamer/Assay	Field of Application	Research or Product State	Refs.
Human influenza virus A/Panama/2007/ 1999(H3N2)	RNA	Medicine, genotyping of viruses, differentiation between closely related strains of influenza viruses	Research	Gopinath et al., 2006
Tumor cells, tenascin-C	RNA (TTA1), fluorescence/radioactive labeled	Medicine, tumor imaging, therapy	Research (animal model)	Hicke et al., 2006
MUC1-peptide (tumor marker for breast, stomach, lung, prostate, colorectal, and other cancers)	DNA	Medicine, tumor imaging, therapy	Research (cell line)	Ferreira et al., 2006
Angiopoietin-2	RNA (11-1.41)	Medicine, antiangiogenic therapeutic agent	Research	White et al., 2003
Leukemia cells (CCRF-CEM)	Aptamer–nanoparticles for collection/detection	Medical diagnostics	Research	Herr et al., 2006
Osteoblasts	Aptamer O-7	Medicine, immobilization on implant material (Ti alloy), accelerates the attachment of seeded cells	Research ("proof of concept")	Guo et al., 2005; Guo et al., 2007
Cytokines	Aptamer-based assays, overview	Medical diagnostic	Research	Guthrie et al., 2006
IgE	D17.4ext/ELONA, screen-printed electrode	Medicine, allergy detection, detect. limit: 23 ng/mL	Research	Papamichael et al., 2007

IgE	D17.4ext/CNT-FET	Medicine, allergy detection, detect. limit: 250 pM	Research	Maehashi et al., 2007
L-2-Phenoxypropionic acid	RNA, produced by Integrated DNA Technologies (Coralville, Iowa), L-RNA chiral stationary phase	Capillary electrochromatography, enantiomer separation of acid herbicides	Research	Andre et al., 2006
Ricin	RNA-aptamer	Capillary electrophoresis with laser-induced fluorescence detection, detect. limit: 500 pM, assay time < 10 min	Research	Hesselberth et al., 2000; Kirby et al., 2004; Haes et al., 2006
Abrin toxin	DNA/molecular luminescence switch $[Ru(phen)_2(dppz)]^{2+}$	Medicine/safety, detect. limit: 1 nM, lin. range: 1 to 400 nM	Research	Tang et al., 2007a
Adenosine (substrate of adenosine deaminase)	DNA signaling aptamer	Bioanalysis, solid-phase enzyme activity assay	Research ("proof of concept")	Rupcich et al., 2006

degeneration (see Section 2.5), has been on the market in the United States for more than three years and in Europe for more than two years. Further therapeutics based on aptamers are in clinical studies: for example, anticoagulants and cancer treatment drugs (see Table 2.1).

Beyond the therapeutic market, aptamers are established in the development of innovative analytical systems and assays (e.g., for medical diagnostics, but also for environmental analysis). The advantages of aptamers over antibodies, such as the chemical synthesis of aptamers, their easy chemical modification without loss of function, and their easier immobilization, should bring about the replacement of antibody assays by aptamer-based assays. A drawback for further applications of aptamers is the limited number of aptamers available. Meanwhile, comprehensive proof of principles of assay formats and sensors are described, based mostly on only a few individual aptamers (e.g., the thrombin aptamer in different biosensor configurations). By use of the newly developed easy-to-use and automated SELEX procedures, work is in progress for the development of additional aptamers for different targets. This is also necessary for a better understanding of the optimum aptamer performance in assays and detection systems, which will help to close the gap between theory and practice (Ito and Fukusaki, 2004). The fascinating world of aptamers should be carried forward out of the research laboratories into the analytical industry (Mukhopadhyay, 2005).

Acknowledgments

We thank Christine Reinemann for valuable assistance and Nadia Nikolaus for helpful discussions and critical reading of the manuscript.

REFERENCES

Andre, C., Berthelot, A., Thomassin, M., Guillaume, Y. C. (2006). Enantioselective aptameric molecular recognition material: design of a novel chiral stationary phase for enantioseparation of a series of chiral herbicides by capillary electrochromatography. *Electrophoresis* 27, 3254–3262.

Andreola, M. L., Calmels, C., Michel, J., Toulmé, J-J., Litvak, S. (2000). Towards the selection of phosphorothioate aptamers: optimizing in vitro selection steps with phosphorothioate nucleotides. *Eur J Biochem* 267, 5032–5040.

Asai, R., Nishimura, S. I., Aita, T., Takahashi, K. (2004). In vitro selection of DNA aptamers on chips using a method for generating point mutations. *Anal Lett* 37, 645–656.

Baker, B. R., Lai, R. Y., Wood, M. S., Doctor, E. H., Heeger, A. J., Plaxco, K. W. (2006). An electronic, aptamer-based small-molecule sensor for the rapid, label-free detection of cocaine in adulterated samples and biological fluids. *J Am Chem Soc* 128, 3138–3139.

Berezovski, M., Musheev, M., Drabovich, A., Krylov, S. N. (2006). Non-SELEX selection of aptamers. *J Am Chem Soc* 128, 1410–1411.

Bianchini, M., Radrizzani, M., Brocardo, M. G., Reyes, G. B., Gonzalez, S. C., Santa-Coloma, T. A. (2001). Specific oligobodies against ERK-2 that recognize both the native and the denatured state of the protein. *J Immunol Methods* 252, 191–197.

Bini, A., Minunni, M., Tombelli, S., Centi, S., Mascini, M. (2007). Analytical performances of aptamer-based sensing for thrombin detection. *Anal Chem* 79, 3016–3019.

Bittker, J. A., Le, B. V., Liu, D. R. (2002). Nucleic acid evolution and minimization by nonhomologous random recombination. *Nat Biotechnol* 20, 1024–1029.

Blank, M., Weinschenk, T., Priemer, M., Schluesener, H. (2001). Systematic evolution of a DNA aptamer binding to rat brain tumor microvessels: selective targeting of endothelial regulatory protein pigpen. *J Biol Chem* 276, 16464–16468.

Bloquel, C., Bourges, J. L., Touchard, E., Berdugo, M., BenEzra, D., Behar-Cohen, F. (2006). Non-viral ocular gene therapy: potential ocular therapeutic avenues. *Adv Drug Deliv Rev* 58, 1224–1242.

Brody, E. N., Willis, M. C., Smith, J. D., Jayasena, S., Zichi, D., Gold, L. (1999). The use of aptamers in large arrays for molecular diagnostics. *Mol Diagn* 4, 381–388.

Burgstaller, P., Girod, A., Blind, M. (2002). Aptamers as tools for target prioritization and lead identification. *Drug Discov Today* 7, 1221–1228.

Burke, D. H., Willis, J. H. (1998). Recombination, RNA evolution, and bifunctional RNA molecules isolated through chimeric SELEX. *RNA* 4, 1165–1175.

Cassiday, L. A., Maher, L. J. (2003). Yeast genetic selections to optimize RNA decoys for transcription factor NF-kappa B. *Proc Natl Acad Sci U S A* 100, 3930–3935.

Cerchia, L., Ducongé, F., Pestourie, C., Boulay, J., Aissouni, Y., Gombert, K., Tavitian, B., de Franciscis, V., Libri, D. (2005). Neutralizing aptamers from whole-cell SELEX inhibit the RET receptor tyrosine kinase. *PLOS Biology* 3, 697–704.

Chapman, J. A., Beckey, C. (2006). Pegaptanib: a novel approach to ocular neovascularization. *Ann Pharmacother* 40, 1322–1326.

Cheng, J., Teply, B. A., Sherifi, I., Sung, J., Luther, G., Gu, F. X., Levy-Nissenbaum, E., Radovic-Moreno, A. F., Langer, R., Farokhzad, O. C. (2007). Formulation of functionalized PLGA-PEG nanoparticles for in vivo targeted drug delivery. *Biomaterials* 28, 869–876.

Coulter, L. R., Landree, M. A., Cooper, T. A. (1997). Identification of a new class of exonic splicing enhancers by in vivo selection. *Mol Cell Biol* 17, 2143–2150.

Cox, J. C., Ellington, A. D. (2001). Automated selection of anti-protein aptamers. *Bioorg Med Chem* 9, 2525–2531.

Cox, J. C., Rajendran, M., Riedel, T., Davidson, E. A., Sooter, L. J., Bayer, T. S., Schmitz-Brown, M., Ellington, A. D. (2002). Automated acquisition of aptamer sequences. *Comb Chem High Throughput Screen* 5, 289–299.

Daniels, D. A., Chen, H., Hicke, B. J., Swiderek, K. M., Gold, L. (2003). A tenascin-C aptamer identified by tumor cell SELEX: systematic evolution of ligands by exponential enrichment. *Proc Natl Acad Sci U S A* 100, 15416–15421.

Davis, K. A., Abrams, B., Lin, Y., Jayasena, S. D. (1996). Use of a high affinity DNA ligand in flow cytometry. *Nucleic Acids Res* 24, 702–706.

Dobbelstein, M., Shenk, T. (1995). In-vitro selection of RNA ligands for the ribosomal L22 protein associated with Epstein–Barr virus-expressed RNA by using randomized and CDNA-derived RNA libraries. *J Virol* 69, 8027–8034.

Dougan, H., Lyster, D. M., Vo, C. V., Stafford, A., Weitz, J. I., Hobbs, J. B. (2000). Extending the lifetime of anticoagulant oligodeoxynucleotide aptamers in blood. *Nucl Med Biol* 27, 289–297.

Drabovich, A. P., Berezovski, M., Okhonin, V., Krylov, S. N. (2006). Selection of smart aptamers by methods of kinetic capillary electrophoresis. *Anal Chem* 78, 3171–3178.

Dyke, C. K., Steinhubl, S. R., Kleiman, N. S., Cannon, R. O., Aberle, L. G., Lin, M., Myles, S. K., Melloni, C., Harrington, R. A., Alexander, J. H., Becker, R. C., Rusconi, C. P. (2006). First-in-human experience of an antidote-controlled anticoagulant using RNA aptamer technology: a phase 1a pharmacodynamic evaluation of a drug–antidote pair for the controlled regulation of factor IXa activity. *Circulation* 114, 2490–2497.

Ellington, A. D., Szostak, J. W. (1990). In vitro selection of RNA molecules that bind specific ligands. *Nature* 346, 818–822.

Ellington, A. D., Szostak, J. W. (1992). Selection in vitro of single-stranded DNA molecules that fold into specific ligand-binding structures. *Nature* 355, 850–852.

Eulberg, D., Klussmann, S. (2003). Spiegelmers: biostable aptamers. *ChemBioChem* 4, 979–983.

Eulberg, D., Buchner, K., Maasch, C., Klussmann, S. (2005). Development of an automated in vitro selection protocol to obtain RNA-based aptamers: identification of a biostable substance P antagonist. *Nucleic Acids Res* 33, e45.

Famulok, M., (1999). Oligonucleotide aptamers that recognize small molecules. *Curr Opin Struct Biol* 9, 324–329.

Famulok, M., Blind, M., Mayer, G. (2001). Intramers as promising new tools in functional proteomics. *Chem Biol* 8, 931–939.

Fattal, E., Bochot, A. (2006). Ocular delivery of nucleic acids: antisense oligonucleotides, aptamers and siRNA. *Adv Drug Deliv Rev* 58, 1203–1223.

Faulhammer, D., Eschgfaller, B., Stark, S., Burgstaller, P., Englberger, W., Erfurth, J., Kleinjung, F., Rupp, J., Vulcu, S. D., Schroder, W., Vonhoff, S., Nawrath, H., Gillen, C., Klussmann, S. (2004). Biostable aptamers with antagonistic properties to the neuropeptide nociceptin/orphanin FQ. *RNA-A* 10, 516–527.

Ferreira, C. S. M., Matthews, C. S., Missailidis, S. (2006). DNA aptamers that bind to MUC1 tumour marker: design and characterization of MUC1-binding single-stranded DNA aptamers. *Tumor Biol* 27, 289–301.

Geiger, A., Burgstaller, P., von der Eltz, H., Roeder, A., Famulok, M. (1996). RNA aptamers that bind L-arginine with sub-micromolar dissociation constants and high enantioselectivity. *Nucleic Acids Res* 24, 1029–1036.

Gold, L., Brown, D., He, Y. Y., Shtatland, T., Singer, B. S., Wu, Y. (1997a). From oligonucleotide shapes to genomic SELEX: novel biological regulatory loops. *Proc Natl Acad Sci U S A* 94, 59–64.

Gold, L., Singer, B., He, Y. Y., Brody, E. (1997b). SELEX and the evolution of genomes. *Curr Opin Genet Dev* 7, 848–851.

Golden, M. C., Collins, B. D., Willis, M. C., Koch, T. H. (2000). Diagnostic potential of photoSELEX-evolved ssDNA aptamers. *J Biotechnol* 81, 167–178.

Gopinath, S. C., (2007). Methods developed for SELEX. *Anal Bioanal Chem* 387, 171–182.

Gopinath, S. C. B., Misono, T. S., Kawasaki, K., Mizuno, T., Imai, M., Odagiri, T., Kumar, P. K. R. (2006). An RNA aptamer that distinguishes between closely related human influenza viruses and inhibits haemagglutinin-mediated membrane fusion. *J Gen Virol* 87, 479–487.

Göringer, H. U., Homann, M., Lorger, M. (2003). In vitro selection of high-affinity nucleic acid ligands to parasite target molecules. *Int J Parasitol* 33, 1309–1317.

Green, L. S., Jellinek, D., Jenison, R., Ostman, A., Heldin, C. H., Janjic, N. (1996). Inhibitory DNA ligands to platelet-derived growth factor B-chain. *Biochemistry* 35, 14413–14424.

Guo, K., Wendel, H. P., Scheideler, L., Ziemer, G., Scheule, A. M. (2005). Aptamer-based capture molecules as a novel coating strategy to promote cell adhesion. *J Cell Mol Med* 9, 731–736.

Guo, K. T., Scharnweber, D., Schwenzer, B., Ziemer, G., Wendel, H. P. (2007). The effect of electrochemical functionalization of Ti-alloy surfaces by aptamer-based capture molecules on cell adhesion. *Biomaterials* 28, 468–474.

Guthold, M., Cubicciotti, R., Superfine, R., Taylor, R. M. (2002). Novel methodology to detect, isolate, amplify and characterize single aptamer molecules with desirable target-binding properties. *Biophys J*. Supplement, Abstract, 163C.

Guthrie, J. W., Hamula, C. L. A., Zhang, H. Q., Le, X. C. (2006). Assays for cytokines using aptamers. *Methods* 38, 324–330.

Haes, A. J., Giordano, B. C., Collins, G. E. (2006). Aptamer-based detection and quantitative analysis of ricin using affinity probe capillary electrophoresis. *Anal Chem* 78, 3758–3764.

Haller, A. A., Sarnow, P. (1997). In vitro selection of a 7-methyl-guanosine binding RNA that inhibits translation of capped mRNA molecules. *Proc Natl Acad Sci U S A* 94, 8521–8526.

Held, D. M., Greathouse, S. T., Agrawal, A., Burke, D. H. (2003). Evolutionary landscapes for the acquisition of new ligand recognition by RNA aptamers. *J Mol Evol* 57, 299–308.

Hermann, T., Patel, D. J. (2000). Adaptive recognition by nucleic acid aptamers. *Science* 287, 820–825.

Herr, J. K., Smith, J. E., Medley, C. D., Shangguan, D. H., Tan, W. H. (2006). Aptamer-conjugated nanoparticles for selective collection and detection of cancer cells. *Anal Chem* 78, 2918–2924.

Hesselberth, J. R., Miller, D., Robertus, J., Ellington, A. D. (2000). In vitro selection of RNA molecules that inhibit the activity of ricin A-chain. *J Biol Chem* 275, 4937–4942.

Hicke, B. J., Stephens, A. W., Gould, T., Chang, Y. F., Lynott, C. K., Heil, J., Borkowski, S., Hilger, C. S., Cook, G., Warren, S., Schmidt, P.G. (2006). Tumor targeting by an aptamer. *J Nucl Med* 47, 668–678.

Homann, M., Göringer, H. U. (1999). Combinatorial selection of high affinity RNA ligands to live African trypanosomes. *Nucleic Acids Res* 27, 2006–2014.

Huizenga, D. E., Szostak, J. W. (1995). A DNA aptamer that binds adenosine and ATP. *Biochemistry* 34, 656–665.

http://www.archemix.com.

http://www.isis-innovation.com.

http://www.noxxon.net.

http://www.somalogic.com/.

Hwang, B., Lee, S. W. (2002). Improvement of RNA aptamer activity against myasthenic autoantibodies by extended sequence selection. *Biochem Biophys Res Commun* 290, 656–662.

Hybarger, G., Bynum, J., Williams, R. F., Valdes, J. J., Chambers, J. P. (2006). A microfluidic SELEX prototype. *Anal Bioanal Chem* 384, 191–198.

Ireson, C. R., Kelland, L. R. (2006). Discovery and development of anticancer aptamers. *Mol Cancer Ther* 5, 2957–2962.

Ito, Y., Fukusaki, E. (2004). DNA as a 'nanomaterial'. *J Mol Catal B* 28, 155–166.

James, W. (2000). Aptamers. In *Encyclopedia of Analytical Chemistry*, R. A. Meyers, ed., Wiley, Chichester, UK, pp. 4848–4871.

Jayasena, S. D. (1999). Aptamers: an emerging class of molecules that rival antibodies in diagnostics. *Clin Chem* 45, 1628–1650.

Jellinek, D., Green, L. S., Bell, C., Janjic, N. (1994). Inhibition of receptor binding by high-affinity RNA ligands to vascular endothelial growth factor. *Biochemistry* 33, 10450–10456.

Jenison, R. D., Gill, S. C., Pardi, A., Polisky, B. (1994). High-resolution molecular discrimination by RNA. *Science* 263, 1425–1429.

Jensen, K. B., Atkinson, B. L., Willis, M. C., Koch, T. H., Gold, L. (1995). Using in vitro selection to direct the covalent attachment of human immunodeficiency virus type 1 Rev protein to high-affinity RNA ligands. *Proc Natl Acad Sci U S A* 92, 12220–12224.

Jeon, S. H., Kayhan, B., Ben-Yedidia, T., Arnon, R. (2004). A DNA aptamer prevents influenza infection by blocking the receptor binding region of the viral hemagglutinin. *J Biol Chem* 279, 48410–48419.

Jhaveri, S., Olwin, B., Ellington, A. D. (1998). In vitro selection of phosphorothiolated aptamers. *Bioorg Med Chem Lett* 8, 2285–2290.

Jhaveri, S., Rajendran, M., Ellington, A. D. (2000). In vitro selection of signaling aptamers. *Nat Biotechnol* 18, 1293–1297.

Johnson, L., Gershon, P. D. (1999). RNA binding characteristics and overall topology of the vaccinia poly(A) polymerase-processivity factor-primer complex. *Nucleic Acids Res* 27, 2708–2721.

Kawakami, J., Imanaka, H., Yokota, Y., Sugimoto, N. (2000). In vitro selection of aptamers that act with Zn2+. *J Inorg Biochem* 82, 197–206.

Kikuchi, K., Umehara, T., Fukuda, K., Hwang, J., Kuno, A., Hasegawa, T., Nishikawa, S. (2003). RNA aptamers targeted to domain II of hepatitis C virus IRES that bind to its apical loop region. *J Biochem* 133, 263–270.

King, D. J., Bassett, S. E., Li, X., Fennewald, S. A., Herzog, N. K., Luxon, B. A., Shope, R., Gorenstein, D. G. (2002). Combinatorial selection and binding of phosphorothioate aptamers targeting human NF-kappa B RelA(p65) and p50. *Biochemistry* 41, 9696–9706.

Kirby, R., Cho, E. J., Gehrke, B., Bayer, T., Park, Y. S., Neikirk, D. P., McDevitt, J. T., Ellington, A. D. (2004). Aptamer-based sensor arrays for the detection and quantitation of proteins. *Anal Chem* 76, 4066–4075.

Klussmann, S. (2006). *The Aptamer Handbook: Functional Oligonucleotides and Their Applications*, Wiley-VCH, Weinheim, Germany.

Klussmann, S., Nolte, A., Bald, R., Erdmann, V. A., Furste, J. P. (1996). Mirror-image RNA that binds D-adenosine. *Nat Biotechnol* 14, 1112–1115.

Kopylov, A. M., Spiridonova, V. A. (2000). Combinatorial chemistry of nucleic acids: SELEX. *Mol Biol* 34, 940–954.

Kusser, W. (2000). Chemically modified nucleic acid aptamers for in vitro selections: evolving evolution. *J Biotechnol* 74, 27–38.

Lai, R. Y., Plaxco, K. W., Heeger, A. J. (2007). Aptamer-based electrochemical detection of picomolar platelet-derived growth factor directly in blood serum. *Anal Chem* 79, 229–233.

Lee, J. F., Stovall, G. M., Ellington, A. D. (2006). Aptamer therapeutics advance. *Curr Opin Chem Biol* 10, 282–289.

Lee, Y. J., Lee, S. W. (2006). In vitro selection of cancer-specific RNA aptamers. *J Microbiol Biotechnol* 16, 1149–1153.

Li, W., Wang, K. M., Tan, W. H., Ma, C. B., Yang, X. H. (2007). Aptamer-based analysis of angiogenin by fluorescence anisotropy. *Analyst* 132, 107–113.

Liss, M., Petersen, B., Wolf, H., Prohaska, E. (2002). An aptamer-based quartz crystal protein biosensor. *Anal Chem* 74, 4488–4495.

Liu, J. J., Stormo, G. D. (2005). Combining SELEX with quantitative assays to rapidly obtain accurate models of protein–DNA interactions. *Nucleic Acids Res* 33, e141.

Liu, J. W., Lu, Y. (2006). Smart nanomaterials responsive to multiple chemical stimuli with controllable cooperativity. *Adv Mater* 18, 1667–1671.

Liu, J. W., Mazumdar, D., Lu, Y. (2006). A simple and sensitive "dipstick" test in serum based on lateral flow separation of aptamer-linked nanostructures. *Angew Chem Int Ed Engl* 45, 7955–7959.

Lupold, S. E., Hicke, B. J., Lin, Y., Coffey, D. S. (2002). Identification and characterization of nuclease-stabilized RNA molecules that bind human prostate cancer cells via the prostate-specific membrane antigen. *Cancer Res* 62, 4029–4033.

Maberley, D. (2005). Pegaptanib for neovascular age-related macular degeneration. *Issues Emerg Health Technol* 1–4.

Maehashi, K., Katsura, T., Kerman, K., Takamura, Y., Matsumoto, K., Tamiya, E. (2007). Label-free protein biosensor based on aptamer-modified carbon nanotube field-effect transistors. *Anal Chem* 79, 782–787.

Marro, M. L., Daniels, D. A., McNamee, A., Andrew, D. P., Chapman, T. D., Jiang, M. S., Wu, Z. N., Smith, J. L., Patel, K. K., Gearing, K. L. (2005). Identification of potent and selective RNA antagonists of the IFN-gamma-inducible CXCL10 chemokine. *Biochemistry* 44, 8449–8460.

Marshall, K. A., Ellington, A. D. (2000). In vitro selection of RNA aptamers. *Methods Enzymol* 318, 193–214.

Martell, R. E., Nevins, J. R., Sullenger, B. A. (2002). Optimizing aptamer activity for gene therapy applications using expression cassette SELEX. *Mol Ther* 6, 30–34.

Mendonsa, S. D., Bowser, M. T. (2004). In vitro selection of high-affinity DNA ligands for human IgE using capillary electrophoresis. *Anal Chem* 76, 5387–5392.

Misono, T. S., Kumar, P. K. R. (2005). Selection of RNA aptamers against human influenza virus hemagglutinin using surface plasmon resonance. *Anal Biochem* 342, 312–317.

Morris, K. N., Jensen, K. B., Julin, C. M., Weil, M., Gold, L. (1998). High affinity ligands from in vitro selection: complex targets. *Proc Natl Acad Sci U S A* 95, 2902–2907.

Mosing, R. K., Mendonsa, S. D., Bowser, M. T. (2005). Capillary electrophoresis-SELEX selection of aptamers with affinity for HIV-1 reverse transcriptase. *Anal Chem* 77, 6107–6112.

Mukhopadhyay, R. (2005). Aptamers are ready for the spotlight. *Anal Chem* 77, 114A–118A.

Murphy, M. B., Fuller, S. T., Richardson, P. M., Doyle, S. A. (2003). An improved method for the in vitro evolution of aptamers and applications in protein detection and purification. *Nucleic Acids Res* 31, e110.

Ng, E. W. M., Adamis, A. P. (2006). Anti-VEGF aptamer (pegaptanib) therapy for ocular vascular diseases. *Oligonucleotide Ther* 1082, 151–171.

Nimjee, S. M., Rusconi, C. P., Harrington, R. A., Sullenger, B. A. (2005a). The potential of aptamers as anticoagulants. *Trends Cardiovasc Med* 15, 41–45.

Nimjee, S. M., Rusconi, C. P., Sullenger, B. A. (2005b). Aptamers: an emerging class of therapeutics. *Annu Rev Med* 56, 555–583.

Nitsche, A., Kurth, A., Dunkhorst, A., Panke, O., Sielaff, H., Junge, W., Muth, D., Scheller, F., Stocklein, W., Dahmen, C., Pauli, G., Kage, A. (2007). One-step selection of vaccinia virus binding DNA-aptamers by MonoLEX. *BMC Biotechnol* 7, doi:10.1186/1472-6750-7-48.

Ohuchi, S. P., Ohtsu, T., Nakamura, Y. (2006). Selection of RNA aptamers against recombinant transforming growth factor-beta type III receptor displayed on cell surface. *Biochimie* 88, 897–904.

Pan, W., Craven, R. C., Qiu, Q., Wilson, C. B., Wills, J. W., Golovine, S., Wang, J. F. (1995). Isolation of virus-neutralizing RNAs from a large pool of random sequences. *Proc Natl Acad Sci U S A* 92, 11509–11513.

Papamichael, K. I., Kreuzer, M. P., Guilbault, G. G. (2007). Viability of allergy (IgE) detection using an alternative aptamer receptor and electrochemical means. *Sens Actuat B Chem* 121, 178–186.

Patel, D. J. (1997). Structural analysis of nucleic acid aptamers. *Curr Opin Chem Biol* 1, 32–46.

Patel, D. J., Suri, A. K., Jiang, F., Jiang, L. C., Fan, P., Kumar, R. A., Nonin, S. (1997). Structure, recognition and adaptive binding in RNA aptamer complexes. *J Mol Biol* 272, 645–664.

Purschke, W. G., Eulberg, D., Buchner, K., Vonhoff, S., Klussmann, S. (2006). An L-RNA-based aquaretic agent that inhibits vasopressin in vivo. *Proc Natl Acad Sci U S A* 103, 5173–5178.

Que-Gewirth, N. S., Sullenger, B. A. (2007). Gene therapy progress and prospects: RNA aptamers. *Gene Ther* 14, 283–291.

Radrizzani, M., Broccardo, M., Solveyra, C. G., Bianchini, M., Reyes, G. B., Cafferata, E. G., Santa-Coloma, T. A. (1999). Oligobodies: bench made synthetic antibodies. *Medicina (B Aires)* 59, 753–758.

Rajendran, M., Ellington, A. D. (2003). In vitro selection of molecular beacons. *Nucleic Acids Res* 31, 5700–5713.

Ravelet, C., Grosset, C., Peyrin, E. (2006). Liquid chromatography, electrochromatography and capillary electrophoresis applications of DNA and RNA aptamers. *J Chromatogr A* 1117, 1–10.

Rhie, A., Kirby, L., Sayer, N., Wellesley, R., Disterer, P., Sylvester, I., Gill, A., Hope, J., James, W., Tahiri-Alaoui, A. (2003). Characterization of 2′-fluoro-RNA aptamers that bind preferentially to disease-associated conformations of prion protein and inhibit conversion. *J Biol Chem* 278, 39697–39705.

Rimmele, M., (2003). Nucleic acid aptamers as tools and drugs: recent developments. *ChemBioChem* 4, 963–971.

Roulet, E., Busso, S., Camargo, A. A., Simpson, A. J. G., Mermod, N., Bucher, P. (2002). High-throughput SELEX-SAGE method for quantitative modeling of transcription-factor binding sites. *Nat Biotechnol* 20, 831–835.

Rupcich, N., Nutiu, R., Li, Y. F., Brennan, J. D. (2006). Solid-phase enzyme activity assay utilizing an entrapped fluorescence-signaling DNA aptamer. *Angew Chem Int Ed Engl* 45, 3295–3299.

Rusconi, C. P., Roberts, J. D., Pitoc, G. A., Nimjee, S. M., White, R. R., Quick, G., Scardino, E., Fay, W. P., Sullenger, B. A. (2004). Antidote-mediated control of an anticoagulant aptamer in vivo. *Nat Biotechnol* 22, 1423–1428.

Schneider, D., Gold, L., Platt, T. (1993). Selective enrichment of RNA Species for tight-binding to *Escherichia coli* Rho-factor. *FASEB J* 7, 201–207.

Shangguan, D., Li, Y., Tang, Z. W., Cao, Z. H. C., Chen, H. W., Mallikaratchy, P., Sefah, K., Yang, C. Y. J., Tan, W. H. (2006). Aptamers evolved from live cells as effective molecular probes for cancer study. *Proc Natl Acad Sci U S A* 103, 11838–11843.

Shimada, T., Fujita, N., Maeda, M., Ishihama, A. (2005). Systematic search for the Cra-binding promoters using genomic SELEX system. *Genes Cells* 10, 907–918.

Singer, B. S., Shtatland, T., Brown, D., Gold, L. (1997). Libraries for genomic SELEX. *Nucleic Acids Res* 25, 781–786.

Smith, D., Kirschenheuter, G. P., Charlton, J., Guidot, D. M., Repine, J. E. (1995). In-vitro selection of RNA-based irreversible inhibitors of human neutrophil elastase. *Chem Biol* 2, 741–750.

Stojanovic, M. N., de Prada, P., Landry, D. W. (2000). Fluorescent sensors based on aptamer self-assembly. *J Am Chem Soc* 122, 11547–11548.

Stoltenburg, R., Reinemann, C., Strehlitz, B. (2005). FluMag-SELEX as an advantageous method for DNA aptamer selection. *Anal Bioanal Chem* 383, 83–91.

Stoltenburg, R., Reinemann, C., Strehlitz, B. (2007a). SELEX-A (r)evolutionary method to generate high-affinity nucleic acid ligands. *Biomol Eng* 24, 381–403.

Stoltenburg, R., Reinemann, C., Strehlitz, B. (2007b). Development of aptamers as new receptors for environmental biosensors. In *Environmental Biotechnology*, C. S. K. Mishra and A. A. Juwarkar, eds., APH Publishing, New Delhi, India, pp. 83–112.

Tang, J. J., Xie, J. W., Shao, N. S., Guo, L., Yan, Y. (2006a). Capillary electrophoresis as a tool for screening aptamer with high affinity and high specificity to ricin. *Chem J Chin Univ Chin* 27, 1840–1843.

Tang, J. J., Xie, J. W., Shao, N. S., Yan, Y. (2006b). The DNA aptamers that specifically recognize ricin toxin are selected by two in vitro selection methods. *Electrophoresis* 27, 1303–1311.

Tang, J., Yu, T., Guo, L., Xie, J., Shao, N., He, Z. (2007a). In vitro selection of DNA aptamer against abrin toxin and aptamer-based abrin direct detection. *Biosens Bioelectron* 22, 2456–2463.

Tang, Z., Shangguan, D., Wang, K., Shi, H., Sefah, K., Mallikratchy, P., Chen, H. W., Li, Y., Tan, W. (2007b). Selection of aptamers for molecular recognition and characterization of cancer cells. *Anal Chem* 79(13), 4900–4907.

Tombelli, S., Minunni, A., Luzi, E., Mascini, M. (2005). Aptamer-based biosensors for the detection of HIV-1 Tat protein. *Bioelectrochemistry* 67, 135–141.

Tsai, R. Y. L., Reed, R. R. (1998). Identification of DNA recognition sequences and protein interaction domains of the multiple-Zn-finger protein Roaz. *Mol Cell Biol* 18, 6447–6456.

Tucker, C. E., Chen, L. S., Judkins, M. B., Farmer, J. A., Gill, S. C., Drolet, D. W. (1999). Detection and plasma pharmacokinetics of an anti-vascular endothelial growth factor oligonucleotide-aptamer (NX1838) in rhesus monkeys. *J Chromatogr B* 732, 203–212.

Tuerk, C., Gold, L. (1990). Systematic evolution of ligands by exponential enrichment: RNA ligands to bacteriophage T4 DNA polymerase. *Science* 249, 505–510.

Ulrich, H., Martins, A. H. B., Pesquero, J. B. (2004). RNA and DNA aptamers in cytomics analysis. *Cytometry* 59A, 220–231.

Ulrich, H., Trujillo, C. A., Nery, A. A., Alves, J. M., Majumder, P., Resende, R. R., Martins, A. H. (2006). DNA and RNA aptamers: from tools for basic research towards therapeutic applications. *Comb Chem High Throughput Screen* 9, 619–632.

Vater, A., Klussmann, S. (2003). Toward third-generation aptamers: spiegelmers and their therapeutic prospects. *Curr Opin Drug Discov Dev* 6, 253–261.

Vater, A., Jarosch, F., Buchner, K., Klussmann, S. (2003). Short bioactive spiegelmers to migraine-associated calcitonin gene-related peptide rapidly identified by a novel approach: Tailored-SELEX. *Nucleic Acids Res* 31, e130.

Wang, C., Zhang, M., Yang, G., Zhang, D., Ding, H., Wang, H., Fan, M., Shen, B., Shao, N. (2003). Single-stranded DNA aptamers that bind differentiated but not parental cells: subtractive systematic evolution of ligands by exponential enrichment. *J Biotechnol* 102, 15–22.

Wen, J. D., Gray, D. M. (2004). Selection of genomic sequences that bind tightly to Ff gene 5 protein: primer-free genomic SELEX. *Nucleic Acids Res* 32, e182.

White, R., Rusconi, C., Scardino, E., Wolberg, A., Lawson, J., Hoffman, M., Sullenger, B. (2001). Generation of species cross-reactive aptamers using "toggle" SELEX. *Mol Ther* 4, 567–574.

White, R. R., Shan, S., Rusconi, C. P., Shetty, G., Dewhirst, M. W., Kontos, C. D., Sullenger, B. A. (2003). Inhibition of rat corneal angiogenesis by a nuclease-resistant RNA aptamer specific for angiopoietin-2. *Proc Natl Acad Sci U S A* 100, 5028–5033.

Wiegand, T. W., Williams, P. B., Dreskin, S. C., Jouvin, M. H., Kinet, J. P., Tasset, D. (1996). High-affinity oligonucleotide ligands to human IgE inhibit binding to Fc epsilon receptor I. *J Immunol* 157, 221–230.

Williams, K. P., Liu, X. H., Schumacher, T. N., Lin, H. Y., Ausiello, D. A., Kim, P. S., Bartel, D. P. (1997). Bioactive and nuclease-resistant L-DNA ligand of vasopressin. *Proc Natl Acad Sci U S A* 94, 11285–11290.

Wilson, D. S., Szostak, J. W. (1999). In vitro selection of functional nucleic acids. *Annu Rev Biochem* 68, 611–647.

Wlotzka, B., Leva, S., Eschgfaller, B., Burmeister, J., Kleinjung, F., Kaduk, C., Muhn, P., Hess-Stump, H., Klussmann, S. (2002). In vivo properties of an anti-GnRH spiegelmer:

an example of an oligonucleotide-based therapeutic substance class. *Proc Natl Acad Sci U S A* 99, 8898–8902.

Wochner, A., Cech, B., Menger, M., Erdmann, V. A., Glökler, J. (2007). Semi-automated selection of DNA aptamers using magnetic particle handling. *Biotechniques* 43, 344–353.

Wu, L. H., Curran, J. F. (1999). An allosteric synthetic DNA. *Nucleic Acids Res* 27, 1512–1516.

Xiao, Y., Lubin, A. A., Heeger, A. J., Plaxco, K. W. (2005a). Label-free electronic detection of thrombin in blood serum by using an aptamer-based sensor. *Angew Chem Int Ed Engl* 44, 5456–5459.

Xiao, Y., Piorek, B. D., Plaxco, K. W., Heeger, A. J. (2005b). A reagentless signal-on architecture for electronic, aptamer-based sensors via target-induced strand displacement. *J Am Chem Soc* 127, 17990–17991.

Yamamoto, R., Katahira, M., Nishikawa, S., Baba, T., Taira, K., Kumar, P. K. R. (2000). A novel RNA motif that binds efficiently and specifically to the Tat protein of HIV and inhibits the trans-activation by Tat of transcription in vitro and in vivo. *Genes Cells* 5, 371–388.

Yan, A. C., Bell, K. M., Breeden, M. M., Ellington, A. D. (2005). Aptamers: prospects in therapeutics and biomedicine. *Front Biosci* 10, 1802–1827.

Yang, X. B., Li, X., Prow, T. W., Reece, L. M., Bassett, S. E., Luxon, B. A., Herzog, N. K., Aronson, J., Shope, R. E., Leary, J. F., Gorenstein, D. G. (2003). Immunofluorescence assay and flow-cytometry selection of bead-bound aptamers. *Nucleic Acids Res* 31, e54.

Yang, Y., Yang, D., Schluesener, H. J., Zhang, Z. (2007). Advances in SELEX and application of aptamers in the central nervous system. *Biomol Eng.*

Yoshida, W., Yokobayashi, Y. (2007). Photonic boolean logic gates based on DNA aptamers. *Chem Commun* 2, 195–197.

Zolotukhin, A. S., Michalowski, D., Smulevitch, S., Felber, B. K. (2001). Retroviral constitutive transport element evolved from cellular TAP(NXF1)-binding sequences. *J Virol* 75, 5567–5575.

Zuker, M. (2003). Mfold web server for nucleic acid folding and hybridization prediction. *Nucleic Acids Res* 31, 3406–3415.

PART II

BIOSENSORS

CHAPTER 3

ELECTROCHEMICAL APTASENSORS

ITAMAR WILLNER and MAYA ZAYATS

3.1 INTRODUCTION

Aptamers are nucleic acids (DNA or RNA) that bind selectively to low-molecular-weight organic or inorganic substrates or to macromolecules such as proteins (Jayasena, 1999; Luzi et al., 2003; You et al., 2003). The affinity constant of aptamers toward their substrates is in the micromolar to nanomolar range, comparable to the binding constants of antibodies to antigens (Jenison et al., 1994). The interest in aptamers as specific binding agents originates from their relative ease of preparation by an evolutionary selection procedure that eliminates the need for structural design of the receptor sites. Selection of the aptamers for the specific target is based on the SELEX (systematic evolution of ligands by exponential enrichment) procedure (Ellington and Szostak, 1990; Tuerk and Gold, 1990; Wilson and Szostak, 1999; Famulok et al., 2000), that is outlined schematically in Figure 3.1.

The process is initiated with a random library of linear oligonucleotides (usually, 10^{15} to 10^{16}) consisting of linear nucleic acids comprising a random sequence embraced by a $5'$ and a $3'$ nucleic acid sequence of defined composition. An RNA-searched aptamer involves the primary transcription of the DNA library into an RNA pool followed by passing the library through a separating matrix that includes the target substrate. The few nucleic acids that reveal affinity toward the substrate (or some nonspecific nucleic acid adsorbents) bind to the separation matrix, while most of the library components are washed off. The elution of surface-bound nucleic acids followed by their polymerase chain reaction (PCR) amplification yields a mixture of nucleic acids

Aptamers in Bioanalysis, Edited by Marco Mascini
Copyright © 2009 John Wiley & Sons, Inc.

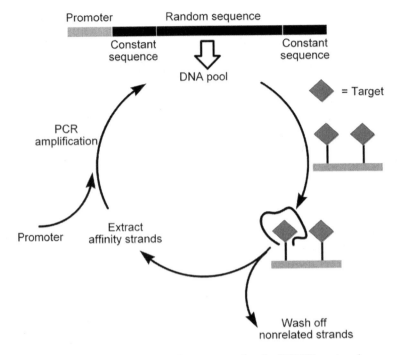

Figure 3.1   Preparation of an aptamer by the SELEX protocol.

of variable affinities toward the target. Repeated separation of the mixture on the ligand-modified surface results in enrichment of the mixture with nucleic acids that reveal a high affinity for the substrate. Usually, 8 to 15 repeated separation and amplification cycles yield aptamers exhibiting high affinities toward the target substrate. The selection of DNA-based aptamers proceeds by an identical selection mechanism that eliminates the primary transcription step. Single-stranded DNA/RNA aptamers associate to their targets by generating three-dimensional structures that involve the formation of loops and hybridized regions using complementary Watson–Crick base pairing (Ellington and Szostak, 1992). Elucidation of the aptamer composition by sequencing, and determination of the tertiary aptamer–substrate structure by nuclear magnetic resonance (Feigon et al., 1996; Patel, 1997), then lead to the selection of a high-affinity binding material from a random oligonucleotide mixture library. The PCR amplification then provides a limitless synthetic source for the aptamers.

Aptamers find growing interest as active separation materials in chromatography (Clark and Remcho, 2002; Deng et al., 2003; Michaud et al., 2003) and electrophoresis (Clark and Remcho, 2003; Mendonsa and Bowser, 2004), as therapeutic and diagnostic agents (Brody and Gold, 2000; White et al., 2002), and as active materials for biosensing (O'Sullivan, 2002; Tombelli et al., 2005a; Navani and Li, 2006). The use of aptamers for biosensing is particularly attractive, since aptamers could substitute antibodies in bioanalytical sensing and reveal advantages over immunosensors:

1. The in vitro selection of aptamers eliminates the need for the in vivo immunization of animals that is required to elicit antibodies. This enables the fabrication of aptamer-binding ligands for toxic materials (e.g., toxins) that it is impossible to obtain by the immune system.

2. In contrast to antibodies that undergo mutation and structural perturbation during the eliciting process, the chemical synthesis of aptamers usually leads to highly reproducible structures of the binding ligands.

3. The chemical modification of aptamers with optical or redox active labels, or with functional groups that enable their tethering to transducers, is usually easier than the manipulation of antibodies.

4. The direct modification of aptamers with imaging labels allows direct readout of the aptamer–substrate complex, without the need for a complex antigen–antibody "sandwich" assay. This is particularly important for low-molecular-weight substrates, where immunosensors are subjected only to competitive assays.

5. Nonspecific adsorption phenomena are usually less pronounced on nucleic acid interfaces than on protein interfaces. Also, the thermal stabilities of nucleic acids compared to antibodies add attractive advantages for using aptamers as the active sensing material: aptasensors.

A fair comparison should also mention, however, the disadvantages of aptamers: DNA and especially RNA are very sensitive to hydrolytic digestion by nucleases, thus requiring highly pure environments for their applications. Several solutions to transform aptamers into nuclease-resistant moieties by the modification of a ribose ring at the $2'$-position (Pieken et al., 1991) or by the specific modification of the pyrimidine nucleotide were reported (Heidenreich and Eckstein, 1992; Kusser, 2000), and such hybrid nucleic acid are anticipated to provide the stabilities requested.

Numerous studies have reported on the development of optical aptasensors (Lee and Walt, 2000; Li et al., 2002; Kirby et al., 2004; Liu and Lu, 2004; Levy et al., 2005; Pavlov et al., 2005;) and progress in the field was summarized in several review articles (O'Sullivan, 2002; Tombelli, 2005a; Navani and Li, 2006). Different optical aptasensors were developed using fluorescent dyes (Nutiu and Li, 2004; Yang et al., 2005), semiconductor quantum dots (Choi et al., 2006), aggregation and deaggregation of metallic nanoparticles (Liu and Lu, 2006), and the use of surface plasmon resonance spectroscopy (Tombelli et al., 2005b). Similarly, rapid progress in the development of electronic aptasensor systems (electrochemical, field-effect transistor, and piezoelectric) was accomplished in recent years. In this chapter we summarize advances in the development of electronic aptasensors.

3.2   ELECTROCHEMICAL APTASENSOR BASED ON REDOX-ACTIVE APTAMER MONOLAYERS LINKED TO ELECTRODES

The conformational changes in nucleic acids upon hybridization have been the basis for the development of different electrochemical DNA sensors (Fan et al., 2003; Mao et al., 2003; Immoos et al., 2004): for example, a molecular hairpin-like DNA stem–loop structure labeled with a redox-active reporter revealed upon immobilization on a conducting support electron-transfer communication with the electrode, due to the close proximity between the redox label in the stem–loop structure and the electrode surface. Hybridization of the stem–loop structure with the complementary DNA, however, resulted in an extended-duplex structure that positioned the redox label in a spatially separated configuration that prevented electrical contact with the electrode. This enabled amperometric detection of the DNA by following depletion of the electrical contact between the redox label and the electrode as a result of hybridization (Fan et al., 2003). This concept was further developed to tailor aptasensors. Single-stranded nucleic acids that act as aptamers for proteins or small molecules change their flexible single-stranded chains into well-defined three-dimensional structures upon complexation with their host substrates. This enabled the tethering of redox-active units to the aptamer nucleic acids and identification of the formation of the aptamer–substrate complex by probing the electrical contact between the redox label and the electrode in the rigidified three-dimensional complex. The thrombin aptamer undergoes a transition into a G-quadruplex structure upon binding the thrombin. Thus, the electrochemical thrombin aptasensor was developed by tethering the redox-active methylene blue label to the aptamer nucleic acids and its immobilization on an electrode (Xiao et al., 2005a) (Figure 3.2). The flexible conformation of the nucleic acid chain enabled electrical contact of the redox label with the electrode, and this resulted in voltammetric response of the methylene blue. Upon the binding of thrombin, the aptamer is assembled in a G-quadruplex structure, and this prohibits the redox label from electron-transfer communication with the electrode, thus enabling the electrical detection of thrombin. This method

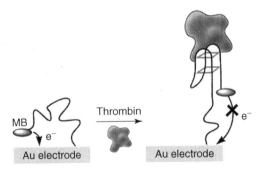

Figure 3.2   Electrochemical aptasensor for thrombin based on the control of electron transfer between a redox-labeled aptamer and the electrode.

enabled the analysis of thrombin with a detection limit corresponding to about 2×10^{-8} M. The electrochemical aptasensors could be regenerated by treatment with 6 M guanidine hydrochloride. The detection of thrombin by following the decrease in amperometric response of the redox label as a result of the association of thrombin is certainly a disadvantage of this sensing scheme, due to the negative readout signal. To overcome this limitation, two alternative approaches were designed. In one approach, a bifunctional thrombin-binding aptamer with a terminal electroactive ferrocene group as the redox label and the thiol group at the second termini of the aptamer strand for the assembly on a gold surface were incorporated through hexamethylene spacers to produce a redox-active nucleic acid hairpin. The long, flexible aptamer chain prevented electrical contact of the ferrocene label with the electrode (Figure 3.3A).

The binding of thrombin to the aptamer domain rigidified the G-quadruplex aptamer configuration, and this oriented the ferrocene units toward the electrode. This led to electron-transfer communication between the electroactive ferrocene units and the electrode and to a positive amperometric readout signal upon the detection of thrombin (Radi et al., 2006) (Figure 3.3B). The sensor revealed a detection limit that corresponds to about 5×10^{-9} M and could be regenerated by unfolding the aptamer in 1 M HCl. This method was applied successfully to develop an electrochemical aptasensor for platelet-derived growth factor (PGDF) (Lai et al., 2007). A thiolated nucleic acid acting as an aptamer for PGDF was functionalized with redox-active methylene blue label. Upon association with PGDF the aptamers folded into a configuration that forced the redox label to a closed position with respect to the electrode, resulting in a voltammetric response. The aptasensor was applied to analyze the PDGF in a serum sample, and in serum diluted twofold by a buffer solution, a sensitivity of 50 pM (1.25 ng mL^{-1}) was reported. This analytical procedure was compared to other PDGF aptasensors and a 10^4-fold enhancement in sensitivity compared to optical aptasensors was claimed. A related approach (Xiao et al., 2005b) shown in Figure 3.3C has employed a DNA duplex assembly on a gold electrode that consisted of two double-stranded domains separated by a noncomplementary nucleic acid bridge. The upper duplex domain included the thrombin aptamer sequence, and its complementary nucleic acid was functionalized with methylene blue. In the presence of thrombin, the thrombin-binding duplex was dissociated to form the G-quadruplex/thrombin complex, and the separated methylene blue–functionalized nucleic acid chain revealed electrical contact with the electrode. The amperometric responses of the redox label enabled amperometric readout of the detection of thrombin (Figure 3.3D). This method enabled the detection of thrombin with a detection limit of about 3×10^{-9} M. A related electrochemical aptasensor for the low-molecular-weight target, cocaine, was also developed (Baker et al., 2006). The thiolated cocaine-binding aptamer–nucleic acid functionalized with methylene blue as a redox label was assembled on a gold electrode. Formation of the aptamer–cocaine complex rigidified the nucleic acid into the cocaine-binding configuration, where the redox-active unit is in proximity to the electrode and yields a voltammetric response. This method

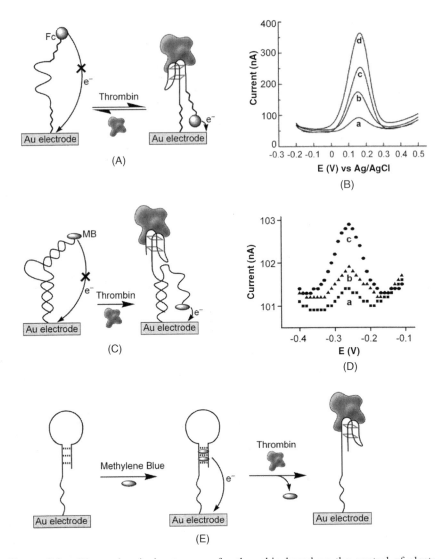

Figure 3.3   Electrochemical aptasensor for thrombin based on the control of electron transfer between redox-labeled aptamer and the electrode. (A) Controlling the orientation of the redox label with respect to the electrode upon formation of a thrombin–aptamer complex. (B) Differential pulse voltammetry corresponding to an analysis of different concentrations of thrombin by a ferrocene-tethered aptamer: (a) 0, (b) 10, (c) 20, and (d) 30 nM. (Reprinted with permission from Radi et al., 2006. Copyright 2006 American Chemical Society.) (C) Activation of the electrical contact of methylene blue–tethered aptamer upon formation of the respective aptamer–thrombin complex. (D) Voltammograms corresponding to analysis of the thrombin by the configuration depicted in part (C): curves (a) no thrombin; (b) thrombin 10 nM; (c) thrombin 256 nM. (Reprinted with permission from Xiao et al., 2005. Copyright 2005 American Chemical Society.) (E) Blocking the electrical response of methylene blue intercalated into the stem of a DNA hairpin as a result of formation of an aptamer–thrombin complex.

was recently modified by the blocking of a ferrocene-tethered aptamer against adenosine triphosphate (ATP) in an extended duplex configuration that prohibited electrical contact between the ferrocene units and the electrode (Zuo et al., 2007). Dissociation of the duplex structure and formation of the ATP–aptamer complex folded the aptamer into a configuration where a short distance is attained between the redox label and the electrode. This resulted in a voltammetric response that was intensified as the concentration of ATP increased. A method employed to follow the aptamer–substrate interaction electrochemically has used redox-active reporter units that intercalate into double-stranded DNA rather than being tethered covalently to the aptamer (Bang et al., 2005). A nucleic acid in a hairpin configuration that includes the thrombin recognition sequence was linked to a gold electrode, and methylene blue was intercalated in the duplex stem of the probe hairpin (Figure 3.3E). The association of thrombin to the aptamer unit opened the hairpin structure, and this process released the intercalated redox-active units. As a result, binding of thrombin to the interface decreased the amperometric response of the system, and the process enabled the detection of thrombin with a detection limit of 11×10^{-9} M.

A reusable electrochemical electrode for the use of low-molecular-weight molecules by its aptamers was demonstrated with the development of an electrochemical aptasensor for adenosine monophosphate (Wu et al., 2007). The anti-adenosine aptamer was functionalized at its end with a redox-active ferrocene unit. An electrode was functionalized with nucleic acid complementary to a region of the aptamer, and this was hybridized to the redox-tethered aptamer, giving rise to a redox response. In the presence of adenosine, an adenosine–aptamer complex was formed, resulting in dissociation of the redox-tethered aptamer from the surface and depletion of the redox signal. The extent of the decrease in electrical response was controlled by concentration of the adenosine analyte.

3.3 ENZYME-BASED AMPLIFIED ELECTROCHEMICAL APTASENSORS

Numerous approaches to amplify the biosensing process were reported in the past decade. These methods were adapted and modified appropriately to design aptasensors. Catalytic labels (e.g., enzymes, inorganic or organic catalysts, or nanoparticles) are often used as catalytic labels for amplifying recognition processes. Enzymes, specifically redox enzymes, are often coupled to antibodies or nucleic acids, and these act as biocatalytic conjugates for the amplified electronic readout of immunocomplexes or DNA analysis. For example, the hydrolysis of p-aminophenyl phosphate by an antibody–alkaline phosphatase conjugate generated p-aminophenol, which acted as an electrochemically detectable redox label (Tang et al., 1988). The analysis of DNA was accomplished by the use of a nucleic acid–functionalized horseradish peroxidase (HRP) as a label for the bioelectrocatalyzed reduction of H_2O_2 (Patolsky et al., 2002). Also, the enzyme-catalyzed precipitation of an insoluble product on electrodes was used for the amplified electrochemical or microgravimetric detection of biorecognition events. For example, biotin-labeled HRP or alkaline phosphatase or nucleic acid–functionalized alkaline phosphatase were used as labels for the analysis of immunocomplexes (Bardea et al., 2000; Alfonta et al., 2001), DNA (Patolsky et al., 2003), or single-base mismatches in DNA (Patolsky et al., 2001) through the biocatalyzed precipitation of insoluble products on the electrode (e.g., an insoluble indigo product) and the detection of biorecognition events by following the insulation of the electrode using faradaic impedance spectroscopy or by following the mass changes on piezoelectric quartz crystals. Catalytic enlargement of nucleic acid–metal nanoparticle conjugates acting as labels for DNA hybridization events was used to develop electrical (conductivity) DNA sensors (Park et al., 2002) and to develop amplified microgravimetric DNA-sensing schemes (Weizmann et al., 2001; Willner et al., 2002).

Two different configurations that employ the biocatalytic properties of enzymes for the amplified analysis of thrombin by its aptamer were reported (Mir et al., 2006) (Figure 3.4). By one approach (Figure 3.4A), the thiolated aptamer, (1), was assembled on a gold electrode, and it acted as a sensing interface for the association of thrombin. The bound thrombin acted as a protease and hydrolyzed a nitroaniline-functionalized peptide, (2), to the redox-active product p-nitroaniline, (3). The latter product was then analyzed electrochemically. The second configuration mode (Figure 3.4B) makes use of the fact that the thrombin includes two different binding sites for aptamers (Bock et al., 1992; Padmanabhan et al., 1993; Tasset et al., 1997). Accordingly, a *sandwich-type assay*, where a second aptamer bound to horseradish peroxidase was linked to the primary aptamer–thrombin complex, acted as an amplifying reporter label for the primary association of thrombin to the sensing aptamer. The enzyme catalyzed relay-mediated reduction of H_2O_2. The amperometric response was then used to quantify the concentration of thrombin, and the protein was analyzed with a detection limit that corresponded to 8×10^{-8} M.

Also, the pyrroloquinoline quinone–dependent glucose dehydrogenase (GDH) was employed as a biocatalytic label for the amplified amperometric detection of thrombin (Figure 3.4C) (Ikebukuro et al., 2005). The fact that two different DNA aptamers bind to thrombin (Bock et al., 1992; Padmanabhan et al., 1993; Tasset et al., 1997) was utilized to develop the aptasensor for thrombin. Thrombin was linked to a 15-mer thiolated aptamer linked to a gold electrode, and the GDH–avidin conjugate was linked to the surface by its association to the biotinylated 29-mer aptamer bound as a secondary aptamer to the thrombin on the surface. The bioelectrocatalyzed oxidation of glucose in the presence of 1-methoxyphenazine methosulfate (m-PMS), as a diffusional mediator, enabled amperometric detection of the thrombin with a linear response in the range 4 to 10×10^{-8} M.

A related approach substituted the enzyme-amplifying agents with platinum nanoparticles (NPs) as catalytic labels (Polsky et al., 2006) (Figure 3.5). Pt-NPs were functionalized with the thiolated aptamer, (1). The stepwise assembly of

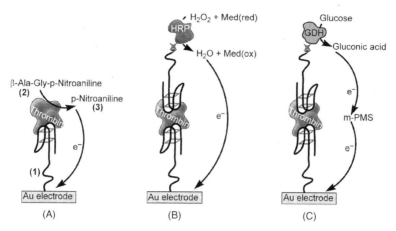

Figure 3.4   Amplified electrochemical analysis of thrombin using (A) thrombin-catalyzed generation of the electroactive substrate (3); (B) and (C) enzyme-tethered anti-thrombin aptasensors as amplifying labels (HRP and GDH, respectively).

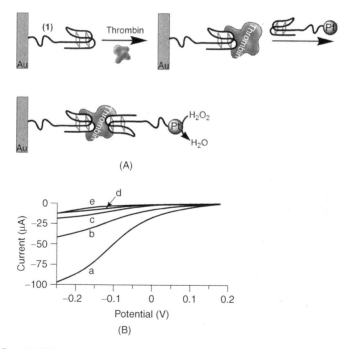

Figure 3.5   (A) Electrochemical analysis of thrombin by aptamer-labeled Pt-NPs acting as electrocatalysts for the reduction of H_2O_2. (B) Voltammograms corresponding to the analysis of various concentrations of thrombin: (a) 1×10^{-6} M; (b) 1×10^{-7} M; (c) 1×10^{-8} M; (d) 1×10^{-9} M; (e) control sample in the absence of thrombin. (Reprinted with permission from Gill et al., 2006. Copyright 2006 American Chemical Society.)

the aptamer on the surface was followed by formation of the aptamer–thrombin complex on the surface and the secondary association of aptamer-functionalized Pt-NPs to the aptamer–thrombin complex bound to the electrode (Figure 3.5A). The NPs catalyzed the electrochemical reduction of H_2O_2, and the resulting cathodic currents (Figure 3.5B) enabled the amplified detection of thrombin with a detection limit that corresponded to 1×10^{-9} M. The Pt-NPs reveal advantages over the enzyme-based amplified aptasensors; in addition to the enhanced stability of the metallic NPs, the Pt-NPs reveal an 80-fold-higher detection limit. The improved sensitivity observed with the NPs was attributed, in part, to the need to activate the biocatalytic functions of the enzymes by a diffusional electron mediator that is weakly coupled with the electrode surface.

Interesting electrochemical aptasensors based on the interaction of ferrocene-functionalized cationic poly(3-alkoxy-4-methylthiophene) [(4) in Figure 3.6A], acting as a polyelectrolyte have been reported (Floch et al., 2006). The interaction of positively charged polythiophene polyelectrolytes with double-stranded DNA and the development of optical nucleic acid sensors were studied extensively (Ho and Leclerc, 2004; Ho et al., 2005). Application of the redox-labeled polyelectrolyte, (4), enabled the voltammetric detection of thrombin by the respective aptamer. In one configuration (Figure 3.6A) the thiolated anti-thrombin aptamer was linked to an electrode surface. Interaction of the polyelectrolyte with the thrombin aptamer yielded a voltammetric response of the ferrocene unit of (4). The association of the thrombin blocked binding of the cationic polymer and depleted its electrochemical response. The disadvantage of the configuration is the appearance of a negative detection signal upon thrombin analysis, and an unsatisfactory detection limit, higher than 1×10^{-6} M. A related improved method, albeit of increased complexity that applies (4) as an electrochemical tracer is depicted in Figure 3.6B. According to this method, the anti-thrombin aptamer and thrombin were interacted in solution. The resulting mixture was then treated with a S1 nuclease that digested all free nucleic acid, leaving the aptamer–thrombin complex intact, due to insulation of the aptamer by the protein shell. The nuclease was then inhibited by EDTA, and the aptamer–thrombin complex was separated by thermal treatment. Hybridization of the dissociated aptamer with the complementary peptide nucleic acid (PNA) linked to a gold surface resulted in the negatively charged duplex structure that binds (4). As coverage of the duplex increased with thrombin concentration, the voltammetric responses of the electrode were enhanced as the concentration of the thrombin was elevated (Figure 3.6C). The method enabled the analysis of thrombin with a detection limit corresponding to 7.5×10^{-8} M.

3.4   AMPLIFIED ELECTROCHEMICAL APTASENSORS BASED ON NANOPARTICLES

Indirect amplified electrochemical analysis of aptamer–protein complexes has employed nanoparticles as labels for the development of electrochemical aptasensors. The use of metal nanoparticles as tracers for the analysis of nucleic acid

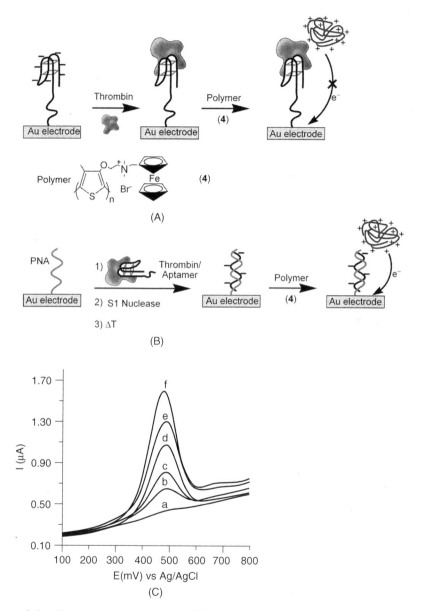

Figure 3.6 Electrochemical detection of thrombin by interaction between nucleic acid and a redox-active oligothiophene polyelectrolyte (4). (A) Blocking the electrical contact between the polyelectrolyte and the electrode by means of a aptamer–thrombin complex. (B) Separation of the aptamer–thrombin complex by the formation of a PNA–opened aptamer duplex on the electrode and its analysis by the redox-active polyelectrolyte. (C) Voltammetric response to the analysis of different concentrations of thrombin according to method B: (a) 0 M; (b) 125 nM; (c) 250 nM; (d) 500 nM, 1 μM; (f) 2 μM. (Reprinted with permission from Floch et al., 2006. Copyright 2006 American Chemical Society.)

hybridization is well established, and numerous DNA-sensing configurations were reported (Wang et al., 2001). For example, the hybridization of a biotinylated target nucleic acid with a complementary nucleic acid associated with magnetic particles was followed by binding of the streptavidine-coated gold nanoparticles to the target DNA. The magnetic separation of the metal tracers labeled magnetic particles, dissolution of the metal nanoparticles, and the electrochemical analysis of the metal ions released by stripping voltammetry enabled indirect amplified analysis of DNA. Similarly, different metal sulfide semiconductor quantum dots (QDs) were used for the parallel analysis of different DNAs (Wang et al., 2003). Magnetic nanoparticles were modified with three different nucleic acids acting as probes for three different DNA targets (Figure 3.7A). The magnetic particles were then interacted with a sample consisting of all three DNA chains complementary to probes linked to the magnetic particles. Semiconductor QDs, comprised of ZnS, CdS, and PbS, each functionalized with nucleic acids complementary to the free chains of the analyte DNAs, were then hybridized with the duplexes linked

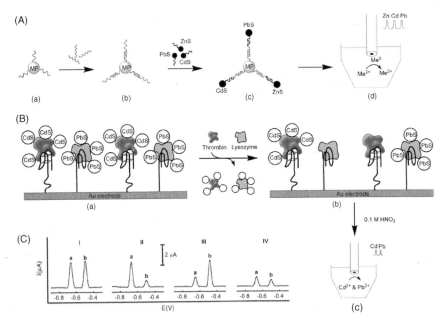

Figure 3.7 (A) Parallel electrochemical analysis of different DNAs using magnetic particles functionalized with probes for the various DNA targets and specific nucleic acid–functionalized metal sulfides as tracers. (B) Simultaneous electrochemical analysis of the two proteins thrombin and lysozyme using a competitive assay where thrombin modified with CdS-QDs and lysozyme modified with PbS-QDs are used as tracers. (C) Square-wave stripping voltammograms corresponding to the simultaneous detection of lysozyme (a) and thrombin (b): (I) no (a); no (b); (II) 1 μg L^{-1} (a), no (b); (III) no (a), 0.5 μg L^{-1} (b); (IV) 1 μg L^{-1} (a), 0.5 μg L^{-1} (b). (Reprinted with permission from Hansen et al., 2006. Copyright 2006 American Chemical Society.)

to the magnetic particles. The semiconductor QDs acted as tracers for detection of the hybridization of the analyte DNA since their binding to the magnetic particles proceeded only upon primary hybridization of the target DNA. Separation of the magnetic particles by an external magnet, followed by dissolution of the various QDs, enabled to trace and quantitatively analyze the respective DNAs by voltammetric assay of the various metal ions at their characteristic potentials. This method to encode biomolecular identity by semiconductor QDs was extended for the parallel analysis of a variety of proteins by their specific aptamers (Hansen et al., 2006). A gold electrode was functionalized using aptamers specific for thrombin and lysozyme (Figure 3.7B). Thrombin and lysozyme were modified with CdS-QDs and PbS-QDs, respectively, and these were bound to the respective aptamers associated with the surface. The QD-functionalized proteins acted as tracing labels for analysis of the proteins. In the presence of nonfunctionalized thrombin or lysozyme, displacement of the respective labeled proteins proceeded, and dissolution of the respective remaining metal sulfide on the surface with the concomitant detection of the released ions by electrochemical stripping then enabled quantitative detection of the two proteins (Figure 3.7C).

3.5 LABEL-FREE ELECTROCHEMICAL APTASENSORS

Whereas the previous electrochemical aptasensors employed redox labels or catalysts as reporter units for the formation of aptamer–substrate complexes, other label-free electrochemical aptasensors began to be developed. Faradaic impedance spectroscopy proved to be an effective method to probe biorecognition events at electrodes by following the changes in electron-transfer resistances at the electrodes that resulted from biosensing events (Katz and Willner, 2003). The association of proteins to electrode supports (e.g., formation of antigen–antibody or receptor–protein complexes) was found to introduce a barrier to electrical contacting of a redox label solubilized in the electrolyte solution and the electrode, resulting in an increase in electron-transfer resistance at the conducting support (Bardea et al., 2000; Pei et al., 2001; Alfonta et al., 2001). Similarly, the formation of double-stranded DNA on a nucleic acid–functionalized electrode increased the negative charge associated with the electrode, thus repelling a negatively charged redox label solubilized in the electrolyte solution. Repulsion of the redox-active probe increased the interfacial–electron transfer resistance, and this enabled quantitative analysis of the hybridization of DNA by following the changes in the interfacial electron-transfer resistances (Bardea et al., 1999; Vagin et al., 2002). The phenomena were extended to develop impedimetric aptasensors. Control of the electrode surface properties by aptamer–substrate interactions led to the development of label-free electrochemical aptasensors. The biotinylated aptamer for the lysozyme was linked to the streptavidin-functionalized ITO electrode, and this acted as a sensing interface for the lysozyme (Rodriguez et al., 2005). The aptamer generated a negatively charged interface that electrostatically repelled an anionic redox probe such as $[Fe(CN)_6]^{3-/4-}$. The electrostatic

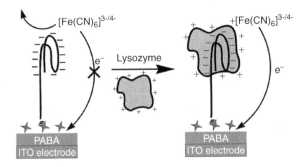

Figure 3.8 Impedimetric analysis of an aptamer–lysozyme complex using [Fe (CN)$_6$]$^{3-/4-}$ as a redox label.

repulsion of the redox probe from the electrode introduced a barrier for electron transfer at the electrode support and resulted in increased electron-transfer resistance. The association of the protein to the aptamer at an appropriate pH resulted in switching of the surface charge, and this provided a positive charge interface that led to a decrease in electron-transfer resistances due to the electrostatic attraction of the redox label (Figure 3.8). Changes in the interfacial electron-transfer resistances at various protein concentrations enabled quantitative analysis of the lysozyme. Faradaic impedance spectroscopy was used to analyze thrombin on a thiolated aptamer-functionalized gold electrode (Radi et al., 2005; Cai et al., 2006). Association of the protein with the aptamer-functionalized surface insulated the electrode surface toward electron transfer with the Fe(CN)$_6^{3-/4-}$ redox probe solubilized in the electrolyte solution, giving rise to an increase in electron-transfer resistances, and this enabled quantitative analysis of the thrombin. The faradaic impedance transduction of aptamer–protein recognition processes was also analyzed in an array configuration of electrodes (Xu et al., 2005). The aptamer for human IgE antibody was assembled on the gold electrodes of an array, and an increase in interfacial electron-transfer resistance was demonstrated upon binding of the immunoglobulin E (IgE) antibody using Fe(CN)$_6^{3-/4-}$ as a redox probe. The interfacial electron-transfer resistance increased as concentration of the antibody was elevated, and the IgE antibody could be detected with a detection limit of 1×10^{-10} M. Although the electrode array could be used for the multiplex analysis of different proteins, this capability has not yet been demonstrated. The electrode array was, however, used to examine the effect of different base mutations in the aptamer sequence on the binding affinities toward the IgE antibody. By the deposition of different aptamer mutants on the electrode array and the readout of interfacial electron-transfer resistance upon the binding of the IgE antibody, the affinity features of the various mutants to the protein were elucidated.

The use of faradaic impedance spectroscopy as a means to identify affinity complexes between the aptamers and small molecules is, however, more difficult since the reorganization of aptamer–small molecule complexes on the electrode yields a minute, usually undetectable change in the interfacial electron-transfer resistance compared to the free aptamer-modified electrode. An approach to overcoming this difficulty included the immobilization of a duplex DNA on an electrode that consisted of an amino-functionalized aptamer for adenosine, (5), and a hybridized nucleic acid fragment, (6), complementary to the aptamer (Figure 3.9A). In the presence of adenosine, the duplex DNA was separated and the aptamer folded into its three-dimensional structure that binds adenosine (Zayats et al., 2006). Separation of the duplex associated with the electrode removed the negative charge from the electrode surface, thus decreasing the interfacial electron-transfer resistance in the presence of the negatively charged redox probe. Separation of the duplex DNA was controlled by the concentration of

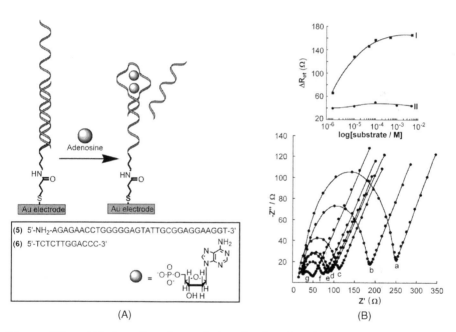

(A)   (B)

Figure 3.9   (A) Impedimetric analysis of adenosine by the formation of an aptamer–adenosine complex through the separation of an aptamer–nucleic acid duplex associated with the electrode. The redox label in solution was $[Fe(CN)_6]^{3-/4-}$. (B) Interfacial electron-transfer resistance on the analysis of various concentrations of adenosine: (a) the monolayer generated by the association of the aptamer–nucleic acid duplex to the electrode; (b) to (f) after treatment with adenosine: 1×10^{-6} M, 1×10^{-5} M, 1×10^{-4} M, 1×10^{-3} M, 1×10^{-2} M, respectively; (g) the bare electrode prior to immobilization of the aptamer–nucleic acid duplex. Inset: (I) calibration curve for the analysis of different concentrations of adenosine; (II) calibration curve corresponding to the control experiment where the respective concentrations of cytidine are analyzed by the aptamer–nucleic acid duplex. (Reprinted with permission from Zayats et al., 2006. Copyright 2006 American Chemical Society.)

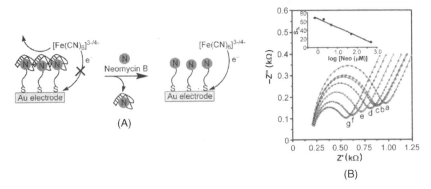

Figure 3.10   (A) Electrochemical analysis of neomycin B by its RNA aptamer using $Fe(CN)_6]^{3-/4-}$ as the redox label and impedance spectroscopy as the readout signal. (B) Faradaic impedance spectra upon analyzing different concentrations of neomycin B: (a) 0; (b) 0.75; (c) 2; (d) 5; (e) 50; (f) 500 μM. Inset: Calibration curve corresponding to changes in the interfacial electron-transfer resistances upon analyzing different concentrations of neomycin B. (Reprinted with permission from De-los-Santos et al., 2007. Copyright 2006 American Chemical Society.)

adenosine, thus enabling the quantitative analysis of the low-molecular-weight substrate (Figure 3.9B).

An alternative competitive impedimetric assay for analyzing neomycin B was demonstrated (de-los-Santos et al., 2007) (Figure 3.10A). A gold electrode was modified with neomycin B, and the respective RNA aptamer was linked to the surface. The resulting neomycin B–aptamer complex introduced an electron-transfer barrier to the solution-solubilized redox label $Fe(CN)_6^{3-/4-}$, and this resulted in high interfacial electron-transfer resistance. In the presence of neomycin B, partial displacement of the aptamer occurred, and this resulted in a decrease in the interfacial electron-transfer resistance. Since the extent of the aptamer displacement was controlled by the concentration of neomycin B, the quantitative analysis of the substrate was achieved (Figure 3.10B). The method was applied successfully to analyze neomycin B in milk.

3.6   FIELD-EFFECT TRANSISTOR–BASED APTASENSORS

An alternative electronic transduction means of biorecognition events involves the use of ion-selective field-effect transistors (ISFETs). Control of the gate potential of FET devices as a result of biorecognition processes that occur on the gate surface became a common principle to develop biosensor devices (Schçning and Poghossian, 2002, 2006; Bergveld, 2003; Dzyadevych et al., 2006). The general configuration of a ISFET-based biosensor includes two electrodes, source and drain, deposited on a semiconductor (e.g., silicon). A gate surface is deposited on the semiconductor by its mounting on a thin insulating layer, and it separates the source and drain electrodes. The gate potential may be affected by chemical transformation occurring on it (e.g., charging or ionization at different pH values). For biosensing purposes, the device is immersed in an electrolyte medium, and a reference electrode linked to the source electrode reflects the gate potentials. Thus, for a certain gate potential, a source-to-drain potential (V_{ds}) can be applied, and this maintains a current, I_{ds}. Upon changing the potential of the gate electrode as a result of a biosensing event, the source-to-drain current is perturbed and its value can be retained either by changing the V_{ds} potential or by altering the potential between the reference electrode and the source, V_{gs}, for compensation of the potential changes on the gate. For example, alteration of the charge (and thus the potential) on the gate of the ISFETs upon hybridization of the complementary nucleic acid to the gate-confined DNA was used for the label-free reagentless detection of DNA (Fritz et al., 2002; Kim et al., 2004; Shin et al., 2004; Uslu et al., 2004; Sakata and Miyahara, 2005). Binding of a low-molecular-weight substrate (e.g., adenosine) to its aptamer was followed by an ion-selective field-effect transistor, and this demonstrated the development of a label-free aptasensor (Zayats et al., 2006). The Al_2O_3 gate surface was functionalized with 3-aminopropylthriethoxysilane, subsequently modified with glutaric dialdehyde, and then amino-functionalized nucleic acid, (5), which acts as an aptamer for adenosine monophosphate, was covalently immobilized on the gate surface (Figure 3.11A). The nucleic acid, (6), was hybridized with the aptamer to enhance the negative charge on the gate surface. The adenosine was analyzed by the displacement of (6) and reorganization of the adenosine–aptamer complex on the surface. The displacement of (6) affected the local charge associated with the gate, thus enabling the transduction of adenosine sensing by the ISFET. Figure 3.11B, curve a, shows the changes in the gate-to-source potential, V_{gs}, upon analyzing different concentrations of adenosine. The detection limit for analyzing adenosine was 5×10^{-5} M. The analysis of adenosine monophosphate by the aptamer revealed high specificity, and the aptamer-modified ISFET did not respond to other nucleotides (e.g., cytidine; Figure 3.11B, curve b). The method represents a label-free, reagentless, analytical procedure to follow aptamer-low-molecular-weight binding reactions.

Aptamer–protein affinity binding was also followed on a single-walled carbon nanotube (SWCNT) FET device (So et al., 2005). Monitoring of protein–protein binding on FET devices is difficult since the recognition occurs outside the electrical double layer associated with the gate, and thus the potential changes on the gate are small. However, the dimensions of aptamers (1 to 2 nm) enable perturbation of the gate potential by proteins that link to the aptamers, since the recognition binding event occurs within the Debye length of a double layer (ca. 3 nm at 10 mM ionic concentrations). Accordingly, the SWCNT was assembled between the source and drain electrode, and the aptamer against thrombin was assembled on the CNTs (Figure 3.11C). The binding of thrombin to the aptamer altered conductance through the device, thus enabling the sensing of the protein (Figure 3.11D). Conductance through the device was specific for thrombin, and another protein (e.g., elastase) showed a minute effect on the conductance. A similar approach was applied to analyze the IgE antibody by its aptamer

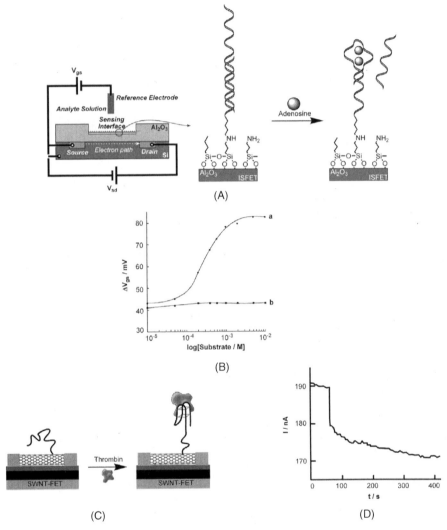

Figure 3.11   (A) Label-free reagentless analysis of adenosine on an ISFET device through separation of the aptamer–nucleic acid complex by adenosine. (B) Changes in gate-to-source potentials upon the analysis of different concentrations of (a) adenosine and (b) cytidine. (Reprinted with permission from Zayats et al., 2006. Copyright 2006 American Chemical Society.) (C) Analysis of thrombin by the anti-thrombin aptamer-functionalized SWCNT acting as a gate on a field-effect transistor. (D) Conductance changes of an aptamer-functionalized SWCNT-FET device upon the analysis of thrombin. (Reprinted with permission from So et al., 2005. Copyright 2005 American Chemical Society.)

(Maehashi et al., 2007). The detection limit for the antibody corresponded to 250 pM.

3.7 CONCLUSIONS AND PERSPECTIVES

Electronic aptasensors reveal certain advantages when compared to optical aptasensors: (1) The possibility of coupling amplifying catalytic or biocatalytic labels to an aptamer–target complex enables amplified detection of the target substrate and thus enhancing the sensitivity of sensing processes; and (2) recent advances demonstrated the electronic label-free detection of the substrate by the aptamers, and thus the elimination of fluorescent labels may be excluded.

Possible variations in the sequences of oligonucleotide enable an enormous diversity of aptamers for almost any molecule or macromolecule. Indeed, DNA or RNA aptamers were developed for hundreds of targets (Gold et al., 1995; Famulok, 1999; Patel and Suri, 2000). The impressive selectivity of aptamers suggests that evolutionary selected binding of oligonucleotides might conquer the selectivity of antibodies. For example, an anti-theophyllin aptamer revealed about a 10^4-fold enhanced binding affinity toward theophyllin compared to caffeine (which differs by only a single methyl group in the molecular structure; Jenison et al., 1994). Even though substantial progress was accomplished in the use of aptamers in analytical chemistry, several exciting opportunities still exist in the field of aptasensors: The use of synthetically modified nucleotides as co-components in the evolution of aptamers will lead to nuclease-resistant oligonucleotides, and eventually to aptamers of higher affinities to substrates. Furthermore, we have emphasized the use of SELEX for the synthesis of aptamers. Nonetheless, similar to scientific efforts to elicit catalytic antibodies, the SELEX procedure was applied to synthesize catalytic DNA/RNA (DNAzymes/RNAzymes) by selecting oligonucleotides to the respective reaction (Lu, 2002; Silverman, 2004; Fiammengo and Jäschke, 2005; Lilley, 2005). Integration of the binding properties of aptamers with DNAzyme units might yield hybrid systems that include a built-in amplifying label. Such molecular hybrids would reveal superior sensing functions by combining evolutionary-selected binding and catalytic properties of nucleic acids.

REFERENCES

Alfonta, L., Bardea, A., Khersonsky, O., Katz, E., Willner, I. (2001). Chronopotentiometry and faradaic impedance spectroscopy as signal transduction methods for the biocatalytic precipitation of an insoluble product on electrode supports: routes for enzyme sensors, immunosensors and DNA sensors. *Biosens Bioelectron* 16, 675–687.

Baker, B. R., Lai, R. Y., Wood, M. S., Doctor, E. H., Heeger, A. J., Plaxco, K. W. (2006). An electronic, aptamer-based small-molecule sensor for the rapid, label-free detection of cocaine in adulterated samples and biological fluids. *J Am Chem Soc* 128, 3138–3139.

Bang, G. S., Cho, S., Kim, B-G. (2005). A novel electrochemical detection method for aptamer biosensors. *Biosens Bioelectron* 21, 863–870.

Bardea, A., Patolsky, F., Dagan, A., Willner, I. (1999). Sensing and amplification of oligonucleotide–DNA interactions by means of impedance spectroscopy: a route to a Tay–Sachs sensor. *Chem Commun* 1, 21–22.

Bardea, A., Katz, E., Willner, I. (2000). Probing antigen–antibody interactions on electrode supports by the biocatalyzed precipitation of an insoluble product. *Electroanalysis* 12, 1097–1106.

Bergveld, P. (2003). Thirty years of ISFETOLOGY: what happened in the past 30 years and what may happen in the next 30 years. *Sens Actuat B* 88, 1–20.

Bock, L. C., Griffin, L. C., Latham, J. A., Vermass, E. H., Toole, J. J. (1992). Selection of single-stranded-DNA molecules that bind and inhibit human thrombin. *Nature* 355, 564–566.

Brody, E. N., Gold, L. (2000). Aptamers as therapeutic and diagnostic agents. *Rev Mol Biotechnol* 74, 5–13.

Cai, H., Lee, T. M-H., Hsing, I-M. (2006). Label-free protein recognition using an aptamer-based impedance measurement assay. *Sens Actuat B* 114, 433–437.

Choi, J. H., Chen, K. H., Strano, M. S. (2006). Aptamer-capped nanocrystal quantum dot: a new method for label-free protein detection. *J Am Chem Soc* 128, 15584–15585.

Clark, S. L., Remcho, V. T. (2002). Aptamers as analytical reagents. *Electrophoresis* 23, 1335–1340.

Clark, S. L., Remcho, V. T. (2003). Electrochromatographic retention studies on a flavin-binding RNA aptamer sorbent. *Anal Chem* 75, 5692–5696.

De-los-Santos Alvarez, N., Lobo-Castanon, M. J., Miranda-Ordieres, A. J., Tunon-Blanco, P. (2007). Modified-RNA aptamer-based sensor for competitive impedimetric assay of neomycin B. *J Am Chem Soc* 129, 3808–3809.

Deng, Q., Watson, C. J., Kennedy, R. T. (2003). Aptamer affinity chromatography for rapid assay of adenosine in microdialysis samples collected in vivo. *J Chromatogr A* 1005, 123–130.

Dzyadevych, S. V., Soldatkin, A. P., El'skaya, A. V., Martelet, C., Jaffrezic-Renault, N. (2006). Enzyme biosensors based on ion-selective field-effect transistors. *Anal Chim Acta* 568, 248–258.

Ellington, A. D., Szostak, J. W. (1990). In vitro selection of RNA molecules that bind specific ligands. *Nature* 346, 818–822.

Ellington, A. D., Szostak, J. W. (1992). Selection in vitro of single-stranded DNA molecules that fold into specific ligand-binding structures. *Nature* 355, 850–852.

Famulok, M. (1999). Oligonucleotide aptamers that recognize small molecules. *Curr Opin Struct Biol* 9, 324–329.

Famulok, M., Mayer, G., Blind, M. (2000). Nucleic acid aptamers-from selection in vitro to application in vivo. *Acc Chem Res* 33, 591–599.

Fan, C., Plaxco, K. W., Heeger, A. J. (2003). Electrochemical interrogation of conformational changes as a reagentless method for the sequence-specific detection of DNA. *Proc Nat Acad Sci U S A* 100, 9134–9137.

Feigon, J., Dieckmann, T., Smith, F. W. (1996). Aptamer structures from A to zeta. *Chem Biol* 3, 611–617.

Fiammengo, R., Jäschke, A. (2005). Nucleic acid enzymes. *Curr Opin Biotechnol* 16, 614–621.

Floch, F. L., Ho, H. A., Leclerc, M. (2006). Label-free electrochemical detection of protein based on a ferrocene-bearing cationic polythiophene and aptamer. *Anal Chem* 76, 4727–4731.

Fritz, J., Cooper, E. B., Gaudet, S., Sorger, P. K., Manalis, S. R. (2002). Electronic detection of DNA by its intrinsic molecular charge. *Proc Nat Acad Sci U S A* 99, 14142–14146.

Gold, L., Polisky, B., Uhlenbeck, O., Yarus, M. (1995). Diversity of oligonucleotide functions. *Annu Rev Biochem* 64, 763–797.

Hansen, J. A., Wang, J., Kawde, A-N., Xiang, Y., Gothelf, K. V., Collins, G. (2006). Quantum dot/aptamer-based ultrasensitive multi-analyte electrochemical biosensor. *J Am Chem Soc* 128, 2228–2229.

Heidenreich, O., Eckstein, F. (1992). Hammerhead ribozyme-mediated cleavage of the long terminal repeat RNA of human immunodeficiency virus type 1. *J Biol Chem* 267, 1904–1909.

Ho, H. A., Leclerc, M. (2004). Optical sensors based on hybrid aptamer/conjugated polymer complexes. *J Am Chem Soc* 126, 1384–1387.

Ho, H. A., Doré, K., Boissinot, M., Bergeron, M. G., Tanguay, R. M., Boudreau, D., Leclerc, M. (2005). Direct molecular detection of nucleic acids by fluorescence signal amplification. *J Am Chem Soc* 127, 12673–12676.

Ikebukuro, K., Kiyohara, C., Sode, K. (2005). Novel electrochemical sensor system for protein using the aptamers in sandwich manner. *Biosens Bioelectron* 20, 2168–2172.

Immoos, C. E., Lee, S. J., Grinstaff, M. W. (2004). Conformationally gated electrochemical gene detection. *ChemBioChem* 5, 1100–1103.

Jayasena, S. (1999). Aptamers: an emerging class of molecules that rival antibodies in diagnostics. *Clin Chem* 45, 1628–1650.

Jenison, R. D., Gill, S. C., Pardi, A., Polisky, B. (1994). High-resolution molecular discrimination by RNA. *Science* 263, 1425–1429.

Katz, E., Willner, I. (2003). Probing biomolecular interactions at conductive and semiconductive surfaces by impedance spectroscopy: routes to impedimetric immunosensors, DNA-sensors, and enzyme biosensors. *Electroanalysis* 15, 913–947.

Kim, D-S., Jeong, Y. T., Lyu, H. K., Park, H-J., Shin, J.-K., Choi, P., Lee, J. H., Lim, G. (2004). An FET-type charge sensor for highly sensitive detection of DNA sequence. *Biosens Bioelectron* 20, 69–74.

Kirby, R., Cho, E. J., Gehrke, B., Bayer, T., Park, Y. S., Neikirk, D. P., McDevitt, J. T., Ellington, A. D. (2004). Aptamer-based sensor arrays for the detection and quantitation of proteins. *Anal Chem* 76, 4066–4075.

Kusser, W. (2000). Chemically modified nucleic acid aptamers for in vitro selections: evolving evolution. *J Biotechnol* 74, 27–38.

Lai, R. Y., Plaxco, K. W., Heeger, A. J. (2007). Aptamer-based electrochemical detection of picomolar platelet-derived growth factor directly in blood serum. *Anal Chem* 79, 229–233.

Lee, M., Walt, D. R. (2000). A fiber-optic microarray biosensor using aptamers as receptors. *Anal Biochem* 282, 142–146.

Levy, M., Cater, S. F., Ellington, A. D. (2005). Quantum-dot aptamer beacons for the detection of proteins. *ChemBioChem* 6, 2163–2166.

Li, J. J., Fang, X., Tan, W. (2002). Molecular aptamer beacons for real-time protein recognition. *Biochem Biophys Res Commun* 292, 31–40.

Lilley, D. M. J. (2005). Structure, folding and mechanisms of ribozymes. *Curr Opin Struct Biol 15*, 313–323.

Liu, J., Lu, Y. (2004). Adenosine-dependent assembly of aptazyme-functionalized gold nanoparticles and its application as a colorimetric biosensor. *Anal Chem* 76, 1627–1632.

Liu, J., Lu, Y. (2006). Fast colorimetric sensing of adenosine and cocaine based on a general sensor design involving aptamers and nanoparticles. *Angew Chem Int Ed Engl* 45, 90–94.

Lu, Y. (2002). New transition-metal-dependent DNA-zymes as efficient endonucleases and as selective metal biosensors. *Chem Eur J* 8, 4588–4596.

Luzi, E., Minunni, M., Tombelli, S., Mascini, M. (2003). New trends in affinity sensing: aptamers for ligand binding. *Trends Anal Chem* 22, 810–817.

Maehashi, K., Katsura, T., Kerman, K., Takamura, Y., Matsumoto, K., Tamiya, E. (2007). Label-free protein biosensor based on aptamer-modified carbon nanotube field-effect transistors. *Anal Chem* 79, 782–787.

Mao, Y., Luo, C., Ouyang, Q. (2003). Studies of temperature-dependent electronic transduction on DNA hairpin loop sensor. *Nucleic Acids Res* 31, e108.

Mendonsa, S. D., Bowser, M. T. (2004). In vitro evolution of functional DNA using capillary electrophoresis. *J Am Chem Soc* 126, 20–21.

Michaud, M., Jourdan, E., Villet, A., Ravel, A., Grosset, C., Peyrin, E. (2003). A DNA aptamer as a new target-specific chiral selector for HPLC. *J Am Chem Soc* 125, 8672–8679.

Mir, M., Vreeke, M., Katakis, I. (2006). Different strategies to develop an electrochemical thrombin aptasensors. *Electrochem Commun* 8, 505–511.

Navani, N. K., Li, Y. (2006). Nucleic acid aptamers and enzymes as sensors. *Curr Opin Chem Biol* 10, 272–281.

Nutiu, R., Li, Y. (2004). Structure-switching signaling aptamers: transducing molecular recognition into fluorescence signalling. *Chem Eur J* 10, 1868–1876.

O'Sullivan, C. K. (2002). Aptasensors: the future of biosensing? *Anal Bioanal Chem* 372, 44–48.

Padmanabhan, K., Padmanabhan K. P., Ferrara, J. D., Sandler, J. E., Tulinsky, A. (1993). The structure of alpha-thrombin inhibited by a 15-mer single stranded-DNA aptamer. *J Biol Chem* 268, 17651–17654.

Park, S-J., Taton, T. A., Mirkin, C. A. (2002). Array-based electrical detection of DNA with nanoparticle probes. *Science* 295, 1503–1506.

Patel, D. J. (1997). Structural analysis of nucleic acid aptamers. *Curr Opin Chem Biol* 1, 32–46.

Patel, D. J., Suri, A. K. (2000). Structure, recognition and discrimination in RNA aptamer complexes with cofactors, amino acids, drugs and aminoglycoside antibiotics. *J Biotechnol* 74, 39–60.

Patolsky, F., Lichtenstein, A., Willner, I. (2001). Detection of single-base DNA mutations by enzyme-amplified electronic transduction. *Nat Biotechnol* 19, 253–257.

Patolsky, F., Katz, E., Willner, I. (2002). Amplified DNA detection by electrogenerated biochemiluminescence and by the catalyzed precipitation of an insoluble product on electrodes in the presence of the doxorubicin intercalator. *Angew Chem Int Ed Engl* 41, 3398–3402.

Patolsky, F., Lichtenstein, A., Willner, I. (2003). Highly sensitive amplified electronic detection of DNA by biocatalyzed precipitation of an insoluble product onto electrodes. *Chem Eur J* 9, 1137–1145.

Pavlov, V., Shlyahovsky, B., Willner, I. (2005). Fluorescence detection of DNA by the catalytic activation of an aptamer/thrombin complex. *J Am Chem Soc* 127, 6522–6523.

Pei, R., Cheng, Z., Wang, E., Yang, X. (2001). Amplification of antigen–antibody interactions based on biotin labeled protein–streptavidin network complex using impedance spectroscopy. *Biosens Bioelectron* 16, 355–361.

Pieken, W., Olsen, D. B., Benseler, F., Aurup, H., Eckstein, H. F. (1991). Kinetic characterization of ribonuclease-resistant 2′-modified hammerhead ribozymes. *Science* 253, 314–317.

Polsky, R., Gill, R., Willner, I. (2006). Nucleic acid–functionalized Pt nanoparticles: catalytic labels for the amplified electrochemical detection of biomolecules. *Anal Chem* 78, 2268–2271.

Radi, A-E., Sánchez, J. L. A., Baldrich, E., O'Sullivan, C. K. (2005). Reusable impedimetric aptasensor. *Anal Chem* 77, 6320–6323.

Radi, A-E., Sánchez, J. L. A., Baldrich, E., O'Sullivan, C. K. (2006). Reagentless, reusable, ultrasensitive electrochemical molecular beacon aptasensors. *J Am Chem Soc* 128, 117–124.

Rodriguez, M. C., Kawde, A-N., Wang, J. (2005). Aptamer biosensor for label-free impedance spectroscopy detection of proteins based on recognition-induced switching of the surface charge. *Chem Commun* 34, 4267–4269.

Sakata, T., Miyahara, Y. (2005). Detection of DNA recognition events using multi-well field effect devices. *Biosens Bioelectron* 21, 827–832.

Schçning, M. J., Poghossian, A. (2002). Recent advances in biologically sensitive field-effect transistors (BioFETs). *Analyst* 127, 1137–1151.

Schçning, M. J., Poghossian, A. (2006). Bio FEDs (field-effect devices): state-of-the-art and new directions. *Electroanalysis* 18, 1893–1900.

Shin, J-K., Kim, D-S., Park, H-J., Lim, G. (2004). Detection of DNA and protein molecules using an FET-type biosensor with gold as a gate metal. *Electroanalysis* 16, 1912–1918.

Silverman, S. K. (2004). Deoxyribozymes: DNA catalysts for bioorganic chemistry. *Org Biomol Chem* 2, 2701–2706.

So, H-M., Won, K., Kim, Y. H., Kim, B-K., Ryu, B. H., Na, P. S., Kim, H., Lee, J-O. (2005). Single-walled carbon nanotube biosensors using aptamers as molecular recognition elements. *J Am Chem Soc* 127, 11906–11907.

Tang, H. T., Lunte, C. E., Halsall, H. B., Heineman, W. R. (1988). *p*-Aminophenyl phosphate: an improved substrate for electrochemical enzyme-immunoassay *Anal Chim Acta* 214, 187–195.

Tasset, D. M., Kubik, M. F., Steiner, W. (1997). Oligonucleotide inhibitors of human thrombin that bind distinct epitopes. *J Mol Biol* 272, 688–698.

Tombelli, S., Minunni, M., Mascini, M. (2005a). Analytical applications of aptamers. *Biosens Bioelectron* 20, 2424–2434.

Tombelli, S., Minunni, M., Luzi, E., Mascini, M. (2005b). Aptamer-based biosensors for the detection of HIV-1 Tat protein. *Bioelectrochemistry* 67, 135–141.

Tuerk, C., Gold, L. (1990). Systematic evolution of ligands by exponential enrichment: RNA ligands to bacteriophage T4 DNA polymerase. *Science* 249, 505–510.

Uslu, F., Ingebrandt, S., Mayer, D., Böcker-Meffert, S., Odenthal, M., Offenhäusser, A. (2004). Label free fully electronic nucleic acid detection system based on a field-effect transistor device. *Biosens Bioelectron* 19, 1723–1731.

Vagin, M. Y., Karyakin, A. A., Hianik, T. (2002). Surfactant bilayers for the direct electrochemical detection of affinity interactions. *Bioelectrochemistry* 56, 91–93.

Wang, J., Xu, D., Kawde, A-N., Polsky, R. (2001). Metal nanoparticle-based electrochemical stripping potentiometric detection of DNA hybridization. *Anal Chem* 73, 5576–5581.

Wang, J., Liu, G., Merkoçi, A. (2003). Electrochemical coding technology for simultaneous detection of multiple DNA targets. *J Am Chem Soc* 125, 3214–3215.

Weizmann, Y., Patolsky, F., Willner, I. (2001). Amplified detection of DNA and analysis of single-base mismatches by the catalyzed deposition of gold on Au-nanoparticles. *Analyst* 126, 1502–1504.

White, R. R., Sullenger, B. A., Rusconi, C. P. (2002). Developing aptamers into therapeutics. *J Clin Invest* 106, 929–934.

Willner, I., Patolsky, F., Weizmann, Y., Willner, B. (2002). Amplified detection of single-base mismatches in DNA using micro gravimetric quartz-crystal-microbalance transduction. *Talanta* 56, 847–856.

Wilson, D. S., Szostak, J. W. (1999). In vitro selection of functional nucleic acids. *Annu Rev Biochem* 68, 611–647.

Wu, Z-S., Guo, M-M., Zhang, S-B., Chen, C-R., Jiang, J-H., Shen, G-L., Yu, R. Q. (2007). Reusable electrochemical sensing platform for highly sensitive detection of small molecules based on structure-switching signaling aptamers. *Anal Chem* 79, 2933–2939.

Xiao, Y., Lubin, A. A., Heeger, A. J., Plaxco, K. W. (2005a). Label-free electronic detection of thrombin in blood serum by using an aptamer-based sensor. *Angew Chem Int Ed Engl* 44, 5456–5459.

Xiao, Y., Piorek, B. D., Plaxco, K. W., Heeger, A. J. (2005b). A reagentless signal-on architecture for electronic, aptamer-based sensors via target-induced strand displacement. *J Am Chem Soc* 127, 17990–17991.

Xu, D., Xu, D., Yu, X., Liu, Z., He, W., Ma, Z. (2005). Label-free electrochemical detection for aptamer-based array electrodes. *Anal Chem* 77, 5107–5113.

Yang, C. J., Jockusch, S., Vicens, M., Turro, N. J., Tan, W. (2005). Light switching excimer probes for rapid protein monitoring in complex biological fluids. *Proc Natl Acad Sci U S A* 102, 17278–17283.

You, K. M., Lee, S. H., Im, A., Lee, S. B. (2003). Aptamers as functional nucleic acids: in vitro selection and biotechnological applications. *Biotechnol Bioprocess Eng* 8, 64–75.

Zayats, M., Huang, Y., Gill, R., Ma, C. A., Willner, I. (2006). Label-free and reagentless aptamer-based sensors for small molecules. *J Am Chem Soc* 128, 13666–13667.

Zuo, X., Song, S., Zhang, J., Pan, D., Wang, L., Fan, C. (2007). A target-responsive electrochemical aptamer switch (TREAS) for reagentless detection of nanomolar ATP. *J Am Chem Soc* 129, 1042–1043.

CHAPTER 4

APTAMERS: HYBRIDS BETWEEN NATURE AND TECHNOLOGY

MORITZ K. BEISSENHIRTZ, EIK LEUPOLD, WALTER STÖCKLEIN, ULLA WOLLENBERGER, OLIVER PÄNKE, FRED LISDAT, and FRIEDER W. SCHELLER

4.1 INTRODUCTION

Among the "specifiers" used in bioanalysis enzymes, antibodies and ribonucleic acids are the most versatile. They interact in a specific manner with small molecules and biomacromolecules to control the complex processes of living cells. These interactions of enzymes with their substrates, inhibitors, activators, and regular macromolecules offer the potential to quantify these substances by appropriate analytical assays. On the other hand, antibodies and carbohydrates, which form a huge repertoire of diverse species, recognise an individual structural motif. Despite the almost unlimited diversity of antibodies, their function is restricted to one process: binding of the respective hapten or antigen. Only a few formats of immunoassays combine the binding event with the signal generation.

The application of nucleic acids as analytical tools mimics their function in transcription: for example, pairing of complementary nucleotides, gene regulation by proteinous transcription factors, and intercalation into the double strand by antibiotics. Aptamers present this broad spectrum of biological interactions of nucleic acids for use in cell biology, therapy, and analytics. This aspect is discussed in several chapters of this book.

Aptamers in Bioanalysis, Edited by Marco Mascini
Copyright © 2009 John Wiley & Sons, Inc.

4.2 SPECIFIC FEATURES OF APTAMERS

The aim of the present contribution is to illustrate important features in the analytical application of aptamers which are unique in relation to other recognition elements.

1. The effective charge transfer by double-stranded DNA should allow for the marker-free indication of base pairing and analyte binding to the aptamer.

2. Based on a defined three-dimensional structure of aptamers which include loops and hybridized sequences, the selectivity may approach that of antibodies. However, the high negative charge density causes a strong dependence on the ionic strength and imposes strong nonspecific interactions with positively charged targets or surfaces. This feature will be discussed for the specificity of cytochrome c binding.

3. Nucleic acids have been designed to realize molecular (LEGO) building sets. In addition to structural elements for molecular recognition, functional assemblies—so-called *molecular machines*—have also been realized. In the future, molecular recognition by nucleic acids should be combined with these technologically introduced molecular functions. In this respect they represent successful hybrids of nature and technology.

4.3 ELECTROCHEMICAL DETECTION OF NUCLEIC ACIDS

Hybridization of complementary oligonucleotide strands is a key feature of nucleic acids, which is also essential in the structure formation of aptamers. When this process occurs at a surface, the surface structure changes because of the duplex formation, which is much more rigid than single-stranded DNA. In addition, the charge distribution is altered because of the accumulation of negative charge connected with the DNA backbone. Both effects can be used for the detection of such a binding event.

The DNA duplex can be considered as a stacked-pi system, and there has been a long debate on the possibility of electron transfer through such a molecular structure (Fink and Schönenberger, 1999; Kelley et al., 1999; Ye and Jiang, 2000). It is clear, meanwhile, that DNA can effectively transport electrons under defined conditions, but it is critical to achieve this situation at an electrode. However, disregarding the pathway of electron transfer, DNA-bound redox labels can be detected sensitively at the electrode and thus used for the indication of hybridization. Alternatively, the charge distribution altered by the hybridization can be detected by impedance spectroscopy analyzing the interfacial properties of an electrode surface.

Recently, we presented a fast and reliable electrochemical method to quantify DNA oligonucleotides and detect single base-pair mismatches within DNA duplexes (Pänke et al., 2007). Using differential pulse voltammetry (DPV),

immobilized   nonlabeled target and   voltammetric detection
probe ssDNA   MB-labeled reporter
                     ssDNA

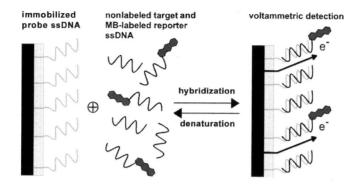

Figure 4.1   Voltammetric detection of DNA oligonucleotides. After immobilization of the ssDNA probe on gold (black), the electrode surface was additionally covered with mercaptobutanol (gray) to prevent unspecific binding. Competitive hybridization of MB-labeled reporter and nonlabeled target ssDNA. The amount of hybridized reporter DNA is gauged by differential pulse voltammetry. [From Pänke (2007).]

label-free detection of the DNA was achieved by the use of a competitive binding assay, as depicted in Figure 4.1.

The sensor electrode was fabricated by immobilizing a single-stranded DNA sequence (ssDNA) as a probe on gold by thiol chemistry. Regeneration of the modified electrode after DNA detection was possible and allowed repeated use. Competitive binding was observed after exposing the sensor to mixtures containing nonlabeled target ssDNA and covalently labeled reporter ssDNA. Methylene blue (MB) was used as a redox indicator, since it is a fast and reversible redox compound. Using the competitive binding protocol, 3 nM of nonlabeled target ssDNA was recognized by the sensor electrode, which is sufficient to detect DNA fragments from an amplification protocol such as polymerase chain reaction (PCR).

Moreover, the DNA-modified electrodes were capable of sensing single base-pair mismatches within the sequence of hybridized and nonlabeled DNA duplexes. After incubating the sensor with mixtures containing equivalent amounts of MB-labeled reporter ssDNA and nonlabeled target ssDNA carrying a single base-pair mismatch in relation to the immobilized probe ssDNA, the surface was analyzed by DPV. The lowest peak current was observed when the nonlabeled target ssDNA was fully complementary to the immobilized probe ssDNA and the MB-labeled reporter ssDNA had effectively been impeded from binding to the electrode surface. Accordingly, the highest peak current was measured when the nonlabeled target ssDNA was noncomplementary. Targets with single base-pair mismatches yielded peak currents between these two limiting currents, which were characteristic for each mismatch. The behavior observed can be explained in terms of reduced binding affinities caused by single base-pair mismatches. The lower the affinity of the nonlabeled ssDNA to

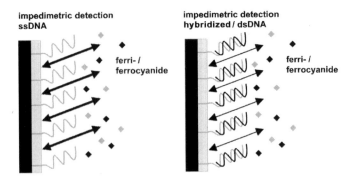

Figure 4.2 Impedimetric detection of DNA oligonucleotides. Sensor preparation as in Figure 4.1. The accessibility of the gold surface for ferri-/ferrocyanide depended on the amount of negative charge accumulated on the electrode surface. DNA hybridization increased the negative charge density and yielded increased charge transfer resistance as measured by impedance spectroscopy.

the sensor surface, the lower the ability to impede the binding of MB-labeled reporter ssDNA.

An alternative approach used the same sensor electrode, but DNA hybridization was detected by electrochemical impedance spectroscopy (EIS), which eliminates the necessity of applying a redox-labeled reporter oligonucleotide (Kafka et al., 2008). The ferri-/ferrocyanide system was used as a redox probe instead. DNA immobilized on the gold electrode affected the electrochemical conversion of the negatively charged ferri- and ferrocyanide ions, as depicted in Figure 4.2.

Hybridization of target ssDNA to immobilized probe ssDNA increases the negative charge at the surface, and ferri-/ferrocyanide conversion at the electrode was hampered, as indicated by an increased charge transfer resistance, R_{CT}, which was deduced from an analysis of the impedance spectra.

Additionally, this method was sensitive enough to detect single base-pair mismatches. The increase in R_{CT} after hybridization was lower than that in fully complementary DNA when a single base-pair mismatch was present in the DNA duplex. The different R_{CT} values suggest that in addition to the accumulation of negative charge after DNA hybridization, the single base-pair mismatch also induces a structural change within the immobilized DNA film. This change was characteristic for each mismatch and altered the accessibility of the electrode surface for the ferri-/ferrocyanide redox reaction.

4.4 CYTOCHROME c BINDING BY APTAMERS

Cytochrome c is an electron-transfer protein which adsorbs easily to negatively charged substrates, due to the presence of a cluster of positive charges on the protein surface. This cluster supports the communication with target proteins for directed electron transfer (Fridman et al., 2000). It also enables the formation of multilayers, formed by anionic polymers and cytochrome *c* (Beissenhirtz et al., 2004). Chinnapen and Sen (2002) reported the selection of aptamers against cytochrome *c*, starting with aptamers previously selected for hemin binding. The new aptamers had two domains, one for hemin and one for cytochrome *c* binding. Although the aptamers were selected under high-salt conditions, the influence of the salt concentration on aptamer–cytochrome *c* binding and the specificity with respect to control aptamers were not examined. We characterized the binding of cytochrome *c* to one of these aptamers, CH6, and to an unrelated control aptamer, 37-1F, using surface plasmon resonance spectroscopy (SPR) at different salt concentrations. Cytochrome *c* bound to CH6 as well as to the control aptamer 37-1F (Figure 4.3).

The 1.9-fold-higher binding of the protein to 37-1F may be due to the 1.9-fold-higher molar density of 37-1F compared with CH6. As the mass of immobilized DNA was approximately the same for both aptamers, the molar density of aptamer 37-1F was higher, due to its lower molecular mass.

To test the influence of ionic strength, cytochrome *c* binding to immobilized CH6 and control aptamer 31-1F was tested in buffers with NaCl concentrations increasing from 120 mM to 228 mM. The results show binding decrease with increasing NaCl concentrations for both aptamers (Figure 4.4). The signal for cytochrome *c* binding to the control aptamer was 1.8-fold higher than for CH6 binding at low NaCl concentrations, in agreement with Figure 4.3. The factor decreased to 1.6 at higher salt concentrations. This means that the cytochrome *c* binding is governed mainly by ionic interactions, while the cytochrome *c*–CH6 binding is slightly less salt sensitive than is control aptamer binding.

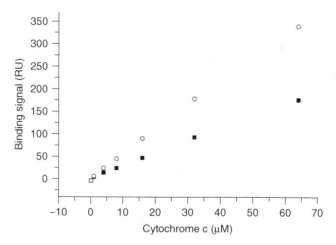

Figure 4.3 Standard curves for cytochrome *c* binding to immobilized aptamer CH6 (squares) and 37-1F (circles). The steady-state signals obtained during injection were corrected for the corresponding signal observed with the control flow cell containing immobilized streptavidin.

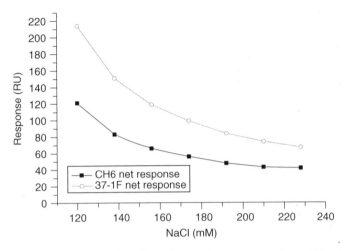

Figure 4.4 NaCl-dependent binding of cytochrome c to immobilized aptamers (steady-state signal; for curve correction, see Figure 4.3).

Further control experiments were performed to find out if other proteins also bind to the DNA aptamers. All proteins used as controls—human serum albumin (pI 4.9), bovine serum albumin (pI 4.7), myoglobin (pI 7.3), and mutarotase, 10 μM each—showed negligible binding to the aptamer compared with cytochrome c (pI 10.65). At least by comparison with these few proteins, cytochrome c binding to DNA aptamers can be regarded as specific, although this specificity could simply reflect the pI value of the protein.

It can be concluded that CH6 as well as other aptamers bind cytochrome c rather unspecifically by ionic interactions of the polyanionic DNA and a protein containing a cationic cluster, at least when the aptamers are immobilized, as in SPR experiments.

4.5 DNA MACHINES AND APTAMERS

Nucleic acids have been developed in nature to faithfully contain, transcribe, and proliferate genetic information, and thus knowledge of life itself, in all living organisms. In these processes, the opening, copying, and regeneration of the Watson–Crick base-paired double helix takes center stage.

In addition to this innate function, modern molecular biology has discovered alternative structural variations of RNA and DNA, which may be formed in a stable and at the same time reversible fashion, both spontaneously and upon induction by the presence of additional, non-nucleic acid molecules.

Over the last two decades, aptamers, stable nucleic acid structures, which have the ability to bind proteins, carbohydrates, and several other non-nucleic acid targets with affinities down to the nanomolar or even picomolar range, have proven to be feasible alternatives to antibodies in biosensorial detections.

However, although countless examples of fluorescence or electrochemical labels have been demonstrated as signal-generating components of aptamer-based sensors, the specific nucleic acid structure of these novel binders has so far not been used sufficiently to improve the sensitivities, and thus the detection limit, of sensorial approaches. A combination of aptamers and DNA machines might take the field to new heights.

Unlike proteins, nucleic acids can easily be designed to undergo several structural permutations without irreversible denaturation. The simple computer-assisted design of DNA sequences and the cheap and reliable automated synthesis of oligonucleotides make possible the concept of active DNA assemblies which can increase the sensorial output by converting a single binding event into a series of steps or a continuous process which might, in turn, be applied to the generation of an output signal, thus amplifying the signal-generating process and reducing the detection limits by orders of magnitude.

DNA machines are now understood to be assemblies of nucleic acids which upon outside stimulation undergo a series of structural permutations which are predesigned in their respective sequences (Beissenhirtz and Willner, 2006). These reactions are powered by outside "fuel" molecules, often nucleic acids themselves, which are converted into accumulating "waste," much like gasoline-driven macroscopic machines in our daily life. Exhaustion of one of the fuel components brings the system to a stop. Figure 4.5 shows the general mode of operation of most DNA machines.

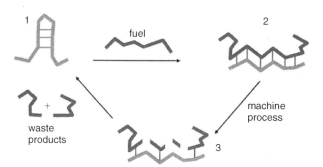

Figure 4.5   General operation mode of DNA machines exemplified by the catalytic scission of a substrate DNA strand. In the absence of the fuel–analyte molecule (blue), the DNA machine is inactive (1). Binding of the fuel and subsequent hybridization start the machine's operations (2), in this case a catalytical DNA cleavage reaction (3). Release of waste products due to reduced base pairing resets the machine to its starting configuration (1) and begins the next round of action. Note that different examples of machines performing a vast array of actions other than DNA cutting are shown. (See insert for color representation.)

In the last decade, several well-designed experiments have demonstrated DNA-based systems that mimic machinelike macroscopic systems such as tweezers (Yurke et al., 2000), gears (Tian and Mao, 2004), and walking systems (Sherman and Seeman, 2004), as well as more compliated novel assemblies utilizing DNAzyme activities or replication systems. The general concept of producing a predefined type of product (the waste) upon encountering a certain target–analyte molecule (such as the fuel) bears a striking resemblance to the definition of biosensors in which the analyte molecule is used to produce some type of signal, often a signal molecule that has no or little structural similarity to the target (such as hydrogen peroxide in electrochemical glucose sensors) but can be used to obtain quantitative or qualitative information about the target molecule's presence. The use of fluorescence labels and their localization in the vicinity of quencher or donor molecules within the waste product has been a successful approach to generate signals. One concept that has reached the market stage is the use of DNA beacons. These are fluorescence donor/acceptor-modified hairpins which upon hybridization with a target nucleic acid change from their stem–loop structure and thereby generate a long distance between donor and acceptor, which prohibits successful energy transfer between the labels. The decrease in fluorescence energy transfer has been applied to quantify the target DNA or RNA concentration, even in in vivo systems such as living cells (Yang et al., 2005; Santangelo et al., 2006).

Most real-time PCR kits are based on such DNA beacons, which upon replication are incorporated into the newly formed strand, or conversely, digested by DNA polymerases, in both cases generating a fluorescence signal in dependence on the target DNA concentration.

The structural changes in redox-modified single-chain DNA have been applied in electrochemical biosensors. In one example, a hairpin structure on an electrode surface places the ferrocene label on its free 5' end in close proximity to the electrode, thus allowing a constant electron flow (Fan et al., 2003). Hybridization with its target DNA leads to formation of the duplex, in which the distance between the redox label and the electrode surface is increased beyond the possibility of electron transfer.

Similarly, an electrode modified with a thrombin aptamer containing a redox label on its free end creates a steady current in the absence of the target (Xiao et al., 2005). When thrombin is added, the aptamer structure forms and removes the label from the surface, thus interrupting the electron transfer (Figure 4.6).

One common trait of many modern biosensors is their ability to translate one input signal into a series of events, thus allowing the production of more than one signal "bit" per input and displaying a catalytical signal generation. The DNA assemblies described earlier fall short of this objective. However, some systems of higher complexity might satisfy such requirements.

The powerful SELEX process allows the identification and formation of catalytically active nucleic acids (Ellington and Szostak, 1990; Tuerk and Gold, 1990). Several DNA and RNA sequences capable of selectively nicking or cutting nucleic acids have been discovered, among them a DNAzyme that is active

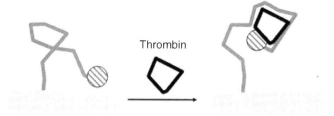

Figure 4.6   Thrombin biosensor based on a ferrocene-labeled aptamer. In the absence of thrombin, the aptamer (green) displays a random coil structure, allowing contact between the ferrocene label (striped ball) and the electrode. Addition of thrombin forces the aptamer into the thrombin-binding structure, thus removing the redox label from the electrode, allowing no further electrochemical contact. (See insert for color representation.)

only in the presence of lead (Pb^{2+}) molecules. Attempts have been made to utilize this capacity for biosensorial purposes. Among them is a sensor for lead ions, which is based on the color changes exhibited by gold nanoparticles when in close proximity or at a greater distance from each other (Liu and Lu, 2003).

The sensor uses DNA-modified gold nanoparticles which are interconnected by single-stranded DNA to form complexes of significant proportions. The DNAzyme can anneal to the linking DNA single strand in between. In the presence of lead ions, the DNAzyme will consecutively sever the bridging units between the nanoparticles and depose of the complex catalytically, leaving only unconnected single nanoparticles. The spectral differences between aggregated and dissolved gold nanoparticles allow the discrimination of these two states, and the time course of the dissolution of the complex can easily be used to quantify the lead ion concentration.

A higher degree of complexity is presented by a DNA machine that used the protein FokI to cleave nucleic acid double strands and was applied to the detection of a mutated human genetic sequence implied in the outbreak of Tay–Sachs disease (Weizmann et al., 2006a). The FokI protein is activated in the presence of double-stranded DNA containing its specific binding site. Upon binding, the protein cleaves the DNA downstream from the recognition site, yielding sticky ends.

A hairpin nucleic acid has been designed to contain the binding site of the protein FokI, but not the scission sequence. Upon hybridization with DNA containing the mutant sequence (responsible for the outbreak of Tay–Sachs), the scission site is formed, and the FokI can cut the double strand. In contrast, hybridization to the healthy wild-type sequence leads to a mismatch in the vicinity of the scission region and therefore inhibits FokI.

After scission of the mutant DNA, the FokI/DNA assembly remains intact as a FokI "cutter unit," which can hybridize to a signaling hairpin containing a fluorophore–quencher pair at its end. This hairpin structure acts as the fuel for the DNA machine. After hybridization, the FokI cutter unit cleaves the hybridized fuel molecule, and upon its spontaneous release the waste product diffuses away, thereby separating flourophore and quencher, allowing simple readout for detection of the Tay–Sachs mutant genetic sequence. This signaling process continues without further stimulation from the outside until eventual exhaustion of the signaling hairpin DNA. Therefore, one target DNA strand activates one cutting unit, which processes countless fuel strands, leading to a catalytically enhanced signal generation process (Figure 4.7).

This FokI/DNA machine has been connected to a thrombin-binding aptamer to form a machine for the controlled release of a protein upon stimulation (Beyer and Simmel, 2006). In contrast to the detection system described above, the cut fuel strand yields no fluorescence but a single-stranded DNA that can hybridize to the thrombin–aptamer sequence. After encountering an aptamer–thrombin couple, this "waste" DNA hybridizes to the aptamer, thereby breaking its structure, forming the DNA duplex, and releasing the thrombin molecule previously bound by the aptamer. Although the triggered release of thrombin serves no purpose in the applied sciences, the availability of dozens of known aptamer sequences might in the future allow this concept to be used to control transport-and-release systems, which might one day facilitate the precise delivery of pharmaceutical compounds inside the human body to their respective sites of action.

Another example of a DNA machine that can easily be triggered by the actions of an aptamer is based on the isothermal strand displacement amplification (Weizmann et al., 2006b). In this case, a single-stranded DNA track contains three distinct domains: domain I for the binding of a primer DNA to start replication; domain II, which can, upon duplex formation, be recognized by a nicking endonuclease (which cuts one strand of a double-stranded structure); and domain

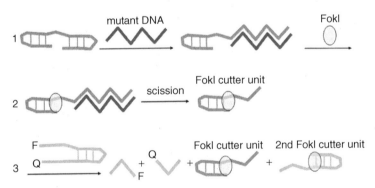

Figure 4.7   FokI/DNA machine for the detection of mutant DNA using catalytical signal amplification. (1) The machine is contained in a double hairpin. Addition of the mutant DNA (the analyte) opens the hairpin and creates binding and scission sites for the FokI protein. (2) the first scission creates the FokI cutter unit. (3) Hybridization of the cutter with a signaling hairpin containing fluorophor (F) and quencher (Q) with the cutter leads to scission of the signaling hairpin and release of fluorophor and quencher on different oligonucleotides. An increase in fluorescence signal can be detected. The remainder of the signaling hairpin binds to free FokI to form additional cutter units. Step 3 is repeated until exhaustion of the signaling hairpin.

III, which encodes an output sequence for the signal generation. In the presence of a strand-displacing DNA polymerase, a nicking enzyme, and a sufficient amount of dNTPs, the binding of a primer oligonucleotide to position I of the track starts polymerization, yielding a DNA duplex. The nicking enzyme can now recognize its binding site (II) and nicks the newly formed product strand. The DNA polymerase can now bind to the nicked position at II and restart polymerization of the primer side of the new duplex, thereby displacing and releasing the product strand formed at position III, and this process continues until the supply of dNTPs is exhausted (Figure 4.8).

This newly formed single-stranded product strand can be used for the generation of biosensorial signals in response to binding of the primer DNA. The formation of a catalytically active peroxidase-mimicking DNAzyme, as well as the controlled bridging of DNA-modified gold nanoparticles (Beissenhirtz et al., 2007) which were interconnected by the product strand, and the formation of photoelectrochemical assemblies (Freeman et al., 2007) have been shown to be model systems for the detection of viral DNA (phage M13).

In these sensors the primer sequence is locked inside a hairpin DNA, which in the absence of viral nucleic acids (the analyte) remains closed and thus prohibits binding of the primer to the DNA machine track. Only in the presence of the

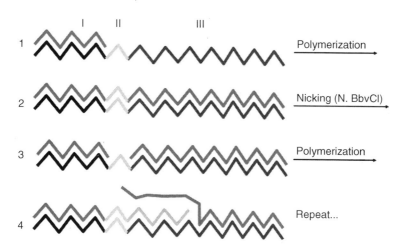

Figure 4.8 DNA machine based on strand displacement amplification. The "track" consists of a primer binding sequence (black, I), a nicking recognition sequence (green, II), and a template for the product DNA (blue, III). (1) Binding of the primer starts replication of the entire track by a DNA polymerase. (2) The fully replicated duplex is nicked on the newly formed strand by the nicking enzyme, regenerating the primer site (3). (4) The start of polymerization at the primer site leads to the formation of a new copy DNA strand (gray), and in the process the product strand (red) is pushed off the track by the polymerase. Subsequent nicking and polymerization steps lead to the accumulation of single-stranded product strands. (See insert for color representation.)

target DNA is the hairpin opened, at which point the primer sequence can attach itself to the track, and replication starts.

Although this concept has so far been applied primarily to the detection of DNA, it is ideal for use in aptamer-based assays. Willner and co-workers (Shlya-hovsky et al., 2007) recently described a system in which a cocaine-binding aptamer was included in the primer region (I) of the DNA track. In the absence of cocaine, the start of replication is inhibited by the presence of a blocking oligonucleotide, which is designed to form an incorrect nicking site (II). The addition of cocaine induces formation of the aptamer structure, which replaces the blocking nucleic acid and triggers the machine's functions to start replication of a product DNA. This oligonucleotide, in turn, opens a molecular beacon, generating a fluorescence signal depending on the amount of cocaine present.

The DNA machines described in this work, particularly those that start a catalytic process upon exposure to their target molecules, may offer new approaches to the design of clever biosensorial strategies which due to their catalytic nature may improve sensitivities for sensors based on the interaction of aptamers with their respective targets.

Acknowledgment

Financial support from the German Ministry of Education and Research (BMBF) through aptamer project 03i1308-B and iPOC Research Group 03 IP 515 is gratefully acknowledged.

REFERENCES

Beissenhirtz, M. K., Willner, I. (2006). DNA-based machines. *Org Biomol Chem* 4, 3392–3401.

Beissenhirtz, M. K., Scheller, F. W., Stöcklein, W. F. M., Kurth, D. G., Möhwald, H., Lisdat, F. (2004). Electroactive cytochrome *c* multilayers within a polyelectrolyte assembly. *Angew Chem Int Ed Engl* 43, 4357–4360.

Beissenhirtz, M. K., Elnathan, R., Weizmann, Y., Willner, I. (2007). The aggregation of Au nanoparticles by an autonomous DNA machine detects viruses. *Small.* 3, 375–379.

Beyer, S., Simmel, F. C. (2006). A modular DNA signal translator for the controlled release of a protein by an aptamer. *Nucleic Acids Res* 34, 1581–1587.

Chinnapen, D. J. F., Sen, D. (2002). Hemin-stimulated docking of cytochrome *c* to a hemin–DNA aptamer complex. *Biochemistry* 41, 5202–5212.

Ellington, A. D., Szostak, J. W. (1990). In vitro selection of RNA molecules that bind specific ligands. *Nature* 346, 818–822.

Fan, C., Plaxco, K. W., Heeger, A. J. (2003). Electrochemical interrogation of conformational changes as a reagentless method for the sequence-specific detection of DNA. *Proc Natl Acad Sci USA* 100, 9134–9137.

Fink, H. W., Schönenberger, C. (1999). Electrical conduction through DNA molecules. *Nature* 398, 407–410.

Freeman, R. Gill, R., Beissenhirtz, M. K., Willner, I. (2007). *Photochem Photobiol Sci.* 6, 416–422.

Fridman, V., Wollenberger, U., Bogdanovskaya, V., Lisdat, F., Ruzgas, T., Lindgren, A., Gorton, L., Scheller, F. W. (2000). Electrochemical investigation of cellobiose oxidation by cellobiose dehydrogenase in the presence of cytochrome *c* as mediator. *Biochem Soc Trans* 28, 63–70.

Kafka, J., Pänke, O., Lisdat, F. (2008). A label-free DNA sensor based on impedance spectroscopy. *Electrochim Acta* 53, 7467–7474.

Kelley, S., Jackson, N., Hill, M., Barton, J. (1999). Long-range electron transfer through DNA films. *Angew Chem* 111, 991–996.

Liu, J., Lu, Y. (2003). A colorimetric lead biosensor using DNAzyme-directed assembly of gold nanoparticles. *J Am Chem Soc* 125, 6642–6643.

Pänke, O., Kirbs, A., Lisdat, F. (2007). Voltammetric detection of single base-pair mismatches and quantification of label-free target ssDNA using a competitive binding assay. *Biosens Bioelectron* 22, 2656–2662.

Santangelo, P., Nittin, N., LaConte, L., Woolums, A., Bao, G. (2006). Live-cell characterization and analysis of a clinical isolate of bovine respiratory syncytial virus, using molecular beacons. *J Virol* 80, 682–688.

Sherman, W. B., Seeman, N. C. (2004). A precisely controlled DNA biped walking device. *Nano Lett* 4, 1203–1207.

Shlyahovsky, B., Li, D., Weizmann, Y., Nowarski, R., Kotler, M., Willner, I. (2007). Spotlighting of cocaine by an autonomous aptamer-based machine. *J Am Chem Soc* 129, 3814–3815.

Tian Y., Mao, C. (2004). Molecular gears: a pair of DNA circles continuously rolls against each other. *J Am Chem Soc* 126, 11410–11411.

Tuerk, C., Gold, L. (1990). Systematic evolution of ligands by exponential enrichment: RNA ligands to bacteriophage T4 DNA polymerase. *Science* 249, 505–510.

Weizmann, Y., Cheglakov, Z., Pavlov, V., Willner, I. (2006a). Autonomous fueled mechanical replication of nucleic acid templates for the amplified optical detection of DNA. *Angew Chem Int Ed Engl* 45, 2238–2242.

Weizmann, Y., Beissenhirtz, M. K., Cheglakov, Z., Nowarski, R., Kotler, M., Willner, I. (2006b). A virus spotlighted by an autonomous DNA machine. *Angew Chem Int Ed Engl* 45, 7384–7388.

Xiao, Y., Lubin, A. A., Heeger, A. J., Plaxco, K. W. (2005). Label-free electronic detection of thrombin in blood serum by using an aptamer-based sensor. *Angew Chem Int Ed Engl* 44, 5456–5459.

Yang, L., Cao, Z. H., Lin, Y. M., Wood, W. C., Staley, C. A. (2005). Molecular beacon imaging of tumor marker gene expression in pancreatic cancer cell. *Cancer Biol Ther* 4, 561–570.

Ye, Y., Jiang, Y. (2000). Electronic structures and long-range electron transfer through DNA molecules. *Int J Quant Chem* 78, 112–130.

Yurke, B., Turberfield, A. J., Mills, A. P., Jr., Simmel, F. C., Neumann, J. L. (2000). A DNA-fuelled molecular machine made of DNA. *Nature* 406, 605–608.

CHAPTER 5

DETECTION OF PROTEIN–APTAMER INTERACTIONS BY MEANS OF ELECTROCHEMICAL INDICATORS AND TRANSVERSE SHEAR MODE METHOD

TIBOR HIANIK

5.1 INTRODUCTION

Biosensors based on DNA or RNA aptamers (aptasensors) are of considerable interest as an alternative to biosensors based on antibodies. Although the first SELEX-related patent was filed in 1989 (Tuerk and Gold, 1990), the potentialities of aptamer-based biosensors have not been fully realized, due to problems with aptamer stability during immobilization and signal registration. Solutions to several problems related to the practical application of aptamers are still being sought: for example, how immobilization of aptamers to the films supported and their microenvironment will affect aptamer structure and aptamer–ligand interactions. Problems are connected with the application of aptamers in complex biological systems in which interference with other molecules could take place. RNA aptamers are especially unstable, due to cleavage by nucleases. So far, mostly radiolabeled aptamers were used: for example, for quantification of protein kinase (Conrad and Ellington, 1996) or in vivo detection of clots (Dougan et al., 1997). However, to be widely employed in clinical practice, aptamers must be detected via a nonradioisotope method with comparable sensitivity; for example, aptamers can be linked covalently to an enzyme (Osborne et al., 1997), or fluorescently labeled aptamers can be exploited (McCauley et al., 2003). Moreover, the most reliable and cost-effective approach would be the exploitation of

Aptamers in Bioanalysis, Edited by Marco Mascini
Copyright © 2009 John Wiley & Sons, Inc.

direct physical methods that do not require labeling of aptamers by additional chemical ligands. This highly promising route is currently a considerable focus of many laboratories, and recent achievements are presented in this book. Among perspective analytical methods for the detection of protein–aptamer interaction, electrochemical indicators can be used. Another popular method is based on mass detection using a quartz crystal microbalance technique. Despite the high analytical value of this relatively simple method, it cannot provide sufficient information about processes at complex surfaces. Changes in the mass of the crystal can be masked by the viscoelastic properties of the layers and the friction between the biolayer and the surrounding liquid. Therefore, the complex impedance of crystal oscillation should be measured and analyzed. This transverse shear mode (TSM) method has recently been used to study protein–aptamer interactions and has already demonstrated high potentiality for detailed study of the mechanisms of molecular interactions at surfaces.

In this chapter the focus is on the use of electrochemical indicators and of the TSM method for the study of protein–aptamer interactions. Aptamers sensitive to thrombin are used for the fabrication of biosensors. Attention is given to methods of immobilization of aptamers to a solid support and how this affects the interaction of thrombin with the aptamer. The effects of aptamer structure, of the presence of the thrombin inhibitor heparin, and of the ions and pH on the binding properties of aptamer are presented. The advantages of electrochemical indicators and the TSM method for the detection of thrombin–aptamer interactions are demonstrated.

5.2   IMMOBILIZATION OF APTAMERS ON A SOLID SUPPORT

Aptamers can be immobilized on a solid support by various methods. In principle these methods are similar to those used previously for immobilization of single- or double-stranded DNA in genosensors or DNA biosensors for the detection of DNA damage (Pividori et al., 2000). Methods of immobilization based on physical adsorption of DNA by means of electrostatic interactions are not suitable, due to the low stability caused by aptamer desorption from the surface. The most effective methods are based on chemisorption of thiol-labeled aptamers to a gold surface (Pavlov et al., 2004) or on the strong affinity of biotin to avidin, streptavidin, or neutravidin. In the latter case, one end of DNA or RNA aptamer is modified by biotin. The solid support is covered by streptavidin, avidin (Liss et al., 2002), or neutravidin (Cavic and Thompson, 2002). In the case of avidin and neutravidin, the proteins are usually linked chemically to the organic layer formed by 3,3′-dithiopropionic acid di(N-succinimidyl ester) (DSP) (Liss et al., 2002). Neutravidin is very convenient because it can be chemisorbed directly on gold and does not require additional chemical modification of the surface (Cavic and Thompson, 2002). Recently, poly(amidoamine) (PAMAM) dendrimers were used for aptamer immobilization (Hianik et al., 2007). PAMAM dendrimers are globular macromolecules with amino terminal groups. Using glutaraldehyde, it is possible to cross-link avidin to a dendrimer surface, so it can then be used for immobilization of biotinylated aptamers. The advantage of dedrimers is their high stability and relatively large surface compared with those of flat electrodes (Svobodová et al., 2004). So and co-workers (2005) reported immobilization of thrombin aptamer onto the surface of carbon nanotubes. The nanotubes were pretreated with carbodiimidazole-activated Tween 20 (CDI-Tween). The 3′ end of the thrombin aptamer was modified by $-NH_2$ groups, which allowed covalent binding of aptamer to CDI-Tween. The use of conducting nanotubes made it possible to fabricate field-effect transistor–based biosensors for the detection of thrombin. In choosing the method of immobilization it is important to provide sufficient conformational freedom of the aptamer. This can also be achieved by using sufficiently long spacers. For example, Liss et al. (2002) showed that substantial improvement of the binding properties of anti-IgE aptamer can be reached by extension of the aptamer length. Aptamer can also be immobilized on glass slides (Potyrailo et al., 1998) or on a silicon surface using ultraviolet radiation (Wei et al., 2003). Some examples of immobilization of aptamers to a solid support are showed schematically in Figure 5.1.

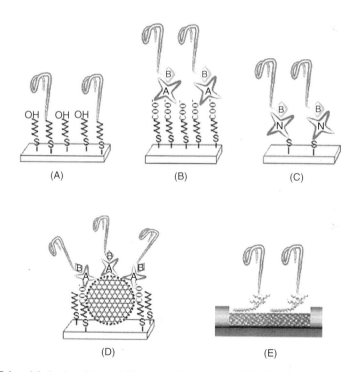

Figure 5.1 Methods of immobilization of aptamers. (A) Chemisorption. Biotinylated aptamer is immobilized on a surface covered by (B) avidin; (C) neutravidin; (D) avidin on a surface of poly(amidoamine) dendrimers (PAMAM). (E) Aptamer is immobilized on carbon nanotubes activated by Tween 20. [(E) Adapted from So et al. (2005), with permission from the American Chemical Society.]

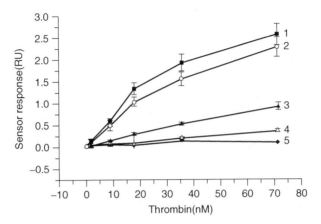

Figure 5.2   SPR sensor response as a function of thrombin concentration for different types of aptamer immobilization and different types of aptamer: APTA immobilized on a (1) streptavidin layer; (2) avidin; (3) dendrimers covered by avidin; (4) SH-APTA. For comparison, nonspecific interaction of HSA with a sensor surface created by APTA immobilized on a streptavidin layer is shown (curve 5). The composition of aptamers was as follows: APTA: 3′–biotin–GGG TTT TCA CTT TTG TGG GTT GGT GTG GTT GG–5′, SH-APTA: SH–$(CH_2)_6$–GGT TGG TGT GGT TGG–5′. Results represent mean ± S.D. obtained from three independent experiments. [Adapted from Ostatná et al. (2008), with permission of Springer.]

The method of immobilization of aptamer to a solid support affects the binding of the protein. This is demonstrated on a plot of the response of aptamer biosensor as a function of thrombin concentration (Figure 5.2). The response represents the shift of resonant wavelength [in relative units (RU)] measured by a multichannel surface plasmon resonance (SPR) method (Piliarik et al., 2005). We can see that a higher response takes place for a sensor composed of APTA immobilized on a streptavidin layer. The nonspecific interaction of aptamers with human serum albumin (HSA) does not cause the resonance angle shift reported above. However, as shown in Figure 5.2, at a relatively low concentration of thrombin (up to 35 nM), nonspecific interaction cannot be distinguished from specific interaction for aptamer modified by a thiol group. Therefore, for analytical purposes the immobilization of biotinylated aptamer to a streptavidin or avidin layer is preferable.

5.3   DETECTION OF APTAMER–LIGAND INTERACTIONS

Aptamer–ligand interactions can be detected by conventional assay based on radioactive labeling, chromatography, capillary electrophoresis, and mass spectrometry. These methods have been reviewed by Tombelli et al. (2005a). A novel approach in detection aptamer–ligand interaction is connected with aptasensors.

In aptasensors the aptamer is immobilized onto a solid support, and the signal following binding of the ligand is detected by various methods: mostly electrochemical, acoustical, and optical. Recent progress in aptasensors has been reviewed (Leca-Bouvier and Blum, 2005; Tombelli et al. 2005a; Hianik, 2007). However, this field of research is progressing so strongly that new achievements have already appeared in the literature, especially those focused on electrochemical methods of detection and on the application of nanotubes and nanoparticles in aptasensors. We therefore describe these new directions in more detail.

5.3.1   Electrochemical Methods

Electrochemical methods of detection affinity interactions at surfaces are rather effective, due to their relative simplicity and low cost. An amperometric aptasensor based on a sandwich assay was proposed by Ikebukuro et al. (2004). They used two aptamers selective for thrombin (Figure 5.3). A 15-mer thiol-labeled aptamer selective to the fibrinogen-binding site of thrombin was immobilized on a gold surface. A second, 29-mer biotinylated aptamer sensitive to the heparin-binding site of thrombin was labeled by enzyme–glucose dehydrogenase from *Burkholdelia cepacia* (GDHPc) (Ikebukuro et al., 2004) or by oxygen-insensitive pyrroquinoline quinone glucose dehydrogenase from *Acinobacter calcoaceticus* [(PQQ)GDH] (Ikebukuro et al., 2005). First, thrombin was added to a measuring cell containing aptasensor immobilized on the surface of a gold electrode. Then enzyme-labeled aptamer was added and attached to the sensor surface, due to the existence of a second binding site at thrombin (Figure 5.3A).

Upon the presence of glucose, the substrate for the enzyme, an electrochemical signal (current) appeared. The current was proportional to the amount of thrombin at the sensor surface. The detection limit was 1 μM for GDHPc and/or 10 nM for the (PQQ)GDH aptasensor. Using fluorescence polarization, authors also proved binding of aptamers to a different binding site of the thrombin. The sensor was selective to thrombin and a negligible response tooks place only in the presence of bovine serum albumin. In this work there were also reported difficulties with sensor regeneration due to strong binding of aptamers to a thrombin.

Electrochemical indicator methods are based on the use of a redox probe that undergoes an oxidation and reduction transition due to electron transfer from an electrode surface to a probe. In 2005, several papers were published that used methylene blue (MB) as an electrochemical indicator. Methylene blue is a positively charged low-molecular-mass compound that can be reduced by two electrons to leucomethylene blue (LB). The reduction process can be monitored effectively by differential pulse voltammetry, cyclic voltammetry, or coulometry. In the presence of a redox probe $Fe(CN)_6^-$, the LB is oxidized to MB and system is regenerated (Boon et al., 2000; Ostatná et al. 2005). In papers of Hianik et al. (2005, 2007) the MB was used as an indicator for the detection of interaction of human thrombin with DNA aptamer. The method of detection is shown

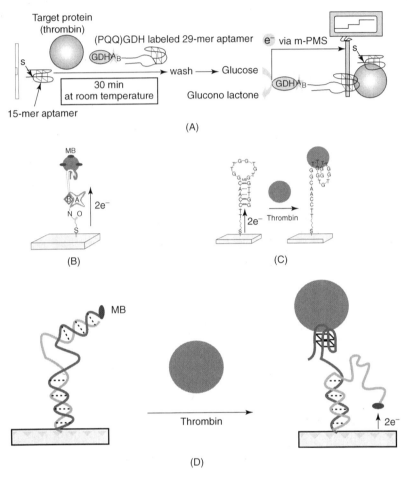

Figure 5.3 Detection of thrombin–aptamer interactions by electrochemical methods. (A) Sandwiched assay according to Ikebukuro et al. (2005). Thiol-modified 15-mer aptamer sensitive to the fibrinogen-binding site of thrombin was immobilized on the surface of a gold electrode. After the addition of thrombin, a second, 29-mer enzyme-labeled aptamer sensitive to the heparin-binding site of thrombin was added. The thrombin was detected by means of the measurement current due to glucose degradation. (B)–(D) Detection of thrombin using the electrochemical indicator methylene blue (MB). (B) MB binds to both aptamer and thrombin. Binding of thrombin with adsorbed MB to aptamer resulted in increased charge transfer from electrode to MB. (C) MB is intercalated between base pairs of aptamer beacon, and electron transfer from electrode to MB is favorable. Addition of thrombin resulted in changes in conformation, release of MB from the double helix, and decrease in electron transport. (D) (MB)-tagged oligonucleotide with aptamer forms a rigid duplex that prevents the MB tag from approaching the electrode surface. The addition of thrombin resulted in quadruplex formation and the approach of MB to a surface of the electrode. In this configuration, reduction of MB is favorable. [(A) From Ikebukuro et al. (2005), with permission from Elsevier; (B) from Hianik et al. (2005); (C) from Bang et al. (2005); (D) adapted from Xiao et al. (2005b), with permission from the American Chemical Society.] (See insert for color representation.)

schematically in Figure 5.3B and was described in the 2005 paper of Hianik et al. Briefly, MB binds to both DNA and to the protein. For charge transfer from the electrode to the MB (i.e., for MB reduction) it is important that MB be close to the electrode surface. Therefore, the charge transfer from the electrode to MB should be more intensive to proteins that bind specifically to the aptamer compared with that in a solution. Because both binding sites at thrombin are positively charged, it is expected that positively charged MB will not bind to these sides. The isoelectric point of thrombin is 7 to 7.6 (Berg et al., 1979); therefore, under experimental conditions the thrombin molecule should be at least 50% negatively charged. Thus, MB could be adsorbed physically to the thrombin, due to electrostatic interactions. The charge consumption can be measured by cyclic voltammetry (CV) or by differential pulse voltammetry (DPV) methods by determination of the area under the reduction peak (Peerce and Bard, 1980). Before determination of the calibration curve, the sensor should be incubated with 2 μM MB for 1 hour. Then the calibration curve can be measured by means of DPV. First, DVP should be determined for a sensor without protein and then in a sensor with an increased amount of protein (Figure 5.4).

An example of the DPV records for an aptamer-based sensor for thrombin is shown in Figure 5.4A. By integration of DPV, the total charge transfer for each record should be determined and the calibration curve can be constructed as a plot of relative changes of charge transfer, $\Delta Q/Q_0 = (Q - Q_0)/Q_0$, versus the concentration of protein, where Q_0 is charge consumption without analyte and Q is that at a particular thrombin concentration. An example of a calibration curve for two independently prepared electrodes is shown in Figure 5.4B. It is seen that results are easily reproducible. Statistical analysis performed earlier by Hianik et al. (2005) revealed that the standard error is approximately 11%. Interference of this aptasensor with other compounds, human serum albumin (HSA), and human IgG are relatively low. An example is shown in Figure 5.5, where the calibration curve for thrombin is compared with those for HSA and IgG. Note that concentrations of HSA and IgG are much higher than that of thrombin.

We have shown (Hianik et al., 2007) that this sensor has a sensitivity and detection limit (around 5 nM) comparable to that in the mass detection (QCM) and SPR methods (Ostatná et al., 2008) and is sufficiently selective compared with nonspecific binding of HSA or human immunoglobulin G (IgG). The detection limit obtained is sufficient for detection thrombin in real blood samples (the physiological concentration of thrombin is in the range low nanomolar to low micromolar (Lee and Walt, 2000).

The affinity of thrombin to the aptamer depends on the ionic strength, on the composition of buffer, and on pH. To check this we used an electrochemical indicator to detect the binding of thrombin to aptamer in the electrolyte of a different concentration of NaCl as well as for three different pH values.

Figure 5.6 shows a plot of relative changes of charge transfer as a function of thrombin concentration for various concentrations of NaCl (X = 0 to 500 mM) in a buffer. Aptamer was immobilized by avidin–biotin technology (Hianik et al.,

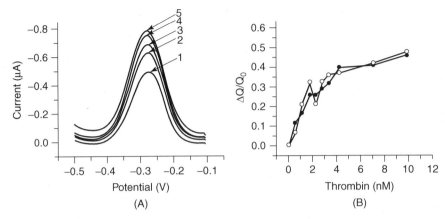

Figure 5.4   (A) Differential pulse voltammogram for (1) an aptamer without thrombin and for an aptamer at different thrombin concentrations: (2) 0.28; (3) 0.57; (4) 1.14; (5) 6.8 nM. (B) A plot of the relative changes of charge consumption during the reduction of MB $\Delta Q/Q_0$ ($\Delta Q = Q - Q_0$, where Q_0 is charge transfer without thrombin and Q that at certain thrombin concentration) as a function of the concentration of thrombin for two independently prepared sensors. A 32-mer DNA aptamer (3'–GGG TTT TCA CTT TTG TGG GTT GGA CGG GAT GG–5') was modified by biotin at the 3' end. This aptamer was sensitive to the heparin-binding site of thrombin. Aptamer was immobilized at the surface of a gold electrode (CH Instruments in the United States) of diameter 2 mm. As an electrolyte, 140 mM NaCl, 5 mM KCl, 1 mM CaCl$_2$, 1 mM MgCl$_2$, and 20 mM TRIS, pH 7.4 containing 2 μmol L^{-1} MB was used. The charge transfer was determined from differential pulse voltammetry measurements performed by means of potentiostat CHI 410 (CH Instruments). [Adapted from Hianik et al. (2005), with permission from Elsevier.]

2007). We can see that increasing the concentration of NaCl decreases the sensitivity of the sensor. This effect could be connected with either the shielding of negative charges at the DNA aptamer and the protein surface or with changes in the conformation of the binding site of the aptamer. For more understanding of the mechanisms of influence of Na$^+$ ions on the binding properties of the thrombin to the aptamer, we compared the effect of the ionic strength on charge transport between an electrode with immobilized aptamer and MB with and without thrombin.

Figure 5.7 shows the dependence of the relative changes in charge transport as a function of the concentration of NaCl in the presence of 2 μM MB without thrombin (curve 1) and in the presence of 10 nM thrombin (curve 2). We can see that in both cases the charge transport decreases with increasing ionic strength. It is likely that the shielding effect is a dominant process that affects the charge transport. The shielding of negative charges of DNA aptamer and protein could result in a decrease in binding of positively charged MB to DNA or to protein molecules, and consequently, the charge transport should decrease. It has been

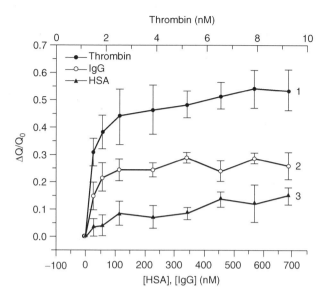

Figure 5.5 Relative changes in charge consumption during reduction of MB as a function of the concentration of (1) thrombin, (2) IgG, or (3) HSA [$\Delta Q/Q_0 = (Q - Q_0)/Q_0$, where Q_0 is charge consumption without an analyte and Q that at a certain thrombin, IgG, or HSA concentration]. Results represent the mean ($\pm$ S.D) obtained for three independent experiments. [Adapted from Hianik et al. (2005), with permission from Elsevier.]

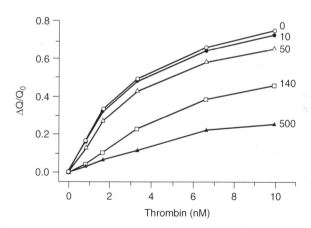

Figure 5.6 Relative changes in charge transfer $\Delta Q/Q_0$ as a function of thrombin concentration and for various concentration of NaCl ($X = 0$ to 500 mM) in a buffer (X NaCl, 5 mM KCl, 1 mM $CaCl_2$, 1 mM $MgCl_2$, 20 mM TRIS, pH 7.4). The concentration of NaCl in mM is showed at the curves. A 32-mer DNA aptamer ($3'$–biotin–GGG TTT TCA CTT TTG TGG GTT GGT GTG GTT GG–$5'$) was immobilized to the gold surface covered by avidin. [Adapted from Hianik et al. (2007), with permission from Elsevier.]

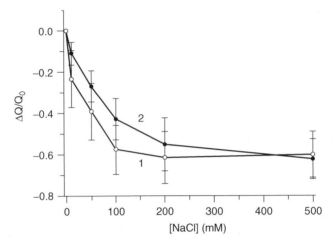

Figure 5.7 Relative changes in charge transfer $\Delta Q/Q_0$ ($\Delta Q = Q - Q_0$, where Q_0 is the charge transfer at $[NaCl] = 0$ and Q at a certain NaCl concentration) as a function of NaCl concentration for an electrode covered by an aptamer and in presence of 2 μM MB: (1) without thrombin and (2) in the presence of 10 nM thrombin. A 32-mer DNA aptamer (3′–biotin–GGG TTT TCA CTT TTG TGG GTT GGT GTG GTT GG–5′) was immobilized to the gold surface covered by avidin. [Adapted from Hianik et al. (2007), with permission from Elsevier.]

shown that the cations Li$^+$, Na$^+$, and Cs$^+$ form weak complexes with thrombin aptamer (Kankia and Marky, 2001) and thus probably weakly influence the aptamer conformation. On the other hand, it may be that at higher concentrations of Na$^+$, thrombin could aggregate, which may cause its lower affinity to the aptamer. The electrolyte pH influences the sensor sensitivity substantially.

A plot of relative changes in charge transfer as a function of thrombin concentration is shown in Figure 5.8 for three different electrolyte pH values. The pH was adjusted by the addition of a small amount of 0.1 M NaOH or 0.1 M HCl to the electrolyte (140 mM NaCl, 5 mM KCl, 1 mM CaCl$_2$, 1 mM MgCl$_2$). We can see that maximal response of the sensor took place for pH 7.5, whereas for lower and higher pH, the sensitivity was weaker. It is likely that pH affects the structure of an aptamer-binding site and is not connected with protonation or deprotonation of the protein. As already mentioned, the isoelectric point of human thrombin is 7.0 to 7.6 (Berg et al., 1979). Therefore, if protonation or deprotonation would be crucial for binding of the thrombin to the aptamer, the sensitivity of the sensor should be different for relatively low and high pH, which was not the case. Thus, ionic strength and pH have substantial influence on the binding effectivity of the thrombin to the aptamer-binding site and should be taken into account in practical application of the sensor.

We also compared how the structural peculiarities of the aptamer affect the binding of thrombin. For this purpose we used biotinylated DNA aptamers

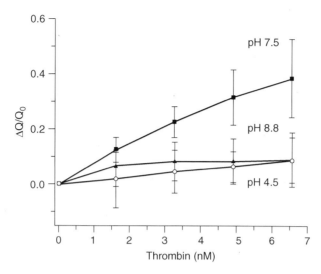

Figure 5.8   Relative changes in charge transfer $\Delta Q/Q_0$ as a function of thrombin concentration and for various values of electrolyte pH. The pH was adjusted by the addition of a small amount of 0.1 M NaOH or 0.1 M HCl to the electrolyte: 140 mM $NaCl + 5$ mM $KCl + 1$ mM $CaCl_2 + 1$ mM $MgCl_2$. The pH values are shown at the curves. A 32-mer DNA aptamer (3′–biotin–GGG TTT TCA CTT TTG TGG GTT GGT GTG GTT GG–5′) was immobilized to the gold surface covered by avidin. Results represent the mean ± SD obtained in three experiments for each pH. [Adapted from Hianik et al. (2007), with permission from Elsevier.]

that have a high affinity to the heparin-binding site of the thrombin: 38-mer (3′–biotin–AAG GTG CTT CAG TGG GGT TGG ACG GGA TGG TGC CTG AC–5′) (APTA A) and 32-mer (3′–biotin–GGG TTT TCA CTT TTG TGG GTT GGA CGG GAT GG–5′) (APTA B). The latter aptamer did not contain a stabilizing structural fragment (Tasset et al., 1997). The aptamers were immobilized to a gold electrode covered by avidin, and the binding of thrombin to the aptamers was monitored by measuring the charge transfer from the electrode to the electrochemical indicator methylene blue (MB) as well as by using a mass detection method.

Figure 5.9A is a plot of the relative changes of charge transfer due to binding of thrombin to aptamers of different structure. We can see that the aptamer containing a stabilizing structural fragment (APTA A) revealed a higher response than that of the aptamer without this fragment (APTA B). Similar results were obtained by a mass detection method using the QCM technique (in this case the aptamer was immobilized on one side of AT-cut quartz crystal of fundamental frequency 8 MHz). The binding of thrombin resulted in an increase in the mass of the crystal, and this caused a decrease in the frequency of the crystal oscillations (see also Section 5.3.2). A plot of changes in frequency as a function of the thrombin concentration for both APTA A and APTA B is shown on Figure 5.9B.

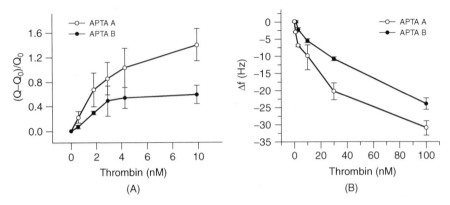

Figure 5.9   Relative changes in charge transfer $\Delta Q/Q_0$ (A) and changes in frequency (B) for biosensors with immobilized aptamers of different structure as a function thrombin concentration. APTA A (open circles), a 38-mer aptamer containing stabilizing structural fragment; APTA B (closed circles), a 32-mer aptamer without a stabilizing fragment (see the text for an explanation).

TABLE 5.1   Binding Constants (K_c) and Number of Binding Sites (N) for an Aptamer-Based Biosensor[a]

Aptamer[b]	K_c (nM^{-1})	N (R.u.)[c]
APTA A (38-mer)	0.028 ± 0.005	14.14 ± 0.83
APTA B (32-mer)	0.012 ± 0.003	12.98 ± 1.20

[a] The biosensor is composed of aptamers of different supporting part structures connected to a sensor surface, but identical in structure to the binding site for thrombin.
[b] APTA A (38-mer) contains a stabilizing structural fragment; APTA B (32-mer) does not.
[c] R.u. = relative units.

Using the Scatchard plot, we determined the binding constants, K_c, and number of binding sites, N, of the thrombin to the aptamers studied (Table 5.1). The binding constant for APTA A is 2.3 times higher than that for APTA B. At the same time, the number of binding sites for APTA A is only slightly higher than that for APTA B. But these differences are statistically insignificant. Therefore, we can conclude that differences in binding constants are due to the higher affinity of thrombin to the APTA A compared with APTA B. Thus, the structural peculiarities connected with the supporting part of the aptamer plays an important role in the affinity of the aptamer-binding site to the protein.

Polyanions such as heparin bind to the thrombin and could affect the binding affinity of proteins to the aptamer. To verify this hypothesis, we studied the interaction of thrombin with DNA aptamer at molar ratios of thrombin to heparin: 1 : 1, 1 : 3, and 1 : 10. The thrombin, heparin, or their complexes were added to the buffer stepwise from a stock solution, and the charge transfer was measured

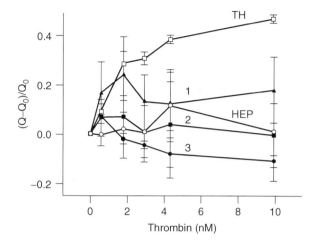

Figure 5.10 Relative changes in charge transfer $\Delta Q/Q_0$ as a function of the concentration of thrombin (TH), heparin (HEP), or their complexes in the molar ratio thrombin/*heparin* = 1:1 (1), 1:3 (2), and 1:10 (3). The sensor was composed of a 38-mer aptamer containing a stabilizing structural fragment (APTA A). The results represent the mean ± SD obtained in three experiments.

by DPV in the presence of 2 μM MB. The addition of heparin in a concentration range up to 9.9 nM had no significant effect on charge transfer from the electrode to MB. This suggests that heparin does not interact or interacts only weakly with the aptamer. However, complexes of thrombin with heparin molar ratio 1:1 substantially reduced the charge transfer, as shown in Figure 5.10. The results obtained suggest that heparin blocks the binding site of thrombin to the aptamer and protects the binding of the thrombin.

The biosensor can be regenerated using high ionic strength (2 M NaCl). This is shown in Figure 5.11, which represents the plot of relative changes in charge transfer from an electrode to MB as a function of thrombin concentration for a biosensor composed of APTA A (a 38-mer aptamer containing a stabilizing fragment) following regeneration of the sensor three times by immersion in 2 M NaCl for 30 minutes. We can see that the charge transfer after regeneration is similar to the calibration curve. However, after the third regeneration, the results at a higher concentration reveal deviation from the calibration curve. Therefore, after calibration, the sensor can be used for analytical purposes twice after corresponding regeneration.

Testing the biosensor in real samples is an important step toward practical application in diagnosis. We therefore tested a DNA aptamer selective for thrombin in blood plasma from healthy individuals. First, a biosensor composed of 38-mer DNA aptamer (APTA A) was immersed in a phosphate buffer and calibrated. The calibration curve is shown in Figure 5.12. Then the sensor was regenerated in 2 M NaCl for 30 minutes. The sensor was then immersed in the

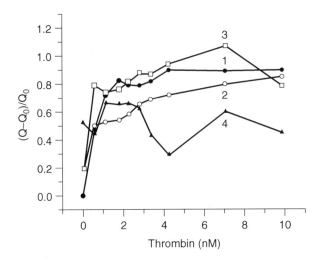

Figure 5.11   Relative changes in charge transfer $\Delta Q/Q_0$ as a function of the concentration of thrombin: (1), calibration curve; (2) to (4), sensor response following first, second, and third regeneration in 2 M NaCl, respectively. The sensor was composed of a 38-mer aptamer containing a stabilizing structural fragment (APTA A).

human blood plasma of a healthy donor and the charge transfer was measured. The value of the relative charge transfer obtained corresponded to a thrombin concentration of 8.3 nM (the first arrow in Figure 5.12). Then we added thrombin in a concentration of 3.5 nM. The value of relative charge transfer obtained (the second arrow in Figure 5.12) corresponds to 16.5 nM of thrombin. In the case of an ideal sensor (i.e., one with ideal selectivity exclusively for thrombin), the final concentration of the thrombin should be 11.8 nM. Thus, considering this result, the sensor recovery was $11.8/16.5 \times 100\% = 71.5\%$. The deviation of sensor recovery from the ideal 100% value could be connected either with interferences of other compounds presented in a blood plasma or with inexact sensor regeneration. Further effort is therefore required for improvement of the sensor recovery in real samples.

Another method of detection of the thrombin–aptamer interaction was proposed by Bang et al. (2005). They immobilized aptamer beacon modified by an amino group to a self-assembled monolayer on a gold surface formed by 11-mercaptoundecanoic acid (MUA). Aptamer beacon without thrombin comprises a short double helix. In the presence of thrombin, the double helix is not stable and the aptamer changes its conformation. The detection method is based on measurement of the electrochemical reduction of MB. At an aptamer hairpin configuration, the MB is intercalated into the double helix and poses a well-resolved reduction peak at— 197 mV versus an Ag/AgCl reference electrode. The conformational changes of aptamer due to binding of thrombin result in release of MB and decrease of the reduction peak. The detection limit of this

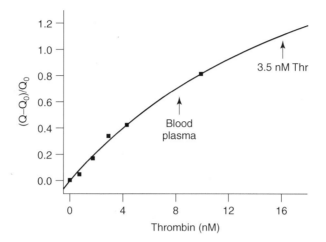

Figure 5.12 Relative changes in charge transfer $\Delta Q/Q_0$ as a function of the concentration of thrombin for an aptamer biosensor. The curve corresponds to calibration of the sensor in a phosphate buffer (PBS). Arrows indicate the values of charge transfer following immersion of the sensor in human blood plasma diluted 10 times and following subsequent addition of 3.5 nM of the thrombin, respectively (for an explanation, see the text). Buffer composition: 50 mM PBS + 140 mM NaCl + 5 mM KCl + 1 mM MgCl$_2$ + 1 mM CaCl$_2$ (pH 7.5). The curve is the fit according to the Langmuir isotherm. The sensor was composed of a 38-mer aptamer containing a stabilizing structural fragment (APTA A).

sensor was 11 nM, and the linear range of the signal was observed between 0 and 50.8 nM of thrombin. This method of detection is shown schematically in Figure 5.3C.

Two different formats of electrochemical detection of thrombin–aptamer interactions using MB were recently developed (Xiao et al., 2005a). The aptamer was modified at one end by a thiol group and at the other end by MB. Without thrombin the MB possesses a reduction signal. The addition of thrombin shifts the equilibrium from an unfolded to a folded aptamer conformation. This resulted in an increase in the distance of the MB from the electrode. As a result, the reduction signal decreased. The limitation of this "signal-off" architecture was improved in further work (Xiao et al., 2005b) in which the aptamer was first allowed to interact with a short, partially complementary nucleotide modified by MB. (MB)-tagged oligonucleotide with aptamer formed a rigid duplex that prevents the MB tag from approaching the electrode surface, so the reduction current is suppressed. The addition of thrombin resulted in quadruplex formation and the approach of MB to an electrode surface. Thus, the amplitude of the reduction signal increased (Figure 5.3D). This method allows the detection of thrombin at concentrations as low as 3 nM.

MB has also been shown useful for the detection of cocaine by means of a specific DNA aptamer (Baker et al., 2006). The MB-tagged aptamer has been immobilized to a gold support via a thiol group. In the absence of cocaine, the aptamer was partially unfolded. The addition of cocaine resulted in the folding of aptamer into a three-way junction, moving MB into close proximity with the electrode surface. This resulted in an increase in the reduction peak measured by ac voltammetry. The sensor was regenerable and allowed to detect cocaine within several seconds with a sensitivity below 10 μM.

Recently, ferrocene (Fc)-modified aptamers were used for the detection of adenosine (Wu et al., 2007). The thiolated aptamers were chemisorbed on a gold electrode modified with polytyramine and gold nanoparticles. An Fc-labeled aptamer probe was prehybridized with a complementary capture DNA sequence that specifically recognizes adenosine. The introduction of adenosine resulted in switching of the hybridized complex into a structure with a specific binding site to the adenosine. As a result, an Fc-labeled aptamer probe dissociated from the sensing interface, resulting in a decrease in the redox current. The decrement of peak current was proportional to the adenosine concentration. A similar strategy has been used for the detection of ATP (Zuo et al., 2007).

Direct detection of thrombin using a single-walled carbon nanotube field-effect transistor (SWNT-FET) was reported by So et al. (2005). In this work, aptamer was covalently immobilized on carbon nanotubes modified by Twin 20. Addition of thrombin resulted in a drop in the conductance of the device. This has been explained by screening of negative charge of DNA by thrombin, which was positively charged under experimental conditions (pH 5.4). The lowest detection limit (LOD) of the sensor was 10 nM. The sensor was selective to thrombin. This was approved by experiment in which elastase instead of thrombin was used. Elastase is serine protease with a similar isoelectric point and a molecular mass similar to that of thrombin. The addition of elastase resulted in considerable lower changes in the current compared with thrombin. The sensor was regenerable by treatment with 6 M guanidine hydrochloride solution, which removed bounded thrombin molecules. Label-free protein biosensors based on aptamer-modified carbon nanotube field-effect transistors (CNT-FETs) for the detection of immunoglobulin E (IgE) have been reported (Maehashi et al., 2007). It has been shown that aptamer-modified CNT-FETs provided better results than those obtained using IgE-mAb-modified CNT-FETs under similar conditions. The detection limit for aptasensor was determined as 250 pM.

Label-free detection of ligand–aptamer interaction was also demonstrated by means of an impedance spectroscopy technique. Simultaneously, Radi et al. (2005) and Rodrigez et al. (2005) reported the application of faradaic impedance spectroscopy (FIS) in the detection interaction of proteins with DNA aptamers. The detection method is based on measurement of resistance in the presence of redox mediator $Fe(CN)_6$. In the absence of the target protein the negatively charged aptamer repulsed the redox mediator molecules from the sensor surface. In Radi et al. (2005), thiol-modified thrombin aptamer was immobilized on a gold electrode together with 2-mercaptoethanol. Thus, the surface was blocked against nonspecific adsorption. This sensor allowed the detection of thrombin at as low a value as 2 nM, with a linear range of 5 to 35 nM. The sensor was regenerable using 2 M NaCl. Rodrigez et al. (2005) detected lysozyme with biotinylated DNA aptamer immobilized on an indium oxide layer coated with polymer and streptavidin. Binding of positively charged protein (at pH < pI) resulted in enhancement of electron transfer, and as a result, the resistance decreases. The method allows detection of the lysozyme with high sensitivity. It was possible to detect 14 nM of the protein reproducibly. The sensor was considerably less sensitive to other proteins: bovine serum albumin, cytochrome c, and thrombin.

Recently, a new electrochemical method of detection of aptamer–protein interactions based on quantum dot (QD) semiconductor nanocrystals has been reported (Hansen et al., 2006). In a competitive binding assay the proteins (thrombin or lysozyme) were conjugated with cadmium or lead nanocrystals. The sensor was prepared by immobilization onto a gold support of thiolated DNA aptamers that selectively bind thrombin and/or lysozyme. After competitive binding with the sensor surface and after the washing step, the conjugated proteins were detected on a surface of mercury-coated glassy carbon electrode using a square-wave stripping voltammetry method. This method make it possible to reach the extremely low detection limit of approximately 0.5 pM of the protein.

An aptamer-based sandwich assay with electrochemical detection for thrombin analysis in complex matrixes using a target-capturing step by aptamer-functionalized magnetic beads was recently reported (Centi et al., 2007). The aptamer-sensing layers were fabricated on a surface of screen-printed electrodes. The high sensitivity of this sensor was demonstrated in the analysis of thrombin in buffer, spiked serum, and plasma. The concentrations detected by the electrochemical assay were in agreement with simulation software that mimics the kinetics of thrombin formation.

Electrochemical methods have great potential, due to the existing wide market of relatively inexpensive instruments produced by CH Instruments and BAS in the United States, Eco Chemie (in the Netherlands), and others. The advantage of this approach consists of simple, low-cost sensor fabrication as well as simple, easy-to-use evaluation of the binding processes.

5.3.2 Acoustic Methods

Acoustic methods belong to label-free detection techniques. In the transverse shear mode method (TSM), the sensor is fabricated on one side of AT-cut quartz. The other side is free and in contact with air or nitrogen gas. When high-frequency voltage at a typical frequency of 5 to 10 MHz is applied to the quartz through electrodes on its top, the crystal starts to oscillate in a direction parallel to the surface and propagates through transverse shear waves. The biochemical processes at the surface of the crystal can cause changes in wave properties and can be monitored in time. The transverse wave travels through the biolayer at the crystal surface with some dissipation, depending on the layer rigidity. This situation is presented schematically in Figure 5.13.

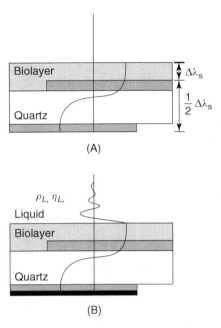

Figure 5.13   Propagation acoustic shear wave through (a) an adsorbed rigid biolayer and (b) in a liquid with density ρ_L and viscosity η_L. In (a) the standing wave is restricted by the substrate and adsorbed rigid layer with a reflection of the shear wave occurring at the upper substrate of the adsorbed layer. For the ideal case, the adsorbate is assumed perfectly rigid and there is no dissipation in the film. However, the biolayer is characterized by viscoelasticity and therefore dissipates the acoustic energy. In (b) the liquid cannot support a shear wave, so it is dissipated viscously, within 300 nm of the surface for a 9-MHz crystal. [Adapted from Wang et al. (2006), with permission from The Royal Society of Chemistry.]

Changes in resonant frequency measured as a function of time can be linked to the adsorption of a molecular layer on a quartz substrate and is related to changes in the thickness of the layer. In the case of pure adsorption, the decrease in resonant frequency is connected with an increase in the layer thickness, Δt_s, as follows: $\Delta f_s = -v\Delta t_s/2t_s^2$, where v is the speed of the transverse wave in quartz ($v = 3318.4$ m s^{-1}). However, when the sensor operates in a liquid, a component of the energy of the standing wave is not reflected at the solid–liquid interface and is dissipated in the liquid. The dissipation is caused by viscous forces (i.e., friction between the biolayer and the surrounding liquid). The most common measure of this loss is motional resistance, R_m (in ohms). R_m represents energy dissipation in the electrical circuit approximation of the resonator device. The energy dissipation causes an increase in the width of the resonant peak, $\Delta\Gamma$, which is proportional to R_m as follows: $\Delta\Gamma = 8K^2C_0f_0^2\,\Delta R_m/\pi$, where C_0 is the parallel capacitance of the TSM device, f_0 is the fundamental resonant frequency, and $K^2 = 7.74 \times 10^{-3}$ is the electromechanical coupling coefficient of quartz (Balantine et al., 1997; Wang et al., 2006).

In the simplest format, usually only the series resonant frequency is measured. This method, known as quartz crystal microbalance (QCM), was developed by Sauerbrey (1959), who derived a relation between changes in resonant frequency, Δf_s, and changes in the surface mass density, ρ_s:

$$\Delta f_s = -\frac{2 f_0^2 \rho_s}{\sqrt{\mu_q \rho_q}} \tag{5.1}$$

where f_0 is the fundamental frequency of the crystal (typically, 5 to 10 MHz) and $\mu_q = 2.947 \times 10^{10}$ Pa and $\rho_q = 2648$ kg m^{-3} are the shear stiffness and mass density of the quartz, respectively. For practical purposes it is convenient to use this equation in the format

$$\Delta f_s = -2.26 \times 10^{-6} f_0^2 (\Delta m / A) \tag{5.2}$$

where Δm are changes in the mass (nanograms) and A is the area of the working electrode (cm^2) (typically, A is about 0.3 cm^2) (Hayward and Chu, 1994).

Equations (5.1) and (5.2) are exactly valid only for dry crystal. In liquid, as we noted earlier, the viscous forces cause damping of the acoustic wave. As a result, observed changes in the series resonant frequency are connected not only with changes in mass but also with shearing viscosity. This has been demonstrated in work by Thompson and co-workers that applied TSM to the study of protein–RNA interactions (Furtado et al., 1999). Detailed study of the interaction of short peptides that model the HIV-1 Tat protein with the HIV-1 mRNA (TAR) showed that for relatively short peptides (12 to 22 amino acid residues), an increase in f_s has been observed. Almost no changes in f_s took place when peptide composed of 27 amino acid residues interacted with TAR. This was in contrast with the mass-response model, which predicts a decrease in f_s [see equation (5.1)]. The f_s value decreased only for longer peptides (Tassew and Thompson, 2002). A decrease in frequency was also observed when native Tat protein (101 amino acid residues) interacted with TAR (Minunni et al., 2004; Tombelli et al., 2005b). The discrepancy observed has been connected with structural changes in RNA due to a binding process. Thus, various layers move with different velocities, which induce changes in acoustic coupling and energy dissipation. This is reflected in changes in series resonant frequency and in motional resistance [see Tassew and Thompson (2002) for a more detailed explanation]. The TSM method has been used further for the study of the interaction of the antibiotics neomycin (Tassew and Thompson, 2002) and streptomycin (Tassew and Thompson, 2003a) with TAR in the presence of Tat. It has been confirmed that neomycin effectively blocked interaction of Tat with TAR. The TSM method is also rather effective for the determination of the kinetics of binding of proteins to RNA (Tassew and Thompson, 2003b) and DNA (Grman et al., 2006) aptamers.

In its simplest, QCM, format, protein–aptamer interactions were analyzed by Liss et al. (2002). They compared the interaction of IgE with DNA aptamer as well as with anti-IgE antibodies. Although the detection limit was similar in the two cases, the advantage of the aptasensor was its possibility of surface regeneration, which was impossible for an antigen-based biosensor. However, recently it has been shown that immobilization of anti-IgE on the dendrimer surface also allows us to regenerate an immunosensor (Svobodová et al., 2006). The QCM method was recently compared with the electrochemical biosensor assay of thrombin detection (Hianik et al., 2005, 2007). It has been shown that the sensitivity of thrombin detection was similar for the two methods. Mascini and co-workers showed that similar results in sensitivity and selectivity in the detection of Tat peptide with RNA aptamer can be obtained by the QCM and SPR methods (Tombelli et al., 2005b).

The sensitivity of QCM for the detection of thrombin by a DNA aptamer was improved using gold nanoparticles (Pavlov et al., 2004). In this work, thiolated DNA aptamer specific to the fibrinogen-binding site was immobilized on an AT-cut quartz surface. The addition of thrombin resulted in a decrease in oscillation frequency. Aptamers of the same structure conjugated with gold nanoparticles were added to the measuring cell and a considerably larger decrease in frequency was observed due to the larger mass of aptamer–nanopraticle complex. However, in contrast with Ikebukuro et al. (2005), Pavlov et al. used the aptamer of the same structure for immobilization to the electrode surface and to the nanoparticles. It should therefore be surprising that they observed changes of the mass following addition of aptamer conjugated with nanoparticles, because the binding site for fibrinogen already was occupied by aptamers immobilized at the crystal. However, as shown by Tasset et al. (1997) in competitive binding studies, the aptamer specific for fibrinogen developed by Bock et al. (1992) and used in the paper by Pavlov et al. (2004) has certain, although low affinity to the heparin-binding site. Therefore, one can assume that when another aptamer, specific to the heparin-binding site of the thrombin, will be used for conjugation with nanoparticles, the sensitivity of the detection could be improved.

A low-wave acoustic sensor (Schlensog et al., 2004) was used to detect the interaction of thrombin with RNA and DNA aptamers (Gronewold et al., 2005). The authors compared the binding of the thrombin to RNA aptamer using a filter binding method utilizing radiolabeling of RNA aptamer by $3'-{}^{32}P$ and also by using the SPR method. They determined the dissociation constants K_d of 294 ± 30 nM for filter binding, 113 ± 20 nM for SPR, and 181 ± 20 nM in the case of an acoustic method. Comparison with the DNA aptamer, which binds thrombin to the fibrinogen-binding site, revealed about a twofold lower effectivity than for RNA aptamers. They also studied the effect of heparin, human antithrombin III, and their complexes on the binding of DNA aptamer to thrombin. The results suggest that binding of thrombin to DNA aptamer affects the binding of heparin–antithrombin III complex to the heparin-binding site. This has been explained by conformational changes in thrombin following binding of DNA aptamer to the fibrinogen-binding site (exocite II), which affect the heparin-binding site (exocite I), and vice versa.

Recently, we applied the TSM sensor for the characterization of all steps in the sensor fabrication and detection of thrombin–aptamer interactions (Grman et al., 2006). For the preparation of aptamer biosensor, we used AT-cut quartz wafers with a fundamental frequency of 8 MHz, covered on both sides by thin gold electrodes. The working area of the crystal was 0.2 cm^2. The crystals were carefully cleaned and mounted between two silicon rubber O-rings in the flow-through cell such that the crystal was exposed to the analyte solution on only one side (Hianik et al., 2008). The DNA aptamer was immobilized on the electrode surface by avidin–biotin technology, as described elsewhere (Hianik et al., 2007). The gold surface was first modified by neutravidin and then by biotinylated aptamer of the following composition: 5′–GGT TGG TGT GGT TGG TTT TTT TTT TTT TTT 3′–biotin. As a buffer, we used 20 mM TRIS, 140 mM NaCl, 5 mM KCl, 1 mM CaCl$_2$, 1 mM MgCl$_2$, pH 7.4. The human thrombin was dissolved in a buffer in a concentration of 1 mg/mL. The impedance spectra were measured by means of a network analyzer (8712ES, Agilent Technology). The analyzer was connected to an IBM Pentium PC. From the impedance spectra the series resonance frequency, f_s, and motional resistance, R_m, were determined.

Figure 5.14 shows changes in the series resonance frequency, Δf_s, and motional resistance, R_m, following various modifications of the quartz crystal. The addition of neutravidin (concentration 0.2 mg mL^{-1}) dissolved in buffer resulted in a substantial decrease in frequency and a slight increase in motional resistance. The decrease in series resonance frequency is a well-known phenomenon and is connected with attachment of the neutravidin to a gold surface. The increase in motional resistance may correspond to an increase in surface viscosity of the layer or be affected by it (Tassew and Thompson, 2003b; Ellis and Thompson, 2004). After the frequency stabilized, the surface of the crystal was washed with buffer, resulting in an increase in the frequency and a slight increase in the motional resistance, due to the desorption of weakly attached neutravidin molecules. The resulting decrease in resonant frequency was on average 150 Hz. Addition of biotinylated aptamer (2 μM) resulted in a sharp decrease in the series resonant frequency (80 Hz) and a substantial increase in the motional resistance (15 Ω), which is caused by binding biotin to the neutravidin layer, due to the high affinity of these two compounds. However, following the initial response, the frequency started to increase and motional resistance decreased, reaching a steady-state value after approximately 45 minutes. The increase in f_s and decrease in R_m may be evidence of an increase in the rigidity of the biolayer. It is likely that aptamers immobilized at the surface do not immediately attain the typical structure of the binding site. The structure of the layer is also developed gradually, due to the fact that aptamers are immobilized at the surface and interact between each other. Thus, any increase in the rigidity of the layer can probably be connected with the ordering of the aptamers.

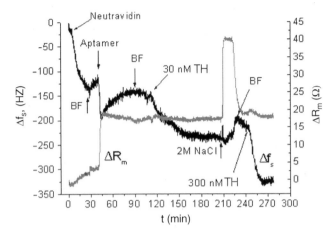

Figure 5.14   Changes in the series resonant frequency Δf_s (black) and motional resistance ΔR_m (gray) following various modifications of the quartz crystal surface and interaction of the aptamer with thrombin. BF, buffer; TH, thrombin. Experiments have been performed at $23 \pm 1°C$. [Adapted from Hianik et al. (2008), with permission of Bentham Science Publishers.]

After flowing the buffer, the weakly adsorbed aptamers were removed from the surface. This resulted in a slight decrease in the frequency and increase in the motional resistance. This may suggest a certain reordering of the aptamer layer. Thus, after binding of the aptamers and reaching steady-state values of f_s and R_m, the sensor was ready to study the interaction with thrombin. For this purpose, thrombin at concentrations of 1 to 300 nM were flowed in a stepwise fashion at the surface of the sensor. Figure 5.14 shows an example of changes of f_s and R_m following the addition of thrombin at 30 and 300 nM.

The addition of thrombin at 30 nM resulted in a decrease in the series resonant frequency by approximately 80 Hz. On the other hand, the motional resistance increase was only 1 Ω. This suggests that attachment of the thrombin does not substantially affect the material properties of the layer, and the changes in frequency are due primarily to the adsorption of thrombin to the crystal surface. Figure 5.14 also shows that the process is reversible using 2 M NaCl. The high ionic strength of the salt solution is able to overcome the weaker electrostatic interaction of the thrombin and the aptamer, so thrombin is removed from the surface. The sensor regeneration is not complete, however, because the frequency is below the value prior to addition of the thrombin. After washing the surface of the sensor with buffer, thrombin in higher concentration (i.e., 300 nM) was added. We can see that in this case a much larger decrease took place in the frequency, accompanied by a slight decrease in the motional resistance.

According to the Sauerbrey equation [equation (5.1)], from changes in f_s it is possible to calculate the change of mass, Δm, at the device surface, and from the known molecular mass (MW) of neutravidin (60 kDa) the number of neutravidin molecules at the gold surface can be estimated. Thus, a change of $\Delta f_s = 150$ Hz corresponds to a change of mass $\Delta m = 208$ ng, and the number of neutravidin molecules at the surface is $N_{neutravidin} = (\Delta m/MW) \times N_A$, where N_A is Avogadro's number (6.02×10^{23} molecules per mole), so $N_{neutravidin} = (2.08 \times 10^{-7}/60,000) \times 6.02 \times 10^{23} = 2.09 \times 10^{12}$. Although this represents the usual invocation of the Sauerbrey effect, it is much more likely that there is a significant contribution from the viscoelasticity of the protein film, which in turn affects coupling between the adsorbed layer and the surrounding electrolyte (Ellis and Thompson, 2004). The effect of viscoelasticity certainly exists, as is evident from the increase in motional resistance R_m (Figure 5.14). It has been suggested that a correction factor of approximately 2 could be used to correct changes in frequency connected with viscoelasticity (Zhou et al., 2001). However, this is an approximation and should not be viewed as a quantitative, analytical factor. With this correction, the number of neutravidin molecules at the surface would be approximately 1.05×10^{12}. Using equation (5.1) with shifts in the series resonant frequency following the addition of biotinylated aptamer, which was approximately 20 Hz, we estimate the change of mass as 28 ng. In analogy with calculation of the number of neutravidin molecules, we can estimate that approximately 8.5×10^{11} molecules of aptamers are immobilized at the surface (here we also take into account a correction factor of 2). Thus, at the surface there are approximately the same number of aptamer molecules as neutravidin molecules. Considering the fact that neutravidin has four binding sites, of which two are unavailable due to orientation on the gold surface, the estimated number of aptamer molecules seems to correspond to the binding ratio of neutravidin to biotin. This is also evidence that full saturation of the binding sites is not possible, due to structural restrictions caused by interaction between aptamers.

From changes in the series resonant frequency of 80 Hz due to binding of the thrombin to the aptamer at 30 nM, it is possible to calculate the surface density of the protein. Using equation (5.1) and the molecular mass of α-thrombin (33.6 kD), we can calculate the number of thrombin molecules attached to the aptamer layer: $N_{thrombin} = 5 \times 10^{11}$ (here again a correction factor of 2 was used). On the other hand, on the base of the crystal structure of the thrombin (Malkowski et al., 1997) the cross-sectional area of this molecule is approximately 20 nm^2. Therefore, approximately 1.0×10^{12} thrombin molecules are required to cover the electrode surface. Thus, the value obtained suggests that at 30 nM the thrombin covers a substantial part of the sensor surface. However, further decrease in the frequency after addition of a higher thrombin concentration may demonstrate the aggregation of thrombin molecules and the formation of a multilayer structure.

The plot of the changes of frequency as a function of concentration of thrombin is presented in Figure 5.15. We can see that by increasing the thrombin concentration, the frequency is decreased. The detection limit for the determination of thrombin was 5 nM.

Thus, acoustic sensors represent a rather useful tool for the study of protein–aptamer interactions without additional labeling (Bini et al., 2007). Using an acoustic sensor in a TSM format, it is also possible to study subtle

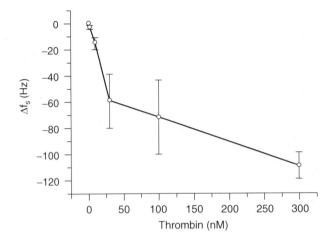

Figure 5.15   Changes in the series resonance frequency of AT-cut crystal covered by neutravidin, with immobilized biotinylated DNA aptamer as a function of the concentration of human thrombin.

structural effects at the sensor surface connected with conformational changes of aptamers induced by protein or low-molecular-mass ligands. These sensors can also be used advantageously for study of the kinetics of ligand–aptamer interactions (Hianik et al., 2007).

5.4   CONCLUSIONS

Aptamer-based biosensors are rather promising. They can be used for the detection of proteins or low-molecular-mass compounds with high sensitivity and selectivity, comparable and even higher than those available so far for antibody-based assay. However, a considerable advantage of aptamers over antibodies is the possibility of their synthesis in vitro without using experimental animals. This allows us to prepare aptamers even for toxic compounds, which is impossible in the case of antibodies. Aptamers can be modified chemically by various ligands and indicator groups. Thus, they can be immobilized on solid support of a different type. Various indicators (optical or electrical) allow us to use various methods for detection of aptamer–ligand interactions, including electrochemical and optical methods. Certain methods (e.g., SPR, QCM, TSM, or FTIR) do not even require aptamer labeling. This detection, especially in the case of acoustic methods, can be amplified substantially using nanoparticles. The TSM method is especially informative, due to the possibility of studying the physical properties of the sensing layer during all steps of sensor fabrication. In contrast with QCM, this method allows us to estimate the contribution of viscoelastic forces into the binding of ligand to the aptamer. Recent achievements in electrochemical detection of aptamer–ligand interactions is rather promising, because it opens new routes for easy-to-use and low-cost assay that may be rather useful in medical and environmental applications of aptasensors. However, further effort is required to optimize sensor properties, especially for application in real samples. For example, the use of RNA aptamers is still in the premature stage, due to problems with their stability in real samples. The problems also involve difficulties of synthesis of longer RNA aptamers (more then 50 bases). DNA aptamers are more stable, but further work is required for the optimization of aptamer immobilization, sensor regenerations, and the detection of ligand–aptamer interactions, especially in the case of small molecules.

Acknowledgments

This work was supported financially by the Slovak Grant Agency (project 1/4016/07), by the Agency for Promotion Research and Development under contract APVV-20-P01705, and by the European Commission under the Noe NeuroPrion (contract FOOD-CT-2004-506579). I would like to thank to Igor Grman and Michaela Sonlajtnerova for technical assistance and Michael Thompson and Jonathan Ellis for valuable discussions and useful comments.

REFERENCES

Baker, B. R., Lai, R. Y., Wood, M. S., Doctor, E. H., Heeger, A. J., Plaxco, K. W. (2006). An electronic aptamer-based small molecule sensor for the rapid, label-free detection of cocaine in adulterated samples and biological fluids. *J Am Chem Soc* 128, 3138–3139.

Balantine, D. S., White, R. M., Martin, S. J., Ricco, A. J., Zellers, E. T., Frye, G. C., Wohlthen, H. In *Acoustic Wave Sensors*, R. Stern and M. Levy, eds., Academic Press, San Diego, CA, 1997.

Bang, G. S., Cho, S., Kim, B-G. (2005). A novel electrochemical detection method for aptamer biosensors. *Biosens Bioelectron* 21, 863–870.

Berg, W., Hillvarn, B., Arwin, H., Stenberg, M., Lundstrom, I. (1979). The isoelectric point of thrombin and its behaviour compared to prothrombin at some solid surfaces. *Thromb Haemost* 42, 972–982.

Bini, A., Minunni, M., Tombelli, S., Centi, S., Mascini, M. (2007). Analytical performances of aptamer-based sensing for thrombin detection. *Anal Chem* 79, 3016–3019.

Bock, L. C., Griffin, L. C., Latham, J. A., Vermaas, E. H., Toole, J. J. (1992). Selection of single-stranded DNA molecules that bind and inhibit human thrombin. *Nature* 355, 564–566.

Boon, E. M., Ceres, D. M., Drummond, T. G., Hill, M. G., Barton, J. K. (2000). Mutation detection by electrocatalysis at DNA-modified electrodes. *Nat Biotechnol* 18, 1096–1100.

Cavic, B. A., Thompson, M. (2002). Interfacial nucleic acid chemistry studied by acoustic shear wave propagation. *Anal Chim Acta* 469, 101–113.

Centi, S., Tombelli, S., Minunni, M., Mascini, M. (2007). Aptamer-based detection of plasma proteins by an electrochemical assay coupled to magnetic beads. *Anal Chem* 79, 1466–1473.

Conrad, R., Ellington, A. D. (1996). Detecting immobilized protein kinase C isozymes with RNA aptamers. *Anal Biochem* 242, 261–265.

Dougan, H., Hobbs, J. B., Weitz, J. I., Lyster, D. M. (1997). Synthesis and radioiodination of a stannyl oligodeoxyribonucleotide. *Nucleic Acid Res* 25, 2897–2901.

Ellis, J. S., Thompson, M. (2004). Slip and coupling phenomena at the liquid–solid interface. *Phys Chem Chem Phys* 6, 4928–4938.

Furtado, L. M., Su, H., Thompson, M. (1999). Interaction of HIV-1 TAR RNA with Tat-derived peptides discriminated by online acoustic wave detector. *Anal Chem* 71, 1167–1175.

Grman, I., Rybar, P., Thompson, M., Hianik, T. (2006). Detection of aptamer–protein interaction using thickness shear mode technique. *Book of Abstracts, Prion 2006*, Torino, Italy.

Gronewold, T. M. A., Glass, S., Quantdt, E., Famulok, M. (2005). Monitoring complex formation in the blood-coagulation cascade using aptamer-coated SAW sensors. *Biosens Bioelectron* 20, 2044–2052.

Hansen, J. A., Wang, J., Kawde, A-N., Xiang, Y., Gothelf, K. V., Collins, G. (2006). Quantum-dot/aptamer-based ultrasensitive multi-analyte electrochemical biosensor. *J Am Chem Soc* 128, 2228–2229.

Hayward, G. L., Chu, G. Z. (1994). Simultaneous measurement of mass and viscosity using piezoelectric crystals in liquid media. *Anal Chim Acta* 288, 179–185.

Hianik, T. (2007). DNA/RNA aptamers: novel recognition structures in biosensing. In *Electrochemical Sensor Analysis*, S. Alegret and A. Merkoçi, eds., Elsevier, Amsterdam.

Hianik, T., Ostatná, V., Zajacová, Z., Stoikova, E., Evtugyn, G. (2005). Detection of aptamer–protein interactions using QCM and electrochemical indicator methods. *Bioorg Med Chem Lett* 15, 291–295.

Hianik, T., Ostatná, V., Sonlajtnerova, M., Grman, I. (2007). Influence of ionic strength, pH and aptamer configuration for binding affinity to thrombin. *Bioelectrochemistry* 70, 127–133.

Hianik, T., Porfireva, A., Grman, I., Evtugyn, G. (2008). Aptabodies—new type of artificial receptors for detection proteins. *Protein Peptide Lett* 15, 799–805.

Ikebukuro, K., Kiyohara, C., Sode, K. (2004). Electrochemical detection of protein using a double aptamer sandwich. *Anal Lett* 27, 2901–2909.

Ikebukuro, K., Kiyohara, C., Sode, K. (2005). Novel electrochemical sensor system for protein using the aptamers in sandwich manner. *Biosens Bioelectron* 20, 2168–2172.

Kankia, B. I., Marky, L. A. (2001). Folding of the thrombin aptamer into a G-quadruplex with Sr^{2+}: stability, heat, and hydratation. *J Am Chem Soc* 123, 10799–10804.

Leca-Bouvier, B., Blum, J. (2005). Biosensors for protein detection: a review. *Anal Lett* 38, 1491–1517.

Lee, M., Walt, D. (2000). A fiber-optic microarray biosensor using aptamers as receptors. *Anal Biochem* 282, 142–146.

Liss, M., Petersen, B., Wolf, H., Prohaska, E. (2002). An aptamer-based quartz crystal protein biosensor. *Anal Chem* 74, 4488–4495.

Maehashi, K., Katsura, T., Kerman, K., Takamura, Y., Matsumoto, K., Tamiya, E. (2007). Label-free protein biosensor based on aptamer-modified carbon nanotube field-effect transistors. *Anal Chem* 79, 782–787.

Malkowski, M. G., Martin, P. D., Guzik, J. C., Edwards, B. F. (1997). The co-crystal structure of unliganded bovine α-thrombin and prethrombin-2: movement of the Tyr-Pro-Pro-Trp segment and active site residues upon ligand binding. *Protein Sci* 6, 1438–1448.

McCauley, G. T., Hamaguchi, N., Stanton, M. (2003). Aptamer-based biosensor arrays for detection and quantification of biological molecules. *Anal Biochem* 319, 244–250.

Minunni, M., Tombelli, S., Gullotto, A., Luzi, E., Mascini, M. (2004). Development of biosensors with aptamers as bio-recognition element: the case of HIV-1 Tat protein. *Biosens Bioelectron* 20, 1149–1156.

Osborne, S. E., Matsumura, I., Elington, A. D. (1997). Aptamers as therapeutic and diagnostic reagents: problems and prospects. *Curr Opin Chem Biol* 1, 5–9.

Ostatná, V., Dolinnaya, N., Andreev, S., Oretskaya, T. Wang, J., Hianik, T. (2005). The detection of DNA deamination by electrocatalysis at DNA-modified electrodes. *Bioelectrochemistry* 67, 205–210.

Ostatná, V., Vaisocherová, H., Homola, J., Hianik, T. (2008). Effect of the immobilisation of DNA aptamers on the detection of thrombin by means of surface plasmon resonance. *Anal Bioanal Chem* 391, 1861–1869.

Pavlov, V., Xiao, Y., Shlyahovsky, B., Willner, I. (2004). Aptamer-functionalized Au nanoparticles for the amplified optical detection of thrombin. *J Am Chem Soc* 126, 11768–11769.

Peerce, P. J., Bard, A. J. (1980). Polymer films on electrodes: III. Digital simulation model for cyclic voltammetry of electroactive polymer film and electrochemistry of poly(vinylferrocene) on platinum. *J Electroanal Chem* 114, 89–115.

Piliarik, M., Vaisocherová, H., Homola, J. (2005). A new surface plasmon resonance sensor for high-throughput screening application. *Biosens Bioelectron* 20, 2104–2110.

Pividori, M. I., Merkoci, A., Alegret, S. (2000). Electrochemical genosensor design: immobilisation of oligonucleotides onto transducer surfaces and detection methods. *Biosens Bioelectron* 15, 291–303.

Potyrailo, R. A., Conrad, R. C., Ellington, A. D., Hieftje, G. M. (1998). Adapting selected nucleid acid ligands (aptamers) to biosensors. *Anal Chem* 70, 3419–3425.

Radi, A. E., Sanchez, J. L. A., Baldrich, E., O'Sullivan, C. K. (2005). Reusable impedimetric aptasensor. *Anal Chem* 77, 6320–6323.

Rodriguez, M. C., Kawde, A-N., Wang, J. (2005). Aptamer biosensor for label-free impedance spectroscopy detection of proteins based on recognition-induced switching of the surface charge. *Chem Commun* 34, 4267–4269.

Sauerbrey, G. (1959). The use of oscillator for weighing thin layers and for microweighing. *Z Phys* 155, 206–210.

Schlensog, M., Schlesog, M. D., Gronewold, T., Tewes, M., Famulok, M., Quandt, E. (2004). A Love-wave biosensor using nucleic acids as ligands. *Sens Actuat B* 101, 308–315.

So, H-M., Won, K., Kim, Y. H., Kim, B-K., Ryu, B. H., Na, P. S., Kim, H., Lee, J-O. (2005). Single-walled carbon nanotube biosensors using aptamers as molecular recognition elements. *J Am Chem Soc* 127, 11906–11907.

Svobodová, L., Šnejdárková, M., Tóth, K., Gyurcsanyi, R. E., Hianik, T. (2004). Properties of mixed alkanethiol–dendrimer layers and their applications in biosensing. *Bioelectrochemistry* 63, 285–289.

Svobodová, L., Šnejdárková, M., Polohová, V., Grman, I., Rybár, P., Hianik, T. (2006). QCM immunosensor based on polyamidoamine dedrimers. *Electroanalysis* 18, 1943–1949.

Tasset, D. M., Kubik, M. F., Steiner, W. J. (1997). Oligonucleotide inhibitors of human thrombin that bind distinct epitopes. *J Mol Biol* 272, 688–699.

Tassew, N., Thompson, M. (2002). RNA-peptide binding and the effect of inhibitor and RNA mutation studied by on-line acoustic wave sensor. *Anal Chem* 74, 5313–5320.

Tassew, N., Thompson, M. (2003a). Binding affinity and inhibitory potency of neomycin and streptomycin on the Tat peptide interaction with HIV-1 TAT RNA detected by on-line acoustic wave sensor. *Org Biomol Chem* 1, 3268–3270.

Tassew, N., Thompson, M. (2003b). Kinetic characterization of TST RNA-Tat peptide and neomycin interactions by acoustic wave biosensor. *Biophys Chem* 106, 241–252.

Tombelli, S., Minunni, M., Mascini, M. (2005a). Analytical applications of aptamers. *Biosens Bioelectron* 20, 2424–2434.

Tombelli, S., Minunni, M., Luzi, E., Mascini, M. (2005b). Aptamer based biosensors for detection of HIV-1 TAT protein. *Bioelectrochemistry* 67, 135–141.

Tuerk, C., Gold, L. (1990). Systematic evolution of ligands by exponential enrichment. *Science* 249, 505–510.

Wang, X., Ellis, J. S., Lyle, E-L., Sundaram, P., Thompson, M. (2006). Conformational chemistry of surface-attached calmodulin detected by acoustic shear wave propagation. *Mol Biosyst* 2, 184–192.

Wei, F., Sun, B., Liao, W., Ouyang, J. H., Zhao, X. S. (2003). FTIR-ATR detection of proteins and small molecules through DNA conjugation. *Biosens Bioelectron* 18, 1149–1155.

Wu, Z. S., Guo, M. M., Zhang, S. B., Chen, C. R., Jiang, J. H., Shen, G. L., Yu, R. Q. (2007). Reusable electrochemical sensing platform for highly sensitive detection of small molecules based on structure-switching signaling aptamers. *Anal Chem* 79, 2933–2939.

Xiao, Y., Lubin, A. A., Heeger, A. J., Plaxco, K. W. (2005a). Label-free electronic detection of thrombin in blood serum by using an aptamer-based sensor. *Angew Chem Int Ed Engl* 44, 5456–5459.

Xiao, Y., Piorek, B. D., Plaxco, K. W., Heeger, A. J. (2005b). A reagentless signal-on architecture for electronic, aptamer-based sensors via target-induced strand displacement. *J Am Chem Soc* 127, 17990–17991.

Zhou, X. C., Huang, L. Q., Li, S. F. Y. (2001). Microgravimetric DNA sensor based on quartz crystal microbalance: comparison of oligonucelotide immobilization methods and the application in genetic diagnosis. *Biosens Bioelectron* 16, 85–95.

Zuo, X., Song, S., Zhang, J., Pan, D., Wang, L., Fan, C. (2007). A target-responsive electrochemical aptamer switch (TREAS) for reagentless detection of nanomolar ATP. *J Am Chem Soc* 129, 1042–1043.

CHAPTER 6

BIOSENSORS USING THE APTAMERIC ENZYME SUBUNIT: THE USE OF APTAMERS IN THE ALLOSTERIC CONTROL OF ENZYMES

KAZUNORI IKEBUKURO, WATARU YOSHIDA, and KOJI SODE

6.1 APTAMERS AS MOLECULAR RECOGNITION ELEMENTS OF BIOSENSORS

6.1.1 Comparing Aptamers to Antibodies

Antibodies are widely used as molecular recognition elements of biosensors because there are many types of high-affinity, high-specificity antibodies to use against various molecules. Aptamers can also recognize their target molecules with high affinity and specificity; therefore, they are expected to be excellent molecular recognition elements. Several cases in which aptamers rather than antibodies were used in biosensing have already been reported. For example, Drolet et al. (1996) showed that human VEGF could be detected by a monoclonal antibody against hVEGF, an anti-hVEGF aptamer labeled with fluorescein and alkaline phosphatase conjugated with anti-fluorescein Fab fragments. We have demonstrated thrombin detection using two antithrombin aptamers, which recognize different epitopes through PQQGDH enzymatic activity (Ikebukuro et al., 2004, 2005). These reports have proven that aptamers can be used in biosensors as molecular recognition elements instead of antibodies.

Aptamers in Bioanalysis, Edited by Marco Mascini
Copyright © 2009 John Wiley & Sons, Inc.

TABLE 6.1   Properties of Aptamers and Antibodies

	Antibodies	Aptamers
Affinity	Excellent	Good/excellent
Selectivity	Excellent	Good/excellent
Cost	Fair (expensive)	Good (inexpensive)
Synthesis	Fair (immunization required)	Good (chemical synthesis)
Stability	Fair (low stability at high temperature)	Good (renaturable)
Modification	Fair (site-specific modification is not easy)	Good (easy)
Construction of conjugate	Fair (difficult)	Good (easy)
Design of structural change	Fair (difficult)	Good (easy)

Antibodies generally have better molecular recognition ability than aptamers, because antibodies are composed of 20 types of amino acid, and aptamers are composed of only four types of nucleotides. However, aptamers have some advantages over antibodies in terms of biosensing: (1) aptamers can be selected in vitro; (2) aptamers can be synthesized chemically; (3) DNA aptamers are chemically stable and their structure is thermally renaturable, even if they are denatured at high temperature; and (4) structural changes in aptamers can be designed by straightforward hybridization rules. Specifically, by designing the structural change of aptamers, biosensor systems that do not require bound–free separation can be developed. Aptamers and antibodies are compared in Table 6.1.

The first advantage, the possibility of screening aptamers in vitro, gives aptamers a strong potential that is difficult to realize using antibodies. Aptamers can be screened in vitro, and if aptamers that recognize unpurified or unidentified target biomolecules are created, biosensors for detecting unpurified or unidentified target biomolecules can be developed. Blank et al. (2001) have reported aptamer screening for cancer cells using whole cells without the purification of biomolecules such as proteins, lipids, or sugars, and we have already reported a method, using aptamer blotting, of selecting aptamers that target unpurified or unidentified proteins (Noma et al., 2006). This method is schematized in Figure 6.1.

Antibodies for unpurified or unidentified target biomolecules can be obtained by the in situ phage screening method (Furuta et al., 2002) using phage-displayed antibodies, but aptamers can be obtained more easily than phage antibodies, considering their respective sizes: A phage is approximately 1 μm long, whereas an aptamer is less than 10 nm. This is an important advantage of aptamers over antibodies in biosensing.

The second advantage of aptamers, the possibility of synthesizing them chemically, is reflected in the cost and the production process, and it allows flexible

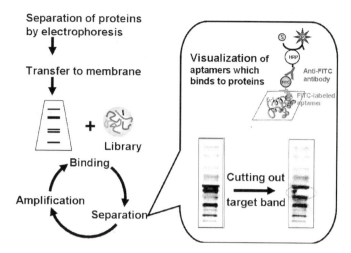

Figure 6.1 Screening using aptamer blotting.

modification and multifunctionalization. Especially in biosensing applications, labeling with fluorophores or enzymes is indispensable, and labeling aptamers is much easier than labeling antibodies. This property of aptamers makes them valuable as molecular recognition elements of biosensors.

The third advantage, the chemical stability of DNA aptamers, can solve the main problem of protein-based biosensors. The chemical and physical instability of protein-based biosensors is always claimed in practical use, and this limits the range of biosensor application. However, DNA is chemically stable. It is stable within the pH range 2 to 12 and is thermally renaturable: Even if it is denatured at $100°C$, it is refolded at room temperature. Even RNA aptamers can gain stability upon $2'$ modification; therefore, aptamers have the potential to enhance the applicability of biosensors in practical contexts. Additionally, aptamers can be immobilized onto substrates using DNA microarray fabrication technology, and aptamer microarrays can be created.

In this chapter we focus mainly on the fourth advantage of aptamers, the possibility of designing aptamers easily due to their simplicity. Aptamers consist of just four types of nucleotides. This simplicity limits their molecular recognition ability compared to that of antibodies, but it simplifies the design of aptamer structures. Moreover, it enables us to design the structural change upon recognition of the target molecule. Many examples have already been reported, and we note them in the next section.

6.1.2 Signaling Aptamers

Hamaguchi et al. (2001) have constructed a thrombin detection system using a designed thrombin aptamer and fluorescence resonance energy transmission (FRET). The oligonucleotides complementary to the nucleotides at the $3'$ end of the thrombin aptamer were connected to the 5' end and modified with a fluorophore at the 5' end and with a quencher at the 3' end. In the absence of thrombin, the thrombin aptamer designed folds into a stem–loop structure. In this structure, the fluorophore is quenched by the quencher. In the presence of thrombin, a conformational change is induced in the aptamer by the binding of thrombin, which causes a change in the distance between the fluorophore and the quencher. Consequently, the fluorophore is no longer quenched and fluorescence is restored. Therefore, thrombin can be detected by measuring the fluorophore's fluorescence intensity. Such aptamers, which generate signals upon binding to the target molecule, are called *signaling aptamers*.

Nutiu and Li (2003) have discussed more generalized designs for signaling aptamers. They used two oligonucleotides: One is an aptamer labeled with a fluorophore, and the other is the complementary strand of the aptamer labeled with a quencher. These two oligonucleotides are able to hybridize with each other and to be in close proximity for fluorescence quenching. In the presence of a target molecule, the aptamer forms an aptamer–target complex, and dehybridization occurs between the aptamer labeled with a fluorophore and the complementary strand labeled with a quencher. Thus, the target molecule can be detected by FRET.

Such signaling aptamers can be designed rationally. However, signaling aptamers also can be selected from a random library. Jhaveri et al. (2000) have selected antiadenosine signaling aptamers using a random library that contains fluoresceinated uridines. Nutiu and Li (2005) have also reported the selection of signaling aptamers by means of a library containing a central 15-nucleotide fixed-sequence domain. The central fixed-sequence domain was designed to be complementary to the biotinylated antisense oligonucleotide. The signaling aptamer species can be eluted by addition of the target molecule, because they are dehybridized from the biotinylated antisense oligonucleotide. After selection, the signaling aptamers were constructed by labeling with a fluorophore and a quencher.

Recently, signaling aptamers have been used in the construction of AND/OR logic gates, which respond to two different aptamer ligands. The AND logic gate works only when both aptamer ligands are present, whereas the OR logic gate works when either one of the aptamer ligands is present. Two types of aptameric logic gates are reported: One is a logic gate that detects aptamer ligands by the aggregation of gold nanoparticles (Liu and Lu, 2006), and the other is a logic gate that yields fluorescence output (Yoshida and Yokobayashi, 2007).

These signaling aptamers have shown that multiple aptamer ligands can be detected without bound–free separation: that is, through homogeneous detection. It is very difficult to realize this type of detection using antibodies, aside from a few exceptions (Ueda et al., 2003). We discuss the problem of homogeneous detection in the next section and introduce our solution: the aptameric enzyme subunit.

6.2 HOMOGENEOUS SENSING

6.2.1 Biosensor Systems That Do Not Require Bound–Free Separation

Sensitivity and accuracy are the most important factors in biosensing. The most critical factor that determines sensitivity is the signal-to-noise ratio; therefore, it is crucial to find ways to reduce the noise. The bound–free (B/F) separation is useful for noise reduction, but the procedure is time consuming and complicated. Apart from high sensitivity and accuracy, rapid and simple detection of the target molecule is always required for diagnosis and is also important for biosensing in general. To detect the target molecule rapidly and simply, a homogeneous assay, which does not require B/F separation, is ideal. However, such an assay often has the disadvantage of lower sensitivity than that of detection with B/F separation. Detection with B/F separation and homogeneous detection are compared in Table 6.2.

To construct a homogeneous detection system, the presence of a molecular recognition element that generates signals upon recognition of the target molecule is essential. Signaling aptamers are one type of good molecular recognition element for homogeneous detection. We propose another candidate for a signal-generating molecular recognition element, described next.

6.2.2 Aptameric Enzyme Subunit

Signaling aptamers are excellent sensing systems, but if we can introduce signal amplification, a more highly sensitive detection can be performed. Allosteric ribozymes and DNAzymes can amplify the detection signals through their catalytic activity, but it takes several minutes to detect their activity (Tang and Breaker, 1997; Levy and Ellington, 2002). Enzymes are the most powerful tools in the biosensing field, because enzymatic reactions can rapidly amplify the sensing signal millions of times. The enzyme-linked immunosorbent assay (ELISA) is widely used because in ELISA, antibodies recognize the target molecules, and enzymes that amplify the detection signals are connected to those antibodies.

TABLE 6.2 Comparison of Detection with B/F Separation and Homogeneous Detection

	Detection with B/F Separation	Homogeneous Detection
Simplicity	Fair (it depends on the separation method)	Excellent
Rapidity	Fair (it depends on the separation method)	Excellent
Sensitivity	Excellent (the noise can be reduced)	Good (the signal can be amplified using enzymes)
Handiness	Fair (it depends on the separation method)	Excellent
Accuracy	Excellent	Fair

If we can find an enzyme that catalyzes the reaction with a target molecule, we can construct effective sensing systems such as the glucose sensor, which is already on the market and is being used daily. However, it is not easy to screen an enzyme that is a good match for a given target molecule. Protein engineers have had numerous successes in changing the substrate specificity, and we have reported such examples (Igarashi et al., 2004), but it is still difficult to change the substrate specificity dramatically.

In the cell, nucleic acids are involved in the storage of genomic information and the production of proteins. However, other functions of nucleic acids, especially RNA, have been discovered. For example, ribosomal RNAs not only fold the structure of the ribosome but play an important role in protein synthesis (Moore and Steitz, 2002), and the RNA-induced silencing complex (RICS) is composed of a guide strand of siRNA or microRNA and proteins to control gene expression through RNA interference (Filipowicz, 2005). Those RNAs and proteins work cooperatively to achieve complicated tasks. This means that if we could engineer a nucleic acid–protein complex, we could construct more functional biosensor systems.

The engineering of an enzyme–aptamer complex is one way to change the substrate specificity of an enzyme. If we can add a substrate-binding site to the aptamer domain and control the corresponding enzymatic activity allosterically, the target molecule would be detected by measuring the enzymatic activity. In other words, we can change the substrate specificity of the enzyme by connecting the target-molecule-binding aptamer to the enzyme-inhibiting aptamer.

We have developed the aptameric enzyme subunit (AES), which enables us to detect the target molecule by measuring the enzymatic activity without B/F separation (Yoshida et al., 2006). The AES is an aptamer composed of an enzyme-inhibiting aptamer and a target molecule–binding aptamer. We designed an AES whose enzyme-inhibiting aptamer site structure changes when the target molecule binds to the target molecule–binding aptamer site. The structural change of the enzyme-inhibiting aptamer site induces a change in the inhibitory activity of the AES. The principle of the AES is schematized in Figure 6.2.

In this study we constructed two AESs that can detect adenosine: One is an AES whose inhibitory activity is increased upon binding to adenosine, and the

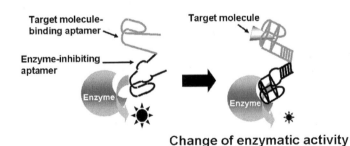

Change of enzymatic activity

Figure 6.2   Aptameric enzyme subunit.

other is an AES whose inhibitory activity is decreased upon binding to adenosine. In this system the adenosine-binding aptamer connects to the thrombin-inhibiting aptamer and the thrombin enzymatic activity is measured.

To construct the AES, which shows increased inhibitory activity upon the binding of adenosine, an adenosine-binding aptamer was designed with a shortened stem at the connection to the thrombin-inhibiting aptamer. The thrombin-inhibiting aptamer is destabilized by insertion of the adenosine-binding aptamer with the shortened stem, and then it becomes stable again when adenosine is bound to the adenosine-binding aptamer domain. Therefore, the thrombin-inhibitory activity of this AES is increased by adenosine. However, this AES design is limited to a special case. We therefore developed a universal AES design.

For the universal design, the thrombin-inhibiting aptamer was split into two oligonucleotides: One oligonucleotide was connected to the adenosine-binding aptamer, and the other was connected to the complementary strand of the adenosine-binding aptamer. These two oligonucleotides hybridize with each other to fold the thrombin-inhibiting aptamer structure and thus inhibit thrombin activity. On the addition of adenosine, the adenosine-binding aptamer site folds into the native structure to dehybridize its complementary strand. Thus, binding the adenosine-binding aptamer to adenosine destabilizes the complex, which leads to the loss of its inhibitory activity. This design, in which the enzyme-inhibiting aptamer is split into two oligonucleotides and the target-binding aptamer is added to one oligonucleotide while the complementary strand to the target-binding aptamer is added to the other oligonucleotide, should be applicable to most aptamers and should therefore be universal (Figure 6.3).

The important step seems to be to split the enzyme-inhibiting aptamer. However, the split aptamer can be folded into its original structure by hybridization of the target-binding aptamer with its complementary strand, and we can choose the split point of the enzyme-inhibiting aptamer freely; therefore, we should be able to find a suitable AES design after optimization.

We have also applied the AES to DNA sensing by inserting the stem–loop structure into the thrombin-inhibiting aptamer (Yoshida et al., 2006). The loop sequence was designed to be complementary to the target DNA sequence. Hybridization of the target DNA with the loop site induces disruption of the entire AES structure, and its inhibitory activity is decreased. Therefore, the target DNA can be detected by measuring the thrombin activity.

As mentioned above, the main advantage of the AES is the signal amplification performed by the enzyme. There is no doubt that enzymes are the best molecular recognition elements for biosensing, since they generate highly amplified signals upon recognition of the target molecule. Some enzymes are used for practical analysis, and the AES can be applied to those detection systems. Therefore, the AES certainly has a great potential for use in the development of rapid and simple homogeneous detection systems that do not rely on the time-consuming and complicated B/F separation step.

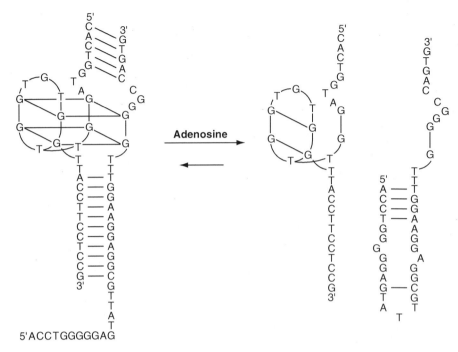

Figure 6.3 Universal AES design. The enzyme-inhibiting aptamer (thrombin aptamer) and the target molecule–binding aptamer (adenosine aptamer) are shown in blue and red, respectively. (See insert for color representation.)

6.3 EVOLUTION-MIMICKING ALGORITHM FOR THE IMPROVEMENT OF APTAMERS

An enzyme-inhibiting aptamer that has high inhibitory activity is necessary for construction of an effective AES. To select an enzyme-inhibiting aptamer, we usually have to carry out the SELEX process. However, selection by SELEX is based on affinity, whereas the aptamer showing the strongest binding activity is not always the best enzyme inhibitor. Ideally, selection should be based on the inhibitory activities of the contending aptamers. We have reported on the evolution-mimicking algorithm (EMA), which is a method of aptamer sequence optimization yielding aptamers with greater inhibitory activities than that of the original aptamer (Ikebukuro et al., 2005, 2006; Noma and Ikebukuro, 2006). The principle of the EMA is illustrated in Figure 6.4.

An EMA consists of selection via an inhibition assay, crossover, and mutation of the oligonucleotide sequence *in silico*. First, a set of 10 DNA sequences, each of which contains a portion of the original aptamer sequence as well as a randomized region, is generated by a computer and synthesized chemically. The inhibitory activity of each aptamer for the target enzyme is measured and

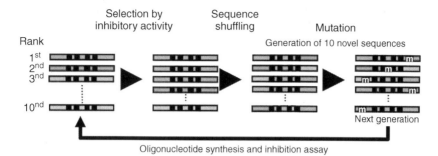

Figure 6.4   Screening using the evolution-mimicking algorithm. (See insert for color representation.)

ranked accordingly. The parent aptamers and their selection ratios are determined manually based on their sequence homologies and activities. Subsequently, the sequences are shuffled or cross-overed among the selected aptamers, followed by generation of the next 10 sequences. Finally, point mutations are introduced into these 10 sequences at a certain constant frequency *in silico*. After several rounds of this process, the enzyme-inhibiting aptamers for thrombin and *Taq* polymerase with greater inhibitory activities than that of the original aptamer are obtained. Although we applied an EMA to obtain aptamers that have high inhibitory activity, this method may also be able to induce evolution of the selectivity and affinity of aptamers.

To summarize, aptamers have several advantages over antibodies in terms of biosensor application and have great potential for use in the development of novel biosensing systems, one that is difficult to realize using antibodies. The AES is especially attractive, since it can realize highly sensitive homogeneous detection without B/F separation and allows rapid and simple detection, which is a crucial requirement in medical diagnosis. We developed a universal AES design and an effective method of selecting the proper enzyme-inhibiting aptamers. Through these methods, the AESs can be improved, and AESs applicable to practical use can be developed in the near future.

REFERENCES

Blank, M., Weinschenk, T., Priemer, M., Schluesener, H. (2001). Systematic evolution of a DNA aptamer binding to rat brain tumor microvessels: selective targeting of endothelial regulatory protein pigpen. *J Biol Chem* 276, 16464–16468.

Drolet, D. W., Moon-McDermott, L., Romig, T. S. (1996). An enzyme-linked oligonucleotide assay. *Nat. Biotechnol.* 14, 1021–1025.

Filipowicz, W. (2005). RNAi: the nuts and bolts of the RISC machine. *Cell*, 122, 17–20.

Furuta, M., Ito, T., Eguchi, C., Tanaka, T., Wakabayashi-Takai, E., Kaneko, K. (2002). Two-dimensional electrophoresis/phage panning (2D-PP): a novel technology for direct antibody selection on 2-D blots. *J Biochem* 132, 245–251.

Hamaguchi, N., Ellington, A., Stanton, M. (2001). Aptamer beacons for the direct detection of proteins. *Anal Biochem* 294, 126–131.

Igarashi, S., Okuda, J., Ikebukuro, K., Sode, K. (2004). Molecular engineering of PQQGDH and its applications. *Arch Biochem Biophys* 428, 52–63.

Ikebukuro, K., Kiyohara, C., Sode, K. (2004). Electrochemical detection of protein using a double aptamer sandwich. *Anal Lett* 37, 2901–2909.

Ikebukuro, K., Okumura, Y., Sumikura, K., Karube, I. (2005a). A novel method of screening thrombin-inhibiting DNA aptamers using an evolution-mimicking algorithm. *Nucleic Acids Res* 33, e108.

Ikebukuro, K., Kiyohara, C., Sode, K. (2005b). Novel electrochemical sensor system for protein using the aptamers in sandwich manner. *Biosens Bioelectron* 20, 2168–2172.

Ikebukuro, K., Yoshida, W., Noma, T., Sode, K. (2006). Analysis of the evolution of the thrombin-inhibiting DNA aptamers using a genetic algorithm. *Biotechnol Lett* 28, 1933–1937.

Jhaveri, S., Rajendran, M., Ellington, A. D. (2000). In vitro selection of signaling aptamers. *Nat Biotechnol* 18, 1293–1297.

Levy, M., Ellington, A. D. (2002). ATP-dependent allosteric DNA enzymes. *Chem Biol* 9, 417–426.

Liu, J., Lu, Y. (2006). Smart nanomaterials responsive to multiple chemical stimuli with controllable cooperativity. *Adv Mater* 18, 1667–1671.

Moore, P. B., Steitz, T. A. (2002). The involvement of RNA in ribosome function. *Nature* 418, 229–235.

Noma, T., Ikebukuro, K. (2006). Aptamer selection based on inhibitory activity using an evolution-mimicking algorithm. *Biochem Biophys Res Commun* 347, 226–231.

Noma, T., Ikebukuro, K., Sode, K., Ohkubo, T., Sakasegawa, Y., Hachiya, N., Kaneko, K. (2006). A screening method for DNA aptamers that bind to specific, unidentified protein in tissue samples. *Biotechnol Lett* 28, 1377–1381.

Nutiu, R., Li, Y. (2003). Structure-switching signaling aptamers. *J Am Chem Soc* 125, 4771–4778.

Nutiu, R., Li, Y. (2005). In vitro selection of structure-switching signaling aptamers. *Angew Chem Int Ed Engl* 44, 1061–1065.

Tang, J., Breaker, R. R. (1997). Rational design of allosteric ribozymes, *Chem Biol* 4, 453–459.

Ueda, H., Yokozeki, T., Arai, R., Tsumoto, K., Kumagai, I., Nagamune, T. (2003). An optimized homogeneous noncompetitive immunoassay based on the antigen-driven enzymatic complementation. *J Immunol Methods* 279, 209–218.

Yoshida, W., Yokobayashi, Y. (2007). Photonic Boolean logic gates based on DNA aptamers. *Chem Commun* 2, 195–197.

Yoshida, W., Sode, K., Ikebukuro, K. (2006a). Aptameric enzyme subunit for biosensing based on enzymatic activity measurement. *Anal Chem* 78, 3296–3303.

Yoshida, W., Sode, K., Ikebukuro, K. (2006b). Homogeneous DNA sensing using enzyme-inhibiting DNA aptamers. *Biochem Biophys Res Commun* 348, 245–252.

CHAPTER 7

NANOMATERIAL-BASED LABEL-FREE APTASENSORS

KAGAN KERMAN and EIICHI TAMIYA

7.1 INTRODUCTION

The code of life imprinted on the nucleic acids has always enchanted researchers and intrigued them into unraveling its secrets at the crossroads of chemistry, physics, and biology. In particular, nucleic acids and the journey that transforms them into proteins have never ceased to provide us with surprises and twists. Right in the center of the crossroads, biosensors emerge as a promising field of research. A biosensor is a device that combines a biological component (a *recognition layer*) and a physicochemical detector component (a *transducer*). The transduction unit can be electrochemical, optical, piezoelectric, magnetic, or calorimetric. The recognition layer can be constructed using enzymes, antibodies, cells, tissues, and in particular, nucleic acids. The aptamers are the latest addition to this amazing catalog of recognition layers. By definition, aptamers are synthetic oligonucleotides that can be generated to recognize certain molecules, such as amino acids, drugs, and proteins, with high affinity (Ellington and Szostak, 1990; Tuerk and Gold, 1990; Jayasena, 1999; Wilson and Szostak, 1999; Cho et al., 2006; Musheev and Krylov, 2006). In this chapter we discuss the voltammetric, electrical, and optical aptasensors from a label-free approach.

7.2 LABEL-FREE ELECTROCHEMICAL APTASENSORS

More than four decades ago, reports regarding electrochemical signals from DNA appeared in the literature (Palecek, 1958). Voltammetry of native and denatured

Aptamers in Bioanalysis, Edited by Marco Mascini
Copyright © 2009 John Wiley & Sons, Inc.

DNA, especially, has served as a milestone for electrochemical monitoring of DNA (Palecek, (1958), (1980); Palecek et al., 1982, 1986). The mechanism for the oxidation of guanine has been studied in detail, because guanine is the most redox active nitrogenous base in DNA (Cai et al., 1996; Jelen et al., (1997); Steenken and Jovanovic, 1997; Fojta et al., 2000). The oxidation of guanine and adenine was shown to follow a two-step mechanism involving the total loss of four electrons and four protons showing current peaks at about 0.9 and 1.2 V, respectively, depending on the pH and ionic strength of the electrolyte and the electrode material. Electrochemical mechanisms for the oxidation and reduction of DNA bases on carbon and mercury electrodes have recently been reviewed in detail (Palecek and Jelen, 2005). The interaction of DNA with drugs, metals, and proteins, as well as the detection of hybridization between two complementary strands with no external redox labels, have been monitored by a large number of groups using the intrinsic oxidation signal of DNA (Wang et al., 1995, 1996, Tomschik et al., 1999; 1999; Kara et al., 2002; Lucarelli, et al., 2002; Ozkan et al., 2002; Kerman et al., 2003, 2005; Ariksoysal et al., 2005; Erdem et al., 2005; Ozsoz et al., 2005; Mascini et al., 2006; Mello et al., 2006). Xie et al. (2007) have recently reported the electrocatalytic oxidation of guanine, guanosine, and guanosine monophosphate. Apilux et al. (2007) reported on the electrochemical behavior of native and thermally denatured fish DNA in the presence of cytosine derivatives and porphyrin using cyclic voltammetry. A comparison study between a disposable electrochemical DNA biosensor and a *Vibrio fischeri*–based luminescent sensor was performed by Tencaliec et al. (Ye and Ju, 2005) for the detection of toxicants in water samples. Potentially toxic compounds present in water were evaluated by changes in the electrochemical signal of guanine. The electrocatalytic oxidation of guanine and DNA on a carbon paste electrode modified with cobalt hexacyanoferate films was also reported (Abbaspour and Mehrgardi, 2004). Koehne et al. (2004) developed a miniaturized multiplex label-free electronic chip for rapid nucleic acid analysis based on carbon nanotube arrays.

Since the middle of the twentieth century, the electrochemical analysis of proteins has been used widely (Brezina and Zuman, 1958). From the early 1970s until today, the major interest of electrochemists has focused on a relatively small group of proteins containing a metal center with reversible redox activity (metalloproteins) (Armstrong, 2002). The fact that most proteins not containing a metal center can show electrochemical activity, depending on their amino acid structure, is gaining the attention of researchers. Since polarography has been a well-established method, the first findings about the label-free electrochemistry of proteins came from mercury electrodes. Peptides and proteins containing cysteine/cystine showed specific electrochemical signals on mercury electrodes with the help of Hg–S bond formation (Havran et al., 2004), reduction of disulfide groups (Tomschik et al., 1998), and the catalytic evolution of hydrogen in cobalt-containing solutions (the Brdicka reaction) (Brdicka, 1933; Heyrovsky, 2004). Hydrogen evolution was also catalyzed at highly negative potentials in the absence of transition-metal ions using mercury electrodes with proteins that contained or lacked sulfur amino acids. In combination with chronopotentiometric stripping analysis, presodium catalysis resulted in a well-defined signal that enabled the detection of several important proteins (Tomschik et al., 2000). The electrooxidation of Tyr residues involves a two-electron and two-proton transfer showing a current peak at about 0.6 V at a carbon electrode in 50 mM phosphate buffer solution (pH 7.4) with an electrode process that is similar to the oxidation of a simple p-substituted phenol (Brabec and Mornstein, 1980; Palecek and Jelen, 2005). Well-developed oxidation peaks of Tyr and Trp in nanomolar concentrations of peptides were obtained by applying voltammetric methods in combination with a sophisticated baseline correction (Cai et al., 1996). Our group has recently reported the electrochemical detection of the aggregation process in amyloid peptides related to Alzheimer's disease by monitoring the electrochemical oxidation signal of Tyr (Vestergaard et al., 2005). Some of the clinically important peptides and proteins, such as telomerase reverse transcriptase (Takata et al., 2006), chorionic gonadotropin hormone (Kerman et al., 2006), metallothioneins (Kizek et al., 2001), bovine serum albumin (Ostatna et al., 2006), α-synuclein (Masarik et al., 2004), avidin (Havran et al., 2004), p53 (Brazdova et al., 2002; Potesil et al., 2006), and MutS (Palecek et al., 2004; Masarik et al., 2007) have been detected using their intrinsic electrochemistry at very low concentrations.

In a similar fashion, aptamer–protein interactions can be monitored using intrinsic DNA and protein oxidation signals on carbon electrodes. The simultaneous monitoring of two signals from aptamer and protein enables us to detect the specific binding event in a rapid and easy format, as illustrated in Figure 7.1.

If the aptamer does not recognize the target protein, no binding event takes place, and the intrinsic oxidation signal of guanine can be monitored. If the target protein exists in the solution, binding takes place between the target protein and the aptamer, resulting in a decrease in the guanine signal and the appearance of Tyr and Trp oxidation signals from the bound protein. The combination of intrinsic DNA and protein oxidation signals provides us with a brand new outlook on the interaction of these proteins with aptamers on a solid surface in the absence of external labels.

Figure 7.2 shows the differential pulse voltammograms for the interaction of an aptamer and immunoglobulin E (IgE) obtained from three different sets of electrodes. The current signal labeled G at about 0.8 V shows the electrochemical oxidation response of guanine residues in the absence of IgE (green line, see color insert). However, after the binding event between 1 nM IgE and the aptamer, the protein oxidation signal appears at about 0.6 V (peak P) and there is a drop in the oxidation of guanine (the blue line, see color insert). When 10 nM of IgE is bound to the aptamers on the surface, peak G drops significantly (red line, see color insert). As demonstrated here, the label-free voltammetry of aptamer and protein interactions promises to provide a new approach with two signals (peaks G and P) at two distinct peak potentials.

In an effort to increase the sensitivity of label-free voltammetry, we can utilize carbon nanotubes (CNTs) as the transducing surface of our aptasensors. It has been well established that the electrochemical responses of DNA are greatly

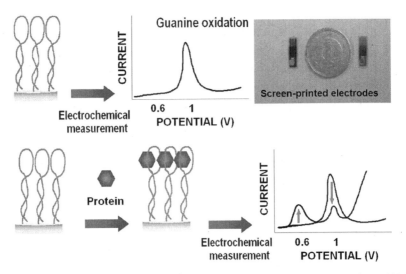

Figure 7.1 Label-free voltammetric detection of aptamer–protein interactions. (A) The electrochemical oxidation response from the guanine bases in the aptamer is observed at about 1 V (peak G) on the surface of the screen-printed electrodes. (The inset shows gold and carbon-based screen-printed electrodes with a three-electrode system.) (B) After the binding event with the target protein, an oxidation signal appears at about 0.6 V (peak P), and the oxidation signal of the aptamer decreases.

enhanced at the CNT-modified electrodes compared with other types of electrode materials (Gooding, 2005; Kerman et al., 2005; Okuno et al., 2007a, 2007b; Pumera et al., 2007). CNTs behave electrically as a metal or as a semiconductor, depending on their atomic structure and therefore are more conductive than graphite (Saito et al., 1992; Britto et al., 1996, 1999; Davis et al., 1997; Odom et al., 1998; Wilder et al., 1998). Moreover, CNTs serve as excellent substrates for the development of biosensors to promote electron-transfer reactions between biomolecules and the solid electrode with the help of its high aspect ratio (Gooding, 2005; Kerman et al., 2005; Okuno et al., 2007a, 2007b; Pumera et al., 2007). In Figure 7.3, the differential pulse voltammetry of 10 nM IgE is displayed on both a CNT-modified (solid line) and a bare screen-printed carbon electrode (dashed line). An enhancement of about 10-fold could be observed in the peak heights of the signals obtained from the CNT-modified aptasensors for IgE detection. Such an enhancement would allow new opportunities for the observation of interactions at trace amounts of proteins.

The label-free voltammetry of DNA and proteins has been known for a long time but has recently started to blossom as an exciting field which new researchers are tempted to pursue with high motivation. With the endless possible combinations of amino acids in peptides and proteins, as well as the nucleobases in DNA and RNA, an immense variety of voltammetric signals can be obtained as a result of aptamer–protein interactions. Since the method does not require any

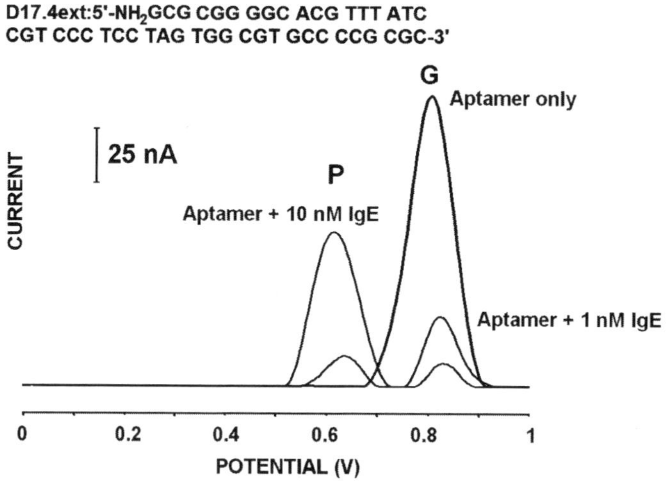
D17.4ext:5'-NH2GCG CGG GGC ACG TTT ATC
CGT CCC TCC TAG TGG CGT GCC CCG CGC-3'
G
Aptamer only
25 nA
P
Aptamer + 10 nM IgE
Aptamer + 1 nM IgE
CURRENT
0
0.2
0.4
0.6
0.8
1
POTENTIAL (V)

Figure 7.2 Differential pulse voltammograms for the label-free voltammetric detection of IgE. The oxidation signal of guanine is observed alone from the 10-μM aptamer on a screen-printed carbon electrode (green line), the oxidation responses from the binding event between 10 nM IgE and the aptamer (red line), and the oxidation responses from the binding event between 1 nM IgE and the aptamer (blue line). Differential pulse voltammetry was performed in blank phosphate buffer solution (PBS, pH 7.4) while scanning from 0.10 to 1.20 V with an amplitude of 25 mV and a step potential of 5 mV. The raw voltammograms were treated using Savitzky–Golay smoothing (level 4) and a moving-average baseline correction with a peak width of 0.005 V. (See insert for color representation.)

external redox labels, the information gathered from these measurements would be valuable for understanding new aspects of the aptamers.

7.3 FIELD-EFFECT TRANSISTOR-BASED APTASENSORS

In this section, field-effect transistors (FETs) (Lieber, 2003) are introduced as another exciting field of developing label-free aptasensors using their electrical properties. In a typical FET, current flows along a semiconductor path called a *channel*. The channel is located between two electrodes, called the *source* and the *drain*. The physical diameter of the channel is fixed, but its effective electrical diameter can be varied by the application of voltage to a control electrode called a *gate*. The electrical diameter of the channels, at any given instant in time, determines the conductivity of the FET. A small change in gate voltage can cause a large variation in the current flowing from the source to the drain. Thus, FET has the ability to amplify small signals.

Field-effect transistors exist in two major types: the junction FET (JFET) and the metal-oxide-semiconductor FET (MOSFET). The junction FET has a

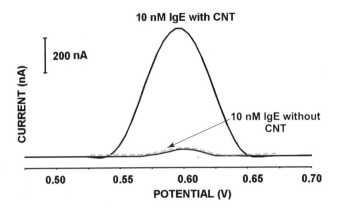

Figure 7.3 Label-free voltammetry of 10 nM IgE on a multiwalled carbon nanotube (MWCNT)-modified screen-printed carbon electrode (solid line) and on a bare screen-printed carbon electrode (dashed line). Experimental conditions for differential pulse voltammetry were as described in Figure 7.2. MWCNTs (1 mg) were dispersed with the aid of ultrasonic agitation in 10 mL of N, N-dimethylformamide to give a 0.1-mg mL^{-1} black solution. MWCNT film was prepared by pipetting a 2-μL drop of MWCNT solution onto the carbon working electrode of the screen-printed electrode and then evaporating the solvent at room temperature.

channel consisting of n-type semiconductor (n-channel) or p-type semiconductor (p-channel) material; the gate is made of the opposite semiconductor type. In p-type material, electric charges are carried mainly in the form of electron deficiencies called *holes*. In n-type material, the charge carriers are primarily electrons. In a JFET, the junction is the boundary between the channel and the gate. Normally, this $p - n$ junction is reverse-biased (a dc voltage is applied to it) so that no current flows between the channel and the gate.

In the MOSFET, the channel can be either an n- or a p-type semiconductor. The gate electrode is usually a piece of metal with an oxidized surface. The oxide layer insulates the gate electrically from the channel. Since the oxide layer acts as a dielectric, there is essentially no current flow between the gate and the channel during any part of the signal cycle, which gives MOSFET a large input impedance. However, MOSFET is susceptible to destruction by electrostatic charges, since the oxide layer is extremely thin. Special precautions are necessary when handling or transporting these devices.

FETs are actually the basic building blocks of integrated circuits. To develop circuits using nanotubes, we first have to design nanotube-based transistors. Silicon nanowires represent one of the best characterized examples of semiconductor nanotubes with the structure, size, and electronic properties controlled reproducibly (Hu et al., 1999; Cui et al., 2001a). In particular, silicon nanowires can be prepared as single-crystal structures with controllable diameters as small as 2 to 3 nm (Cui et al., 2003; Wu, 2004). Both n- and p-type FET devices can be produced with well-defined and reproducible high-performance properties

(Zheng et al., 2004). These reproducible structural and electronic characteristics of the silicon nanowires, which contrast the current state of CNT devices, have provided a strong platform for numerous biosensing schemes (Souteyrand et al., 1997; Dzyadevich, 1999; Cui et al., 2001a; Fritz et al., 2002; Schoning and Poghissian, 2002; Hahm and Lieber, 2004; Wang et al., 2005; Patolsky et al., 2006; Ster et al., 2007). FET-based devices are attractive alternatives to existing biosensor technologies, because they are readily applicable to on-chip integration of arrays with a well-defined signal-processing scheme, and most important, mass production of devices can be made at low cost. Moreover, thermodynamical analysis demonstrated that the surface electrostatics of the FET device would significantly affect DNA and protein detection (Souteyrand et al., 1997; Dzyadevich, 1999; Cui et al., 2001a; Fritz et al., 2002; Schoning and Poghissian, 2002; Hahm and Lieber, 2004; Wang et al., 2005; Patolsky et al., 2006; Gao et al., 2007; Ster et al., 2007). The most attractive property of FET-based biosensors is the ability to control the surface charge, which enables manipulation of the charged biomolecules on the surface (Souteyrand et al., 1997; Dzyadevich, 1999; Cui et al., 2001a; Fritz et al., 2002; Schoning and Poghissian, 2002; Hahm and Lieber, 2004; Wang et al., 2005; Patolsky et al., 2006; Gao et al., 2007; Ster et al., 2007).

As shown in Figure 7.4, we introduce the detection of immunoglobulin E (IgE) using CNT-FETs, in which CNT channels were modified with 45-mer aptamers. Since the aptamer was self-complementary with 12 base pairs, its size was significantly smaller than that of a monoclonal antibody against IgE (about 10 nm). Under similar ionic strength conditions, the antibody-IgE binding event may have occurred outside the electrical double layer in solution, since the antibodies are much larger in size than the Debye length (about 3 nm in 10 mM solutions). Thus, the charges in the bound protein may be "screened" by the double layer, and their effect on the equilibrium carrier distribution would then be vanishingly small. However, they are no more or no less "canceled" in the sense of being paired with an ion from the solution of opposite charge than they would be if much closer to the electrode. On the other hand, aptamers enable sensitive detection possibilities, derived partly from their small size (about 1 to 2 nm). As a result, the binding event between the aptamers and the target proteins can occur within the electrical double layer in buffer solution, and therefore changes in the charge distribution within proximity to the CNT can be detected easily by FETs (Guo et al., 2005; So et al., 2005). Moreover, the density of the immobilized aptamers on the CNT channels can be controlled, and a high density of aptamers can easily be prepared.

Recently, Park et al. (2006) developed a biosensor capable of detecting carcinoembryonic antigen (CEA) markers using single-walled carbon nanotube-FETs (SWCNT-FETs). For biosensor applications, SU-8 negative photoresist patterns were used as an insulation layer. CEA antibodies were employed as recognition elements to specific tumor markers and were immobilized successfully on the sides of a single-walled carbon nanotube using carbodiimidazole-activated Tween 20 (CDI-Tween) linking molecules. Recently, So et al. (2005) developed a

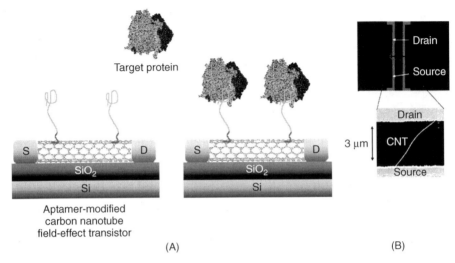

Figure 7.4 Label-free aptasensors based on CNT-FETs. (A) The aptamers are attached onto the sidewalls of a single-walled carbon nanotube which connects the source (S) and drain (D) electrodes. The binding of the target protein changes the electrical properties of the CNT-FETs significantly. (B) CNT-FET devices at high resolution. (See insert for color representation.)

single-walled carbon nanotube FET device for the detection of thrombin. Aptamer immobilization was performed by first modifying the sidewall of the CNT with the same linker, CDI-Tween. While the Tween component was bound to the CNT sidewall through hydrophobic interactions, the carbodiimidazole moiety was used to attach the 3′-amine group of the aptamer covalently. Furthermore, proteins and oligonucleotides have been attached noncovalently to the graphitic sidewalls of carbon nanotubes via a pyrene moiety (Taft et al., 2004). In a similar fashion, we have used 1-pyrenebutyric acid N-hydroxysuccinimidyl ester for the attachment of aptamers on the sidewalls of the SWCNT on our FET device. Figure 7.5 shows the time dependence of source–drain current of CNT-FETs at the source–drain bias of 0.2 V and at the gate bias of 0 V after the introduction on the CNT-FET aptasensor of IgE at various concentrations.

Adding the target IgE caused a sharp drop in the source–drain current and then a gradual saturation at lower values. This drop was attributed to the screening effect of IgE on the negative charge of the aptamers. Moreover, the increase in Schottky barrier height between the metal electrodes (Byon and Choi, 2006) and the CNT channel, which could be due to the adsorption of IgE molecules on CNT channels at the CNT–metal contacts, might have caused the drop in the electrical response (Taft et al., 2004; So et al., 2005; Byon and Choi, 2006; Park et al., 2006).

For the development of FET-based biosensors, aptamers display a significantly better performance than their monoclonal antibody counterparts do under

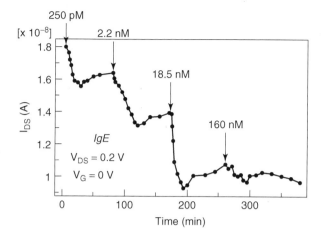

Figure 7.5   Time dependence of source–drain current of the CNT-FET at the source–drain bias of 0.2 V and at the gate bias of 0 V after the introduction of IgE at various concentrations onto the IgE aptamer-modified CNT-FET. Arrows indicate the points of IgE injections.

similar conditions. CNT-FET aptasensor is a promising candidate for the development of an integrated, high-throughput, multiplexed diagnostic device. As the development and optimization studies about FET-based aptasensors continue at a fast rate, we predict that sensitive multiplexed detection of numerous clinically important biomolecules would become possible in a rapid and high-throughput format.

7.4 LABEL-FREE APTASENSORS BASED ON LOCALIZED SURFACE PLASMON RESONANCE

For centuries, metal nanoparticles have never ceased to attract scientists and artists from many diverse cultures. In this section we briefly introduce a phenomenon of metal nanoparticles that still inspires scientists: localized surface plasmon resonance (LSPR) (Hutter and Fendler, 2004). Metal nanoparticles show nonlinear electronic transport (single-electron transport of Coulomb blockade) and nonlinear/ultrafast optical response due to the SPR. Conduction electrons $(-)$ and ionic cores $(+)$ in a metal form a plasma state. When external electric fields (i.e., electromagnetic waves, electron beams etc.) are applied to a metal, electrons move so as to screen perturbed charge distribution, move beyond the neutral states, return to the neutral states, and so on. This collective motion of electrons is called a *plasma oscillation*. SPR is a collective excitation mode of the plasma localized near the surface. Electrons confined in a nanoparticle conform the LSPR mode. The resonance frequency of the surface plasmon is different from an ordinary plasma frequency. The surface plasmon mode arises from the electron confinement in the nanoparticle. Since the dielectric function tends to become continuous at the interface (surface), the oscillation mode shifts from the ordinary plasma resonance and decays exponentially along the depth from the surface.

Since the detection of biomolecular interactions using SPR has been established as a reliable and conventional method (Aslan et al., 2005), numerous new designs toward aptasensors have focused on SPR. Recently, RNA aptamers that bind the opium alkaloid codeine were generated by Win et al. (2006) using an iterative in vitro selection process. The binding properties of these aptamers, including equilibrium and kinetic rate constants, were determined using SPR. Single-stranded DNA aptamers that bind MUC1 tumor marker were selected and characterized using SPR (Ferreira et al., 2006). Tombelli et al. (2005) constructed SPR chips using an RNA aptamer specific for HIV-1 TAT protein. Misono and Kumar (2005) selected RNA aptamers against human influenza virus hemagglutinin using SPR. Murphy et al. (2003) demonstrated the binding ability of aptamers that recognize thyroid transcription factor 1. They characterized their aptamers using SPR and showed them to be useful for enzyme-linked assays, Western blots, and affinity purification.

A methodology for the detection of protein biomarkers at picomolar concentrations that utilizes SPR imaging of RNA aptamer microarrays has recently been developed by Li et al. (2007). The adsorption of proteins onto the RNA microarray was detected by the formation of a surface aptamer–protein–antibody complex. The response signal of SPR imaging was then enhanced using a localized precipitation reaction catalyzed by horseradish peroxidase that was conjugated to the antibody. They detected vascular endothelial growth factor at a biologically relevant concentration of 1 pM. The same group has also developed a surface ligation method to detect protein factor IXa (Li et al., 2006). They determined a Langmuir adsorption coefficient of 1.6×10^{-7} for factor IXa.

LSPR absorption bands are characteristic of the type of nanomaterial, the diameter of nanoparticles, and their distribution (Freeman et al., 1995; Nath and Chilkoti, 2002). LSPR can detect an immediate change in the interfacial refractive index (RI) of the surrounding medium (Mock et al., 2003; Raschke et al., 2003; Nath and Chilkoti, 2004), which is greatly affected by the attachment of biomolecules at the colloid–solution interface (Himmelhaus and Takei, 2000; Mock et al., 2003; Raschke et al., 2003; Nath and Chilkoti, 2004; Endo et al., 2005). LSPR phenomena have been utilized previously to monitor biomolecular interactions by several groups (Haes et al., 2004; Dahlin et al., 2005; Endo et al., 2005, 2006; Sönnichsen et al., 2005; Liu et al., 2006; Shin et al., 2007).

Here we demonstrate the combination of LSPR with interferometry using a gold-coated porous anodic alumina (PAA) layer chip for the detection of aptamer–protein interactions (Figure 7.6). Recently, we have employed this device for the detection of DNA hybridization (Kim et al., 2007). Gold deposition on the surface of the PAA layer chip enables the formation of a novel nanomaterial, which we called gold-capped oxide nanostructures. One of

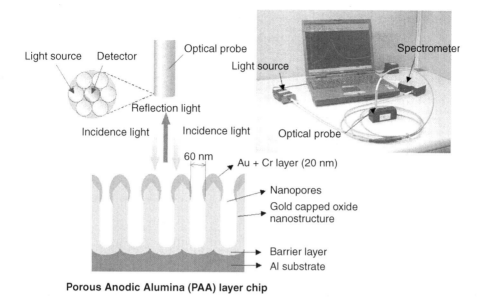

Porous Anodic Alumina (PAA) layer chip

Figure 7.6   Experimental setup and construction of LSPR and interferometry-based label-free optical aptasensor with a porous anodic alumina layer chip.

the major drawbacks of conventional gold nanoparticle-based LSPR chips is the complicated chemistry required to form a self-assembled monolayer (SAM) of gold nanoparticles on a solid surface. During the mass fabrication of numerous chips, small defects in the uniformity of SAMs can cause significant problems in the reproducibility and reliability of the results. It is well defined that the shift in the position of λ_{max} is small in LSPR observations, whereas there is a significant shift in the responses obtained from interferometric experiments.

AFM image of the PAA layer chip after gold deposition is shown in Figure 7.7A. Figure 7.7B shows a sandwich assay with two aptamers and thrombin, and Figure 7.7C shows the spectrum from 790 to 840 nm.

Aptamer I film on a PAA layer substrate resulted in a significant change in the refractive index of the layer medium and was detected as a corresponding shift and an increase in the absorbance intensity. When the bare gold-coated PAA layer was excited with light, we could observe only the interferometric pattern (black line). When 10 μM aptamer I was immobilized on a gold-deposited PAA layer surface, we could observe both the red shift and an increase in the absorbance intensity (red line, see color insert). After stringent washing of the surface, there was no significant change in the aptamer response. However, the binding reaction between 10 μM aptamer I and 1 μM thrombin on the chip surface caused a significant red shift in λ_{max} and an enhancement in the absorbance intensity (blue line, see color insert). After the binding reaction between thrombin and aptamer I, we added aptamer II onto the complex, and observed a further red

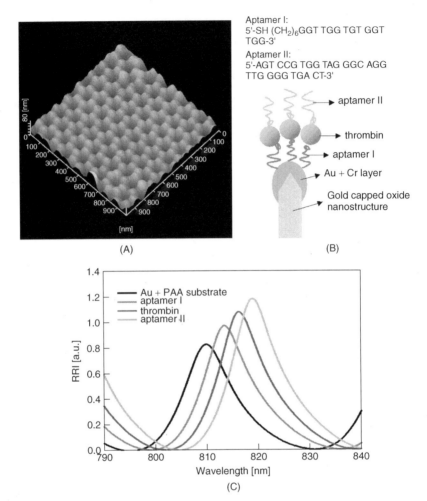

Aptamer I:
5'-SH (CH$_2$)$_6$GGT TGG TGT GGT TGG-3'

Aptamer II:
5'-AGT CCG TGG TAG GGC AGG TTG GGG TGA CT-3'

→ aptamer II

→ thrombin

→ aptamer I

→ Au + Cr layer

→ Gold capped oxide nanostructure

(A)

(B)

Au + PAA substrate
aptamer I
thrombin
aptamer II

(C)

Figure 7.7 (A) AFM analysis of a porous anodic alumina (PAA) layer chip surface after Au deposition. (B) Detection of a sandwich-type binding assay between the aptamers and thrombin using Au-capped oxide nanostructures on the PAA layer chip. (C) Relative reflected intensity characteristics of (black line) a bare PAA layer chip, 10 μM aptamer I immobilized on a gold-deposited PAA layer surface (red line), the binding reaction between 10 μM aptamer I and 1 μM thrombin on the chip surface (blue line), after the binding reaction between aptamer I/thrombin complex and aptamer II (green line). An aliquot (20 μL) of 10 μM thiolated aptamer I solution containing 100 μM 6-mercaptohexanol was introduced to the gold-deposited PAA layer chip surface and incubated for 1 hour. After the immobilization of aptamer I, the PAA layer chip surface was rinsed thoroughly with 20 mM phosphate-buffered solution (PBS, pH 7.4) and dried at room temperature. Then a desired concentration of thrombin solution in PBS (20 μL) was introduced to the aptamer I–immobilized PAA layer chip surface, and the interaction was allowed while incubating for 1 hour at room temperature. Aptamer II (20 μL) at 10 μM was then exposed to the aptamer I/thrombin complex as described above. After a stringent washing of the surface, changes in the absorption spectrum caused by the interaction were observed. (See insert for color representation.)

shift in λ_{max} and an increase in the absorbance intensity (green line, see color insert). Since the epitope of aptamer II was at a different location on thrombin, we could observe formation of an aptamer I/thrombin/aptamer II sandwich structure on the chip surface by means of optical characteristics.

The combination of LSPR with interferometry on a PAA layer chip enabled two important aspects of optical sensing systems: a shift in λ_{max} and an increment in the absorbance intensity in a new and highly sensitive format. The excitation of optical characteristics and the detection were performed using an optical fiber, which made our aptasensor user friendly and suitable for the simple construction of a handheld diagnostic device.

7.5 FORTHCOMING CHALLENGES AND CONCLUDING REMARKS

In this review we described the most commonly used label-free aptasensor strategies. Although aptamers are one of the most promising systems for applications in biosensors, their potential applications using complex matrices as real samples remain the major challenges for point-of-care applications; therefore, complementary strategies involving nanomaterials are a subject of intense study.

We are presently in the early days of the emerging technology of using aptamers and nanomaterials to modify biosensors. So far, thrombin-binding aptamer has been the topic of numerous reports, but although many improvements are required in reproducibility and sensitivity, there is no doubt that a growing number of aptamers and, it is to be hoped, nanomaterial-based label-free aptasensors will soon be used for the diagnosis and therapeutic follow-up of diseases.

REFERENCES

Abbaspour, A, Mehrgardi, M. (2004). Electrocatalytic oxidation of guanine and DNA on a carbon paste electrode modified by cobalt hexacyanoferrate films. A. *Anal Chem* 76, 5690–5696.

Apilux, A., Tabata, M., Chailapakul, O. (2007). Electrochemical behaviors of native and thermally denatured fish DNA in the presence of cytosine derivatives and porphyrin by cyclic voltammetry using boron-doped diamond electrode. *Bioelectrochemistry* 70, 435–439.

Ariksoysal, D. O., Karadeniz, H., Erdem, A., Sengonul, A., Sayiner, A. A., Ozsoz, M. (2005). Label-free electrochemical hybridization genosensor for the detection of hepatitis B virus genotype on the development of lamivudine resistance. *Anal Chem* 77, 4908–4917.

Armstrong, F. A. (2002). Voltammetry of proteins. In *Encyclopedia of Electrochemistry*, G. S. Wilson, ed., Vol. 9, *Bioelectrochemistry*, Wiley-VCH, Weinheim, Germany, pp. 11–29.

Aslan, K., Lakowicz, J. R., Geddes, C. (2005). Metal-enhanced fluorescence: an emerging trend in biotechnology. *Curr Opin Chem Biol* 9, 538–544.

Brabec, V., Mornstein, V. (1980). Electrochemical behaviour of proteins at graphite electrodes: I. Electrooxidation of proteins as a new probe of protein structure and reactions. *Biochim Biophys Acta* 625, 43–50.

Brazdova, M., Kizek, R., Havran, L., Palecek, E. (2002). Determination of glutathione-*S*-transferase traces in preparations of p53C-terminal domain (aa 320–393). *Bioelectrochemistry* 55, 115–118.

Brdicka, R. (1933). Polarographic studies with the dropping mercury method: a new test for proteins in the presence of cobalt salts in ammoniacal solution of ammonium chloride. *Collect Czech Chem Commun* 5, 112–127.

Brezina, M., Zuman, P. (1958). In *Polarography in Medicine, Biochemistry and Pharmacy*, Interscience, New York, p. 713.

Britto, P. J., Santhanam, K. S. V., Ajayan, P. M. (1996). Carbon nanotube electrode for oxidation of dopamine. *Bioelectrochem Bioenerg* 41, 121–125.

Britto, P. J., Santhanam, K. S. V., Alonso, V., Rubio, A., Ajayan, P. M. (1999). Improved charge transfer at carbon nanotube electrodes. *Adv Mater* 11, 154–157.

Byon, H. R., Choi, H. C. (2006). Site-specific delivery of DNA and appended cargo to carbon nanotubes. *J Am Chem Soc* 128, 2188–2189.

Cai, X., Rivas, G., Farias, P. A. M., Shiraishi, H., Wang, J., Palecek, E. (1996). Evaluation of different carbon electrodes for adsorptive stripping analysis of nucleic acids. *Electroanalysis* 8, 753–758.

Cai X., Rivas, G., Farias, M. A. P., Shiraishi, H., Wang, J., Palecek, E. (1996). Potentiometric stripping analysis of bioactive peptides at carbon electrodes down to sub-nanomolar concentrations. *Anal Chim Acta* 332(1), 49–57.

Cho, E. J., Collett, J. R., Szafranska, A. E., Ellington, A. D. (2006). Optimization of aptamer microarray technology for multiple protein targets. *Anal Chim Acta* 564, 82–90.

Cui, Y., Wei, Q., Park, H., Lieber, C. M. (2001a). Nanowire nanosensors for highly sensitive and selective detection of biological and chemical species. *Science* 293, 1289–1292.

Cui, Y., Lauhon, L. J., Gudiksen, M. S., Wang, J., Lieber, C. M. (2001b). Diameter-controlled synthesis of single crystal silicon nanowires. *Appl Phys Lett* 78, 2214–2216.

Cui, Y., Zhong, Z., Wang, D., Wang, W. U., Lieber, C. M. (2003). High performance silicon nanowire field effect transistors. *Nano Lett* 3, 149–152.

Dahlin, A., Zäch, M., Rindzevicius, T., Käll, M., Sutherland, D. S., Höök, F. (2005). Localized surface plasmon resonance sensing of lipid-membrane-mediated biorecognition events. *J Am Chem Soc* 127, 5043–5048.

Davis, J. J., Coles, R. J., Allen, A. O. H. (1997). Protein electrochemistry at carbon nanotube electrodes. *J Electroanal Chem* 440, 279–282.

Dzyadevich, S. V. (1999). Soldatkin application of enzyme field-effect transistors for determination of glucose concentrations in blood serum. *Biosens Bioelectron* 14, 283–287.

Ellington, A. D., Szostak, J. W. (1990). In vitro evolution of new ribozymes with polynucleotide kinase activity. *Nature* 346, 818–822.

Endo, T., Kerman, K., Nagatani, N., Takamura, Y., Tamiya, E. (2005a). Label-free detection of peptide nucleic acid–DNA hybridization using localized surface plasmon resonance based optical biosensor. *Anal Chem* 77, 6976–6984.

Endo, T., Yamamura, S., Nagatani, N., Morita, Y., Takamura, Y., Tamiya, E. (2005b). Localized surface plasmon resonance based optical biosensor using surface modified nanoparticle layer for label-free monitoring antigenantibody reaction. *Sci Technol Adv Mater* 6, 491–500.

Endo, T., Kerman, K., Nagatani, N., Hiep, H. M., Kim, D-K., Yonezawa, Y., Nakano, K., Tamiya, E. (2006). Multiple label-free detection of antigen–antibody reaction using localized surface plasmon resonance-based core–shell structured nanoparticle layer nanochip. *Anal Chem* 78, 6465–6475.

Erdem, A., Kosmider, B., Osiecka, R., Zyner, E., Ochocki, J., Ozsoz, M. (2005). Electrochemical genosensing of the interaction between the potential chemotherapeutic agent, *cis*-bis(3-aminoflavone)dichloroplatinum(II) and DNA in comparison with *cis*-DDP. *J Pharm Biomed Anal* 38, 645–652.

Ferreira, C. S. M., Matthews, C. S., Missailidis, S. (2006). DNA aptamers that bind to MUC1 tumour marker: design and characterization of MUC1-binding single-stranded DNA aptamers. *Tumor Biol* 27, 289–301.

Fojta, M., Havran, L., Fulneckova, J., Kubicarova, T. (2000). Adsorptive transfer stripping ac voltammetry of DNA complexes with intercalators. *Electroanalysis* 12, 926–934.

Fritz, J., Cooper, E. B., Gaudet, S., Sorger, P. K., Manalis, S. R. (2002). Electronic detection of DNA by its intrinsic molecular charge. *Proc Natl Acad Sci U S A* 99, 14142–14146.

Gao, Z., Agarwal, A., Trigg, A. D., Singh, N., Fang, C., Tung, C. H., Fan, Y., Buddharaju, K. D., Kong, J. (2007). Silicon nanowire arrays for label-free detection of DNA. *Anal Chem* 79, 3291–3297.

Gooding, J. J. (2005). Nanostructuring electrodes with carbon nanotubes: a review on electrochemistry and applications for sensing. *Electrochim Acta* 50, 3049–3060.

Guo, X., Huang, L., O'Brien, S., Kim, P., Nuckolls, C. (2005). Directing and sensing changes in molecular conformation on individual carbon nanotube field effect transistors. *J Am Chem Soc* 127, 15045–15047.

Hahm, J., Lieber, C. M. (2004). Direct ultrasensitive electrical detection of DNA and DNA sequence variations using nanowire nanosensors. *Nano Lett* 4, 51–54.

Havran, L., Billova, S., Palecek, E. (2004). Electroactivity of avidin and streptavidin: avidin signals at mercury and carbon electrodes respond to biotin binding. *Electroanalysis* 16, 1139–1148.

Heyrovsky, M. (2004). Early Polarographic studies on proteins. *Electroanalysis* 16, 1067–1073.

Himmelhaus, M., Takei, H. (2000). Cap-shaped gold nanoparticles for an optical biosensor. *Sens Actuat B* 63, 24–30.

Hu, J. T., Odom, T. W., Lieber, C. M. (1999). Chemistry and physics in one dimension: synthesis and properties of nanowires and nanotubes. *Acc Chem Res* 32, 435–445.

Hutter, E., Fendler, J. (2004). Exploitation of localized surface plasmon resonance. *Adv Mater* 16, 1685–1706.

Jayasena, S. D. (1999). Aptamers: an emerging class of molecules that rival antibodies in diagnostics. *Clin Chem* 45, 1628–1650.

Jelen, F., Fojta, M., Palecek, E. (1997). Voltammetry of native double-stranded, denatured and degraded DNAs. *J Electroanal Chem* 427, 49–56.

Kara, P., Kerman, K., Ozkan, D., Meric, B., Erdem, A., Nielsen, P. E., Ozsoz, M. (2002). Label-free and label based electrochemical detection of hybridization by using methylene blue and peptide nucleic acid probes at chitosan modified carbon paste electrodes. *Electroanalysis* 14, 1685–1690.

Kerman, K., Ozkan, D., Kara, P., Erdem, A., Meric, B., Nielsen, P. E., Ozsoz, M. (2003). Label-free bioelectronic detection of point mutation by using peptide nucleic acid probes. *Electroanalysis* 15, 667–670.

Kerman, K., Morita, Y., Takamura, Y., Tamiya, E. (2005). *Escherichia coil* single-strand binding protein–DNA interactions on carbon nanotube–modified electrodes from a label-free electrochemical hybridization sensor. *Anal Bioanal Chem* 381, 1114–1121.

Kerman, K., Nagatani, N., Chikae, M., Yuhi, T., Takamura, Y., Tamiya, E. (2006). Label-free electrochemical immunoassay for the detection of human chorionic gonadotropin hormone. *Anal Chem* 78, 5612–5616.

Kim, D. K., Kerman, K., Saito, M., Sathuluri, R. R., Endo, T., Yamamura, S., Kwon, Y. S., Tamiya, E. (2007). Label-free DNA biosensor based on localized surface plasmon resonance coupled with interferometry. *Anal Chem* 79, 1855–1864.

Kizek, R., Trnkova, T. L., Palecek, E. (2001). Determination of metallothionein at the femtomole level by constant current stripping chronopotentiometry. *Anal Chem* 73, 4801–4807.

Koehne, J. E., Chen, H., Cassell, A. M., Ye, Q., Han, J., Meyyappan M., Li, J. (2004). Miniaturized multiplex label-free electronic chip for rapid nucleic acid analysis based on carbon nanotube nanoelectrode arrays. *Clin Chem* 50, 1886–1893.

Li, Y., Lee, H. J., Corn, R. M. (2006). Fabrication and characterization of RNA aptamer microarrays for the study of protein–aptamer interactions with SPR imaging. *Nucleic Acids Res* 34, 6416–6424.

Li, Y., Lee, H. J., Corn, R. M. (2007). Detection of protein biomarkers using RNA aptamer microarrays and enzymatically amplified SPR imaging. *Anal Chem* 79, 1082–1088.

Lieber, C. M. (2003). Nanoscale science and technology: building a big future from small things. *MRS Bull* 28, 486–491.

Liu, G. L., Yin, Y., Kunchakarra, S., Mukherjee, B., Gerion, D., Jett, S. D., Bear, D. G., Gray, J. W., Alivisatos, P. A., Lee, L. P., Chen, F. F. (2006). A nanoplasmonic molecular ruler for measuring nuclease activity and DNA footprinting. *Nat Nanotechnol* 1, 47–52.

Lucarelli, F., Marrazza, G., Palchetti, I., Cesaretti, S., Mascini, M. (2002). Coupling of an indicator-free electrochemical DNA biosensor with polymerase chain reaction for the detection of DNA sequences related to the apolipoprotein E. *Anal Chim Acta* 469, 93–99.

Masarik, M., Stobiecka, A., Kizek, R., Jelen, F., Pechan, Z., Hoyer, W., Jovin, T. M., Subramaniam, V., Palecek, E. (2004). Sensitive electrochemical detection of native and aggregated a-synuclein protein involved in Parkinson's disease. *Electroanalysis* 16, 1172–1181.

Masarik, M., Cahova, K., Kizek, R., Palecek, E., Fojta, M. (2007). Label-free voltammetric detection of single-nucleotide mismatches recognized by the protein MutS. *Anal Bioanal Chem* 388, 259–270.

Mascini, M., Bagni, G., Di Pietro, M. L., Ravera, M., Baracco, S., Osella, D. (2006). Electrochemical biosensor evaluation of the interaction between DNA and metallo-drugs. *Biometals* 19, 409–418.

Mello, L. D., Hernandez, S., Marrazza, G., Mascini, M., Kubota, L. T. (2006). Investigations of the antioxidant properties of plant extracts using a DNA-electrochemical biosensor. *Biosens Bioelectron* 21, 1374–1382.

Misono, T., Kumar, P. K. R. (2005). Selection of RNA aptamers against human influenza virus hemagglutinin using surface plasmon resonance. *Anal Biochem* 342, 312–317.

Mock, J. J., Smith, D. R., Schultz, S. (2003). Local refractive index dependence of plasmon resonance spectra from individual nanoparticles. *Nano Lett* 3, 485–491.

Murphy, M. B., Fuller, S. T., Richardson, P. M., Doyle, S. A. (2003). An improved method for the in vitro evolution of aptamers and applications in protein detection and purification. *Nucleic Acids Res* 31, e110–e118.

Musheev, M. U., Krylov, S. N. (2006). Selection of aptamers by systematic evolution of ligands by exponential enrichment: addressing the polymerase chain reaction issue. *Anal Chim Acta* 564, 91–96.

Nath, N., Chilkoti, A. (2002). A colorimetric colloidal gold sensor to interrogate biomolecular interactions in real-time on a surface. *Anal Chem* 74, 504–509.

Nath, N., Chilkoti, A. (2004). Label-free biosensing by surface plasmon resonance of nanoparticles on glass: optimization of nanoparticle size. *Anal Chem* 76, 5370–5378.

Odom, T. W., Huang, J. L., Kim, P., Lieber, C. M. (1998). Atomic structure and electronic properties of single-walled carbon nanotubes. *Nature* 391, 62–64.

Okuno, J., Maehashi, K., Matsumoto, K., Kerman, K., Takamura, Y., Tamiya, E. (2007a). *Electrochem Commun* 9, 13–18.

Okuno, J., Maehashi, K., Kerman, K., Takamura, Y., Matsumoto, K., Tamiya, E. (2007b). Label-free immunosensor for prostate-specific antigen based on single-walled carbon nanotube array-modified microelectrodes. *Biosens Bioelectron* 22, 2377–2381.

Ostatna, V., Uslu, B., Dogan, B., Ozkan, S., Palecek, E. (2006). Native and denatured bovine serum albumin: d.c. polarography, stripping voltammetry and constant current chronopotentiometry. *J Electroanal Chem* 593, 172–178.

Ozkan, D., Erdem, A., Kara, P., Kerman, K., Meric, B., Hassmann, J., Ozsoz, M. (2002). Allele-specific genotype detection of factor V Leiden mutation from polymerase chain reaction amplicons based on label-free electrochemical genosensor. *Anal Chem* 74, 5931–5936.

Ozsoz, M., Erdem, A., Ozkan, A., Kara, P., Karadeniz, H., Meric, B., Kerman, K., Girousi, S. (2005). Allele-specific genotyping by using guanine and gold electrochemical oxidation signals. *Bioelectrochemistry* 67, 199–203.

Palecek. E. (1958). Oszillographiche Polarographie der Nucleinsauren und ihrer Bestandteile. *Naturwissenschaften* 45, 186–187.

Palecek. E. (1980). Reaction of nucleic acid bases with the mercury electrode: determination of purine derivatives at submicromolar concentrations by means of cathodic stripping voltammetry. *Anal Biochem* 108, 129–138.

Palecek, E., Jelen, F. (2005). Electrochemistry of nucleic acids. In *Electrochemistry of Nucleic Acids and Proteins: Towards Electrochemical Sensors for Genomics and Proteomics*, E. Palecek, F. Scheller, and J. Wang, eds., Elsevier, Amsterdam, pp. 74–174.

Palecek, E., Osteryoung, J., Osteryoung, R. A. (1982). Interactions of methylated adenine derivatives with the mercury electrode. *Anal Chem* 54, 1389–1394.

Palecek, E., Jelen, F., Trnkova, L. (1986). Cyclic voltametry of DNA at a mercury electrode: an anodic peak specific for guanine. *Gen Physiol Biophys* 5, 315–329.

Palecek, E., Masarik, M., Kizek, R., Kuhlmeier, D., Hassmann, J., Schulein, J. (2004). Sensitive electrochemical determination of unlabeled MutS protein and detection of point mutations in DNA. *Anal Chem* 76, 5930–5936.

Park, D. W., Kim, Y. H., Kim, B. S., So, H. M., Won, K., Lee, J. O., Kong, K. J., Chang, H. (2006). Detection of tumor markers using single-walled carbon nanotube field effect transistors. *J Nanosci Nanotechnol* 6, 3499–3502.

Patolsky, F., Zheg, G., Lieber, C. M. (2006). Fabrication of silicon nanowire devices for ultrasensitive, label-free, real-time detection of biological and chemical species. *Nat Protoc* 1, 1711–1724.

Potesil, D., Mikelova, R., Adam, V., Kizek, R., Prusa, R. (2006). Change of the protein p53 electrochemical signal according to its structural form: quick and sensitive distinguishing of native, denatured, and aggregated form of the "guardian of the genome". *Protein J* 25, 23–32.

Pumera, M., Sanchez, S., Ichinose, I., Tang, J. (2007). Electrochemical nanobiosensors. *Sens Actuat B* 123, 1195–1205.

Raschke, G., Kowarik, S., Franzl, T., Sönnichsen, C., Klar, T. A., Feldmann, J., Nichtl, A., Kürzinger, K. (2003). Biomolecular recognition based on single gold nanoparticle light scattering. *Nano Lett* 3, 935–938.

Saito, R., Fujita, M., Dresselhaus, G., Dresselhaus, M. S. (1992). Electronic structure of chiral graphene tubules. *Appl Phys Lett* 60, 2204–2206.

Schoning, M. J., Poghissian, A. (2002). Recent advances in biologically sensitive field-effect transistors (bioFETs). *Analyst* 127, 1137–1151.

Shin, Y.-B., Lee, J-M., Park, M-R., Kim, M-G., Chung, B. H., Pyo, H-B., Maeng, S. (2007). Analysis of recombinant protein expression using localized surface plasmon resonance (LSPR). *Biosens Bioelectron* 22, 2310–2307.

So, H-M., Won, K., Kim, Y. H., Kim, B-K, Ryu, B. H., Na, P. S., Kim, H., Lee, J. O. (2005). Single-walled carbon nanotube biosensors using aptamers as molecular recognition elements. *J Am Chem Soc* 127, 11906–11907.

Sönnichsen, C., Reinhard, B. M., Liphardt, J., Alivisatos, A. P. (2005). A molecular ruler based on plasmon voupling of single gold and silver nanoparticles. *Nat Biotechnol* 23, 741–745.

Souteyrand, E., Cloarec, J. P., Martin, J. R., Wilson, C., Lawrence, I., Mikkelsen, S., Lawrence, M. F. (1997). Direct detection of the hybridization of synthetic homo-oligomer DNA sequences by field effect. *J Phys Chem B* 101, 2980–2985.

Steenken, S., Jovanovic, S. V. (1997). How easily oxidizable is DNA? One-electron reduction potentials of adenosine and guanosine radicals in aqueous solution. *J Am Chem Soc*, 119, 617–618.

Ster, E., Klemic, J. F., Routenberg, D. A., Wyrembak, P. N., Turner-Evans, D. B., Hamilton, A. D., LaVan, D. A., Fahmy, T. M., Reed, M. A. (2007). Label-free immunodetection with CMOS-compatible semiconducting nanowires. *Nature* 445, 519–522.

Taft, B. J., Lazareck, A. D., Withey, G. D., Yin, A., Xu, J. M., Kelley, S. O. (2004). Site-specific delivery of DNA and appended cargo to carbon nanotubes. *J Am Chem Soc* 126, 12750–12751.

Takata, M., Kerman, K., Nagatani, N., Konaka, H., Namiki, M., Tamiya, E. (2006). Label-free bioelectronic immunoassay for the detection of human telomerase reverse transcrriptase in urine. *J Electroanal Chem* 596, 109–116.

Tombelli, S., Minunni, M., Luzi, E., Mascini, M. (2005). Aptamer-based biosensors for the detection of HIV-1 Tat protein. *Bioelectrochemistry* 67, 135–141.

Tomschik, M., Havran, L., Fojta, M., Palecek, E. (1998). Constant current chronopotentiometric stripping analysis of bioactive peptides at mercury and carbon electrodes. *Electroanalysis* 10, 403–409.

Tomschik, M., Jelen, F., Havran, L., Trnkova, L., Nielsen, P. E., Palecek. E. (1999). Reduction and oxidation of peptide nucleic acids and DNA at mercury and carbon electrodes. *J Electroanal Chem* 476, 71–80.

Tomschik, M., Havran, L., Palecek, E., Heyrovsky, M. (2000). The "presodium" catalysis of electroreduction of hydrogen ions on mercury electrodes by metallothionein: an investigation by constant current derivative stripping chronopotentiometry. *Electroanalysis* 12, 274–279.

Tuerk, C., Gold, L. (1990). Evolution of ligands by exponential enrichment: RNA ligands to bacteriophage T4 DNA polymerase. *Science* 249, 505–510.

Vestergaard, M., Kerman, K., Saito, M., Nagatani, N., Takamura, Y., Tamiya, E. (2005). A rapid label-free electrochemical detection and kinetic study of Alzheimer's amyloid beta aggregation. *J Am Chem Soc* 127, 11892–11893.

Wang, J., Cai, X., Wang, J., Jonsson, C., Palecek, E. (1995). Trace measurements of RNA by potentiometric stripping analysis at carbon paste electrodes. *Anal Chem* 67, 4065–4070.

Wang, J., Cai, X., Jonsson, C., Balakrishnan, M. (1996). Adsorptive stripping potentiometry of DNA at electrochemically pretreated carbon paste electrode. *Electroanalysis* 8, 20–24.

Wang, J., Bollo, S., Paz, J. L. L., Sahlin, E., Mukherjee, B. (1999). Ultratrace measurements of nucleic acids by baseline-corrected adsorptive stripping square-wave voltammetry. *Anal Chem* 71, 1910–1913.

Wang, W. U., Chen, C., Lin, K. H., Fang, Y., Lieber, C. M. (2005). Label-free detection of small-molecule–protein interactions by using nanowire nanosensors. *Proc Natl Acad Sci U S A* 102, 3208–3212.

Wilder, J. W. G., Venema, L. C., Rinzler, A. G., Smalley, R. E., Dekker, C. (1998). Electronic structure of atomically resolved carbon nanotubes. *Nature* 391, 59–62.

Wilson, D. S., Szostak, J. W. (1999). In vitro selection of functional nucleic acids. *Annu Rev Biochem* 68, 611–647.

Win, M. N., Klein, J. S., Smolke, C. D. (2006). Codeine-binding RNA aptamers and rapid determination of their binding constants using a direct coupling surface plasmon resonance assay. *Nucleic Acids Res* 34, 5670–5682.

Xie, H., Yang, D., Heller, A., Gao, Z. (2007). Electrocatalytic oxidation of guanine, guanosine, and guanosine monophosphate. *Biophys J* 92, L70–L72.

Ye, Y., Ju, H. (2005). Rapid detection of ssDNA and RNA using multi-walled carbon nanotubes modified screen-printed carbon electrode. *Biosens Bioelectron* 21, 735–741.

Zheng, G. F., Lu, W., Jin, S., Lieber, C. M. (2004). Synthesis and fabrication of high-performance *n*-type silicon nanowire transistors. *Adv Mater* 16, 1890–1893.

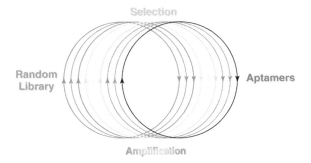

Figure 1.1 Scheme of in vitro selection.

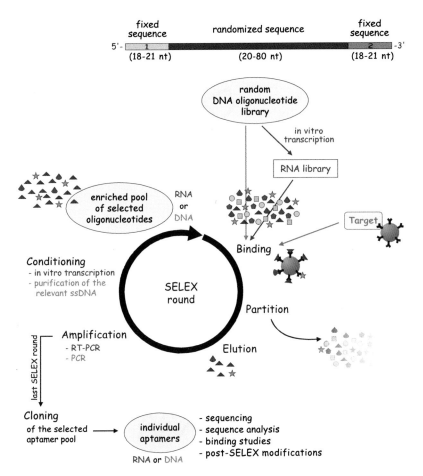

Figure 2.1 In vitro selection of target-specific aptamers using SELEX technology. (*See text for full description.*)

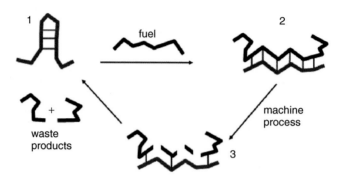

Figure 4.5 General operation mode of DNA machines exemplified by the catalytic scission of a substrate DNA strand. (*See text for full description.*)

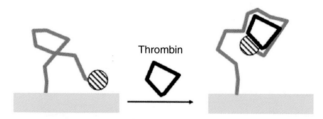

Figure 4.6 Thrombin biosensor based on a ferrocene-labeled aptamer. (*See text for full description.*)

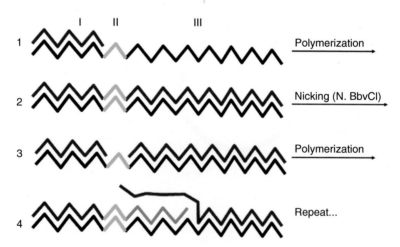

Figure 4.8 DNA machine based on strand displacement amplification. (*See text for full description.*)

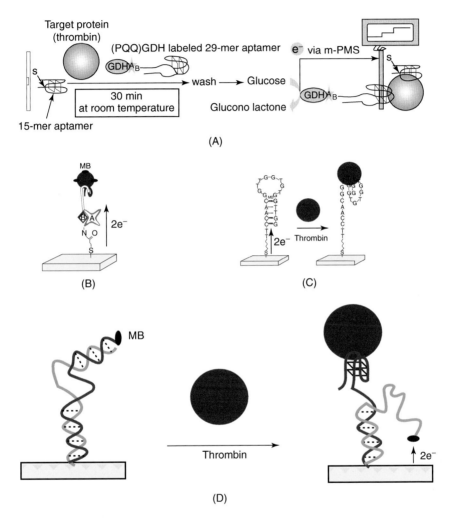

Figure 5.3 Detection of thrombin–aptamer interactions by electrochemical methods. (A) Sandwiched assay according to Ikebukuro et al. (2005). (B)–(D) Detection of thrombin using the electrochemical indicator methylene blue (MB). (*See text for full description.*) [(A) From Ikebukuro et al. (2005), with permission from Elsevier; (B) from Hianik et al. (2005); (C) from Bang et al. (2005); (D) adapted from Xiao et al. (2005b), with permission from the American Chemical Society.]

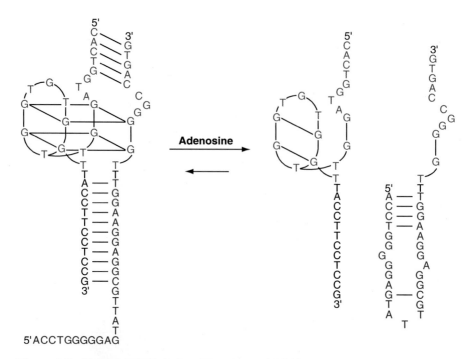

Figure 6.3 Universal AES design. The enzyme-inhibiting aptamer (thrombin aptamer) and the target molecule–binding aptamer (adenosine aptamer) are shown in blue and red, respectively.

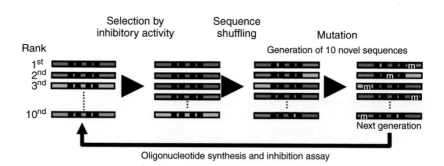

Figure 6.4 Screening using the evolution-mimicking algorithm.

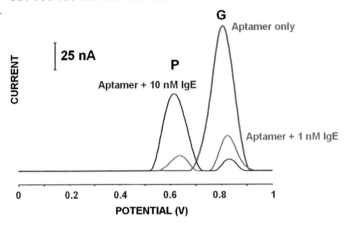

D17.4ext:5'-NH₂GCG CGG GGC ACG TTT ATC CGT CCC TCC TAG TGG CGT GCC CCG CGC-3'

Figure 7.2 Differential pulse voltammograms for the label-free voltammetric detection of IgE. The oxidation signal of guanine is observed alone from the 10-μM aptamer on a screen-printed carbon electrode (green line), the oxidation responses from the binding event between 10 nM IgE and the aptamer (red line), and the oxidation responses from the binding event between 1 nM IgE and the aptamer (blue line). (*See text for full description.*)

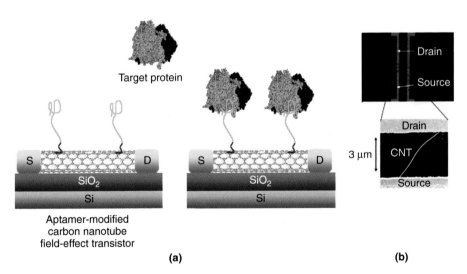

Figure 7.4 Label-free aptasensors based on CNT-FETs. (A) The aptamers are attached onto the sidewalls of a single-walled carbon nanotube which connects the source (S) and drain (D) electrodes. The binding of the target protein changes the electrical properties of the CNT-FETs significantly. (B) CNT-FET devices at high resolution.

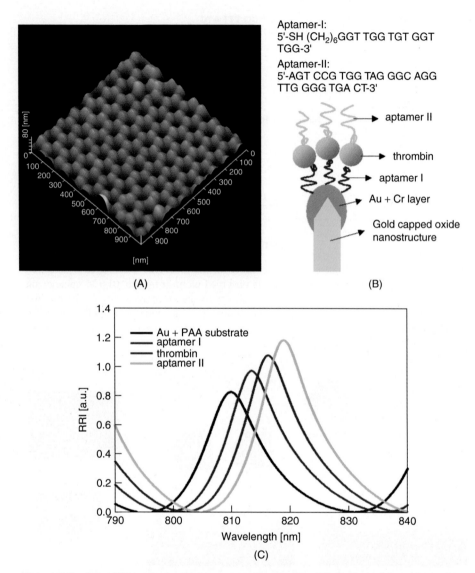

Figure 7.7 (A) AFM analysis of a porous anodic alumina (PAA) layer chip surface after Au deposition. (B) Detection of a sandwich-type binding assay between the aptamers and thrombin using Au-capped oxide nanostructures on the PAA layer chip. (C) Relative reflected intensity characteristics of (black line) a bare PAA layer chip, 10 μM aptamer I immobilized on a gold-deposited PAA layer surface (red line), the binding reaction between 10 μM aptamer I and 1 μM thrombin on the chip surface (blue line), after the binding reaction between aptamer I/thrombin complex and aptamer II (green line). (*See text for full description.*)

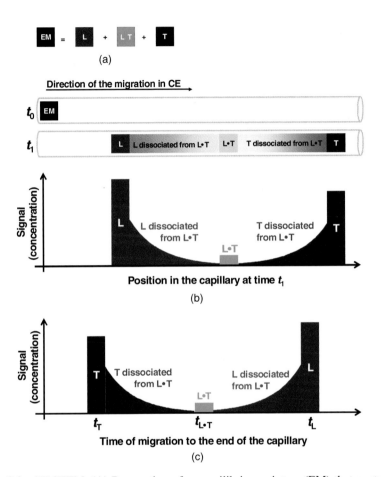

Figure 9.2 NECEEM. (A) Preparation of an equilibrium mixture (EM) that contains L, L · T, and T. (B) NECEEM-based separation of equilibrium fractions of L, L · T, and T. The top part illustrates the spatial distribution of separated components in the capillary at two different times ($t_0 = 0$ and $t_1 > t_0$) from the beginning of separation. The graph at the bottom shows concentrations of the separated components as functions of the position in the capillary at time t_1. (C) Schematic NECEEM electropherogram (signal versus migration time) assuming a point detector at the distal end of the capillary.

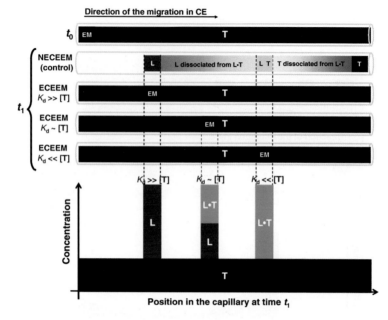

Figure 9.3 ECEEM. The equilibrium mixture (EM) contains L, L · T, and T, as depicted in Figure 9.2. (*See text for full description.*)

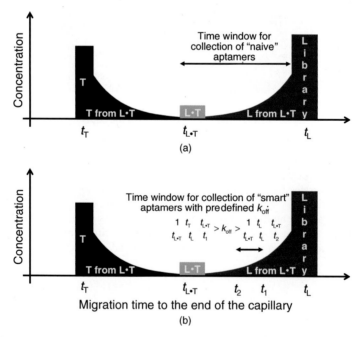

Figure 9.4 NECEEM electropherogram to illustrate (A) the selection of naive aptamers and (B) smart aptamers with predefined k_{off}.

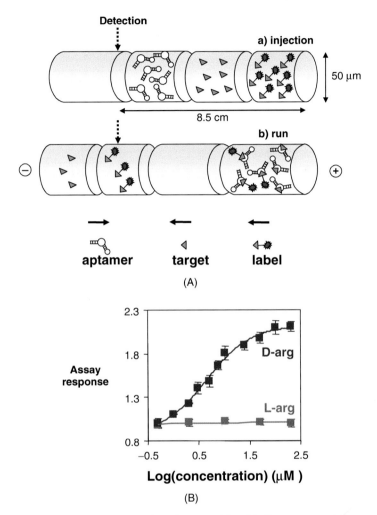

Figure 10.6 (A) Principle of CE-based competitive binding assay using on-capillary mixing of the various species. Arrows indicate the migration direction of the interacting species when an electric field is applied. (B) Standard curves obtained for D- (square) and L- (diamond) arginine enantiomers over the concentration range 0.5 to 200 μM. The relative peak height corresponds to the ratio of free-label peak height for an enantiomeric concentration in the sample plug to the free-label peak height for the blank (no enantiomer in the sample plug). Experimental conditions: running buffer, 25 mM TRIS-HCl, 25 mM KCl, 10 mM MgCl$_2$, (pH 7.0); coated PVA capillary, 50 μm i.d. with an extended light path (total length 64.5 cm); short-end injection method (effective length, 8.5 cm; negative polarity, cathode at the inlet and anode at the outlet; applied voltage, −25 kV); hydrodynamic injection (−50 mbar): 20-s aptamer (100 μM) plug (∼ 24 nL), 32-s sample (∼ 41 nL) plug, 4-s label (400 μM) plug (∼ 5 nL); temperature, 12°C; UV detection at 250 nm.

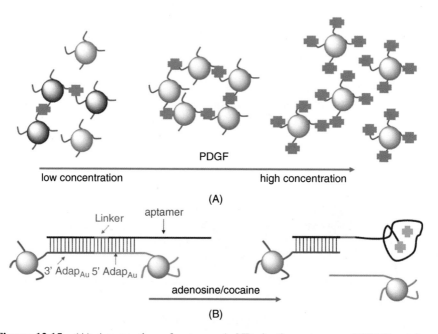

Figure 12.15 (A) Aggregation of aptamer-AuNPs in the presence of PDGFs at low, medium, and high concentrations; (B) colorimetric detection of small molecules. [(A) From Huang et al. (2005); (B) from Liu and Lu (2006).]

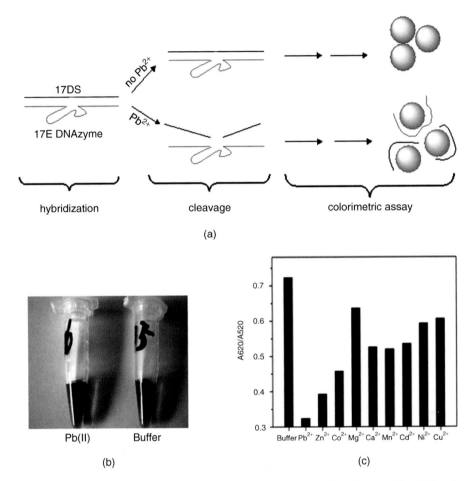

Figure 12.16 (A) AuNP-based colorimetric detection for Pb_{2+}. (B) 200 μL AuNPs/17E-17DS duplex solutions in the absence (right) and in the presence (left) of 50 μM Pb_{2+} after addition of 20 μL of 0.5 M NaCl. (C) Corresponding columns of the absorption ratio (A_{620}/A_{520}) of 200 μL of GNPs/17E-17DS duplex solutions in the presence of various 50 μM metal ions after addition of 20 μL of 0.5 M NaCl. [From H. Wei et al. (submitted).]

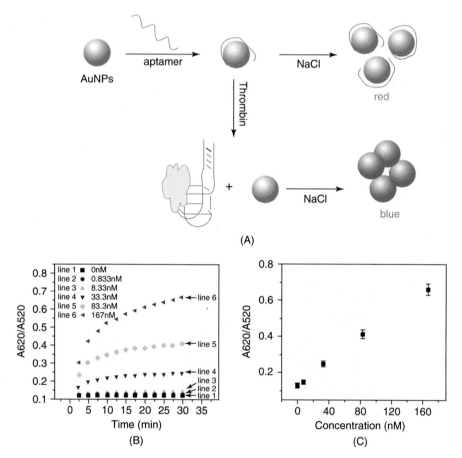

(A)

(B)

(C)

Figure 12.17 (A) AuNP-based colorimetric detection for α-thrombin. (B) 200 μL of AuNPs/17E-17DS duplex solutions in the absence (right) and in the presence (light) of 50 μM Pb$_{2+}$ after addition of 20 μL of 0.5 M NaCl. (C) Corresponding columns of the absorption ratio (A_{620}/A_{520}) of 200 μL of GNPs/17E-17DS duplex solutions in the presence of 50 μM different metal ions after addition of 20 μL of 0.5 M NaCl. [From Wei et al. (2007), with permission. Copyright © 2007 Royal Society of Chemistry.]

CHAPTER 8

APTAMER-BASED BIOANALYTICAL ASSAYS: AMPLIFICATION STRATEGIES

SARA TOMBELLI, MARIA MINUNNI, and MARCO MASCINI

8.1 INTRODUCTION

Aptamers hold great promise as molecular recognition tools for their incorporation into analytical devices, and in particular, they can be used as immobilized ligands in separation technologies (Clark and Remcho, 2002), as affinity probes in capillary electrophoresis-based quantitative assays (Kotia and McGown, 2000), in affinity matrix-assisted laser desorption/ionization mass spectrometry (MALDI-MS)(Dick and McGown, 2004), and as biocomponents in biosensors (Tombelli et al., 2005). Other aptamer-based bioanalytical applications have been reviewed (You et al., 2003; Mukhopadhyay, 2005).

In recent years great progress has been made toward the development of aptamer-based bioanalytical assays using a variety of detection techniques (i.e., optical, electrochemical, piezoelectric, etc.). The high sensitivities requested to aptamer-based assays for the detection of different targets (e.g., thrombin at the pM level) cannot be reached by simple "direct" binding protocols since the affinities of aptamers for their targets is not high enough, ranging from the micro- to the nanomolar level. Recent efforts have been focused on improving the sensitivity of these assays by developing effective amplification paths. With this aim, several methodologies have been employed as signal amplification tools, such as metallic and magnetic nanoparticles (Centi et al., 2007b; Shlyahovsky et al., 2007), enzymatic labels (Li et al., 2007), and quantum dots (Ikanovic et al., 2007). Other amplification techniques, such as polymerase chain reaction, were inspired by molecular biology (Di Giusto et al.,

Aptamers in Bioanalysis, Edited by Marco Mascini
Copyright © 2009 John Wiley & Sons, Inc.

TABLE 8.1 Overview of Aptamer-Based Bioanalytical Assays Developed for the Detection of Thrombin

Detection Method	Amplification Method	Limit of Detection[a]	Ref.
Optical	Enlargement of gold nanoparticles	20 nM (solution); 2 nM (solid phase)	Pavlov et al., 2004
Electrochemical	Catalytic platinum nanoparticles	1 nM	Polsky et al., 2006
Optical	Quantum dots	1 nM	Choi et al., 2006
Electrochemical	Magnetic particles and enzyme	0.5 nM	Centi et al., 2007b
Electrochemical	Quantum dots	0.5 pM	Hansen et al., 2006
	Polymerase chain reaction	Low pM	Di Giusto et al., 2005
Electrochemical and optical	Polymerase chain reaction	~fM	Wang et al., 2004

[a]The limit of detection of the assay is reported with the relative amplification strategies.

2005) or were a combination of DNA–RNA catalytic molecules with aptamers (Hesselberth et al., 2003).

Table 8.1 provides an overview of various aptamer-based bioanalytical assays for the detection of thrombin, in order to evidence the improvements achieved in the assay detection limit by the use of some of the reported amplification tools. The various methods are explained in detail by the use of several examples.

8.2 BIOANALYTICAL ASSAYS BASED ON APTAMER-FUNCTIONALIZED NANOPARTICLES

Biomolecule–nanoparticle systems are used extensively in different bioanalytical assays, the major challenge being to improve assay sensitivity by creating a particular amplification path (Willner, 2005; Willner et al., 2007). In particular, metallic or magnetic nanoparticles have been coupled to different biomolecules, such as DNA or antibodies (Yao et al., 2006; Centi et al., 2007a; Chang et al., 2007). Nanoparticles can be used as electrochemical markers, as catalytic labels for the enlargement of nanoparticles, or as a capturing agent in the case of magnetic nanoparticles coated with biomolecules.

Aptamer-functionalized metallic and magnetic nanoparticles have recently been used for the detection of proteins or for the collection of cancer cells. In particular, gold nanoparticles were used as capturing agents and as an amplification tool for the aptamer-based detection of thrombin (Pavlov et al., 2004). The detection was conducted both in solution and on a solid phase by the use of aptamer-functionalized gold nanoparticles. In the case of the solution phase assay (Figure 8.1A), the gold nanoparticles were first functionalized with the thiolated aptamer for thrombin.

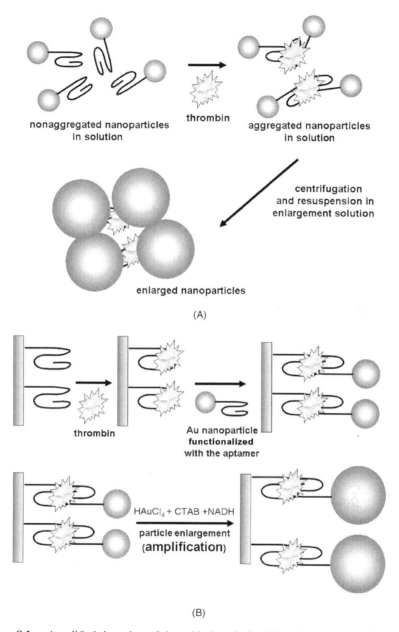

nonaggregated nanoparticles
in solution

thrombin

aggregated nanoparticles
in solution

centrifugation
and resuspension in
enlargement solution

enlarged nanoparticles

(A)

thrombin

Au nanoparticle
functionalized
with the aptamer

HAuCl₄ + CTAB +NADH

particle enlargement
(amplification)

(B)

Figure 8.1 Amplified detection of thrombin in solution (A) and on surfaces (B) by the catalytic enlargement of gold nanoparticles. The assay has a sandwich-like format and is based on the capture of thrombin by immobilized aptamer. A gold nanoparticle-labeled secondary aptamer is then added. Finally, the gold nanoparticles are enlarged in a solution containing HAuCl₄ and CTAB as a surfactant in the presence of NADH. [From Pavlov et al. (2004).]

During interaction of functionalized nanoparticles with the protein, an aggregation of the nanoparticles was observed, due mainly to the binding of the aptamer to the two binding sites present on thrombin. After centrifugation, the absorbance of the solution from which the aggregated nanoparticles are precipitated was recorded, and a decrease in absorbance was observed with the increase in thrombin concentration. The precipitated gold nanoparticles could further be collected, redissolved in aqueous solution, and used as seed for their catalytic enlargement in the presence of $HAuCl_4$. The catalyzed growth of gold nanoparticles in the presence of NAD(P)H cofactors has been reported (Xiao et al., 2004). The solution for growth of the particles consisted of citrate-stabilized gold NPs, $HAuCl_4$, and CTAB as surfactants in the presence of NADH. The same enlargement solution has been used in the aptamer-based method for thrombin.

With this amplification system a detection limit for thrombin of 20 nM could be reached, improving the sensitivity about five times with respect to other traditional colorimetric aptamer-based assays. A further increase in sensitivity was obtained by developing the assay on a solid surface (Figure 8.1b). The aptamer was attached to a maleimide-functionalized siloxane monolayer and the interaction with thrombin was observed. A sandwich assay was then conducted by adding gold nanoparticles functionalized with the aptamer, which binds to the second binding site on thrombin. The resulting gold nanoparticles interface was then enlarged in the growth solution containing $HAuCl_4$. The absorbance of the glass slides presenting the enlarged nanoparticles was then measured and the detection limit of the assay was decreased to 2 nM of thrombin.

An interesting approach based on the use of aptamer-functionalized platinum nanoparticles has recently been optimized (Polsky et al., 2006). In this case the metallic nanoparticles were used not as mere labels for electrochemical detection, but as catalytic molecules. This with the aim of overcoming some of the drawbacks associated with the thermal and environmental instability of traditional biological redox catalytic labels, such as enzymes. By following this approach the amplified electrochemical detection of thrombin was achieved by functionalizing platinum nanoparticles with the specific aptamer (Figure 8.2).

The assay was conducted by immobilizing the thiolated thrombin aptamer onto a gold electrode and using, after interaction with thrombin, the same aptamer fixed onto platinum nanoparticles as a secondary ligand. In this way the method developed has a sandwich assay format by using the same aptamer as both primary and secondary ligand, demonstrating that this aptamer can bind thrombin in two different binding sites.

The platinum nanoparticles were then used to perform the electrocatalytic reduction of added H_2O_2 detected by linear sweep voltammetry. A detection limit of 1 nM for thrombin was achieved with this method, improving the sensitivity 100 times over other aptamer-based thrombin detection schemes.

A different approach based on the use of an enzyme as a traditional amplifying label and aptamer-functionalized magnetic beads has been developed and applied to the detection of thrombin directly into human plasma (Centi et al., 2007b). The assay is based on electrochemical transduction coupled to magnetic particles,

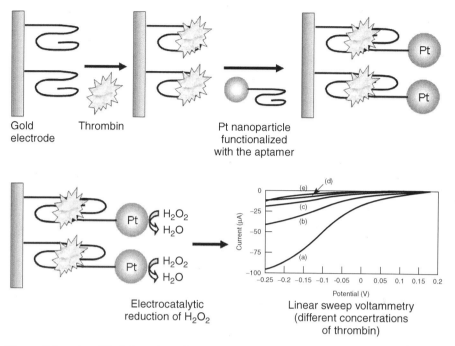

Figure 8.2   Amplified detection of thrombin by a sandwich-like assay based on a platinum nanoparticle-labeled secondary aptamer. Thrombin is captured by an immobilized aptamer and then reacted with the nanoparticle-labeled secondary aptamer. The platinum nanoparticles are then used to perform the electrocatalytic reduction of added H_2O_2 detected by linear sweep voltammetry. [From Polsky et al. (2006).]

modified with the aptamer. Similar to immunoassays, with the addition of a second aptamer after interaction of the analyte with the modified beads, this approach can be thought of as a sandwich method (Figure 8.3).

Two selected aptamers binding thrombin in two different, nonoverlapping sites were used. The protein captured by the first aptamer is detected after the addition of the second biotinylated aptamer and of streptavidin labeled with an enzyme (alkaline phosphatase). Detection of the product generated by the enzymatic reaction was achieved by differential pulse voltammetry onto screen-printed electrodes. A very good analytical performance was obtained, with a detection limit of 0.45 nM, a reproducibility in terms of the average coefficient of variation (CV) of 8%, and high specificity, as demonstrated by the negligible signal obtained with human serum albumin at high concentration (72 μM). The possibility of the real application of the assay was demonstrated by testing plasma samples spiked with different concentrations of thrombin. Comparable responses were found for buffer, serum, and plasma: addition of thrombin to the samples resulted in a protein concentration-dependent signal (Figure 8.4). Using this novel design,

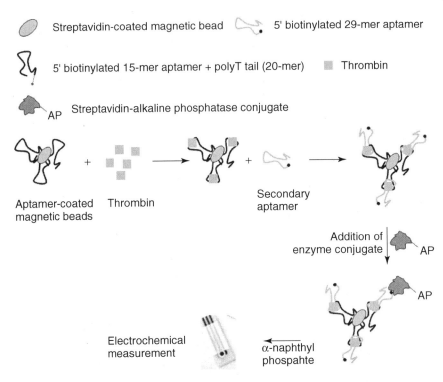

Figure 8.3   Sandwich assay developed for thrombin using two different aptamers, magnetic particles, and enzymatic amplification. Two selected aptamers binding thrombin in two different, nonoverlapping sites are used. The protein captured by the first aptamer fixed onto magnetic particles is detected after addition of the second biotinylated aptamer and of streptavidin labeled with an enzyme (alkaline phosphatase). Detection of the product generated by the enzymatic reaction is achieved by differential pulse voltammetry onto screen-printed electrodes onto which the magnetic nanoparticles are deposited and kept in contact through a magnet. [From (Centi et al., 2007b).]

an electrochemical biosensor recognizing thrombin with high affinity, sensitivity, and specificity was obtained, opening the possibility of a real application to diagnostics or medical investigation.

8.3   APTAMERS AND QUANTUM DOT–BASED ASSAYS

Quantum dots (QDs) are nearly spherical fluorescent nanocrystals composed of semiconductor materials that bridge the gap between individual atoms and bulk semiconductor solids. Owing to the quantum-size effect, they exhibit size-tunable bandgaps and luminescence energies. They can be fabricated in a variety of materials, such as small regions of one material being buried in another material

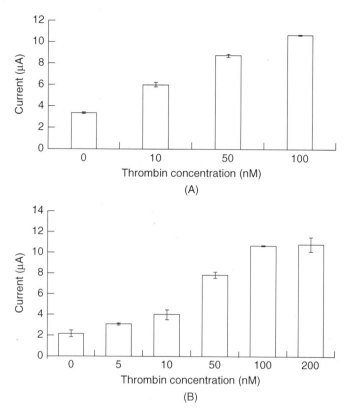

Figure 8.4   Analysis of thrombin-spiked serum (A) and plasma (B) samples by aptamer-based electrochemical sandwich assay coupled to magnetic particles. [From Centi et al. (2007b).]

that has a larger bandgap. These can be so-called *core–shell structures*: for example, with CdSe in the core and ZnS in the shell, in which the ZnS passivates the core surface, protects it from oxidation, and prevents leeching of the Cd/Se into the surrounding solution. Colloidal quantum dots made of ZnS, CdS, ZnSe, PbSe, and PbS have also been prepared (Medintz et al., 2005).

The colloidal quantum dots are now widely employed as targeted fluorescent labels in biomedical research applications (Yin and Alivisatos, 2005). Compared to the traditional organic fluorescent labels, they provide long-term stability and have the advantage of allowing the detection of multiple signals (Medintz et al., 2005). Owing to these properties QD nanocrystals have generated considerable interest for optical detection of different biomolecules, including DNA (Han et al., 2001). Recently, their utility for enhanced electrochemical detection has also been evidenced (Wang et al., 2003; Liu et al., 2004), with special emphasis on the possibility of having multiplexed bioanalysis.

Similar to DNA-based applications, QDs have been coupled to aptamers for the recognition of proteins (Levy et al., 2005; Choi et al., 2006; Hansen et al., 2006; Ikanovic et al., 2007). Most of these very recent applications rely on optical detection, whereas in only a very few cases (Hansen et al., 2006) an electrochemical transduction have been employed.

Among optical-based detection methods, different strategies were followed and critical points were evaluated. When using QDs modified with aptamer beacons (Levy et al., 2005) two important features have to be taken into consideration. The design of the aptamer beacon and the way it is attached to the QD must ensure the correct conformational change in the aptamer after its binding to the target molecule. In addition, coupling of the aptamer to the QD has to ensure correct quenching of the QD itself. When these two features are considered in the assay design, a QD-based aptamer biosensor can be developed (Levy et al., 2005). An antithrombin aptamer, hybridized to a complementary quencher oligonucleotide that disrupts the aptamer structure, was conjugated to a QD. In the presence of the target protein, thrombin, the structure of the aptamer is stabilized, resulting in displacement of the quencher oligonucleotide with a consequent increase of fluorescence (Figure 8.5a). With this format the aptamer-modified QD exhibited a 19-fold increase in fluorescence in the presence of thrombin, a higher increase with respect to traditional organic fluorophores (12-fold) (Nutiu, 2003).

An innovative detection scheme, still based on optical transduction with QDs and aptamers, was presented by Choi et al. (2006). The authors discovered that PbS QDs could be passivated by a DNA aptamer becoming water soluble and

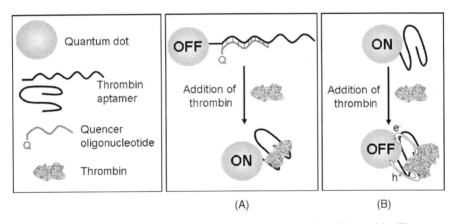

Figure 8.5 (A) Quantum dot aptamer assay for the detection of thrombin. The quantum dots are functionalized with the thrombin aptamer (black) hybridized to a quencher oligonucleotide (gray). The binding of thrombin induces displacement of the quencher oligonucleotide with the activation of quantum dot fluorescence. (B) Novel assay based on aptamer-modified quantum dots. When the aptamer target protein (thrombin) binds to the aptamer fixed onto the QD, a quenching of the QD photoluminescence is observed, due to a charge transfer (white arrows) from functional groups on the protein to the QD. [(A) From Levy et al. (2005); (B) from Choi et al. (2006).]

stable against aggregation. A novel one-step synthesis of PbS QDs in aqueous solution at room temperature has been proposed based on mixing of the thrombin-binding aptamer with lead acetate, followed by the addition of sodium sulfide. The resulting aptamer-coated QDs are stable for several months and have a diameter of 2 to 6 nm.

When the target protein (thrombin) binds to the aptamer fixed onto the QD, a quenching of the QD photoluminescence is observed due to a charge transfer from functional groups on the protein to the QD (Figure 8.5b). The innovative aspect of the assay is the possibility of selectively modulating the QD photoluminescence. The detection of thrombin at low concentrations (1 nM) was also achieved in the presence of high concentrations of nonspecific negatively or positively charged proteins which adsorb on the QD but do not modulate its photoluminescence.

Another optical assay based on QDs modified with aptamers was reported for the detection of *Bacillus thuringiensis* spores (Ikanovic et al., 2007). In this case the background fluorescence of the bacteria was attenuated by choosing the appropriate excitation wavelength. With zinc-capped cadmium selenide QDs, the detection limit of the assay (1000 CFU mL^{-1}) was six times lower than traditional detection methods based on classical fluorophores (Hoile et al., 2006). The QDs here were modified with an aptamer specific for *Bacillus thuringiensis*, which is a bacterium similar to *B. cereus*, a food-poisoning agent, and *B. anthracis*, the causative agent of anthrax. The QDs were coated with the aptamer and were then reacted with the bacteria. The spores attached to the aptamer-modified QDs at the end were collected on membrane filters (0.45 μm) and resuspended in buffer, and the fluorescence signal was read.

Apart from these optical-based methods, electrochemical detection has also been demonstrated to be applicable in QD/aptamer-based assays (Hansen et al., 2006). Two important advantages were achieved by the use of QDs in this type of application: The nanocrystals offer an electrodiverse population of electrical tags that enable multiplexed bioanalysis. In addition, this new multiple-protein aptamer-based biosensing is coupled to enormous signal amplification by nanoparticle-based electrochemical stripping measurements, resulting in subpicomolar (attomole) detection limits. In the work, a very interesting and simple displacement assay was developed by co-immobilizing different thiolated aptamers specific for thrombin and lysozyme onto a gold substrate. PbS QD-modified lysozyme and CdS QD-modified thrombin were then bound to the corresponding aptamer, and their displacement by nonmodified proteins was observed. The remaining nanocrystals captured were then dissolved and detected by electrochemical stripping analysis (Figure 8.6). The position and size of the corresponding metal peaks (cadmium and lead) reflect the identity of a concentration of the corresponding protein (Figure 8.7).

These examples of the coupling of aptamers with the coding and amplification characteristics of QDs demonstrate the high potentialities of these methods, with particular attention to the simultaneous detection of different proteins or the detection of barely optically measurable molecules such as bacteria.

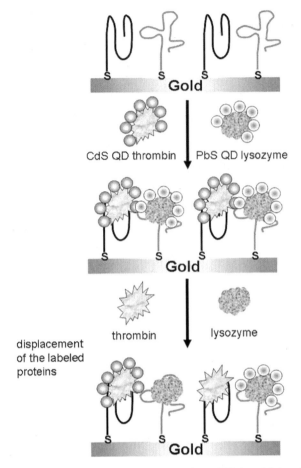

Figure 8.6 Displacement assay developed by co-immobilizing thiolated aptamers specific for thrombin and lysozyme onto a gold substrate. PbS QD-modified lysozyme and CdS QD-modified thrombin first bind to the corresponding aptamer and then displacement by the nonmodified proteins is achieved. [From Hansen et al. (2006).]

8.4 APTAZYMES AND APTAMER-BASED MACHINES

The "discovery" of aptamers in 1990 has opened a new scenario in the development of RNA/DNA-based assays for different molecules (Tuerk and Gold, 1990; McGown et al., 1995), but a further advance in nucleic acid nanotechnology was made in 1994 with the first application of a DNA with a catalytic function (DNAzyme) (Breaker and Joyce, 1994). An example of DNAzyme is a single-stranded guanine-rich nucleic acid that has been reported for binding a hemin to yield a DNAzyme possessing peroxidase-like activity. It was proposed that the intercalation of hemin into

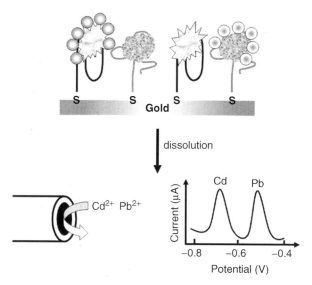

Figure 8.7   Captured PbS and CdS quantum dots attached to aptamer-bound proteins are dissolved and detected by electrochemical stripping analysis. The position and size of the corresponding metal peaks (cadmium and lead) reflect the identity of a concentration of the corresponding protein. [From Hansen et al. (2006).]

G-quadruplex resulted in the formation of a DNAzyme that catalyzes the oxidation of 2,2'-azino-bis(3-ethylbenzthiazoline)-6-sulfonic acid (ABTS) and the chemiluminescence of luminol by hydrogen peroxide (Travascio et al., 2001).

Aptamers and DNAzymes have finally been combined to form allosteric DNAzymes (aptazymes) in which the enzymatic function is activated or inhibited by an effector that is the target molecule for the aptamer (Soukup and Breaker, 1999; Knudsen and Ellington, 2006; Lu and Liu, 2006). In particular, when the aptamer target molecule is binding, conformational changes in the aptamer domain generate a change in the catalytic core of the aptazyme, modulating its catalytic activity (Knudsen and Ellington, 2006). The design and engineering of aptazymes have been facilitated by the simplicity of nucleic acid secondary structures, and aptazymes have simply been created by integrating preexisting aptamers and catalytic RNAs with known specificity and catalytic properties, respectively. In this way, the relative characteristics are combined in one resulting molecule that possesses the desired affinity for a specific target and, at the same time, the expected catalytic effect. This coupling is obtained by *modular rational design* and in vitro selection techniques (Breaker, 2002). A different approach starts directly from an aptazyme in which the receptor site is represented by a random-sequence RNA (library) and the catalytic region is an existing ribozyme. The random sequence region undergoes selection (allosteric selection) to rise with the desired binding specificity (Soukup and Breaker, 1999; Breaker, 2002).

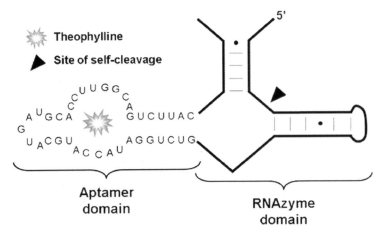

Figure 8.8   Sequence and secondary structure of a theophylline-sensitive allosteric ribozyme. The line represents the ribozyme domain. The arrow identifies the site of self-cleavage. [From Seetharaman et al. (2001).]

Aptazyme-based assays have been developed demonstrating the advantages of these tools in array systems (Seetharaman et al., 2001) or as amplification systems to reach lower limits of detection (Hesselberth et al., 2003; Chelyapov et al., 2006).

Seetharaman et al. (2001) optimized an aptazyme array that discriminated between a variety of metabolites and drug analytes. They generated seven different aptazymes in which the RNAzyme exhibited self-cleavage activity. The aptazymes, each labeled with ^{32}P and triggered specifically by its effector, were then immobilized on a gold surface via a $5'$-thiotriphosphate moiety. The resulting aptazyme array was able to detect different analytes in complex mixtures (e.g., metal ions, enzyme cofactors, metabolites, and drugs) and also to characterize different *Escherichia coli* strains by examining the radioactivity of the $3'$-cleavage labeled fragment released. An example illustrating an aptazyme activated by theophylline is reported in Figure 8.8. This work opened new frontiers in the use of aptazymes in a single-chip format with application to real sample analysis.

Aptazymes can present not only an activity of self-cleavage as in the work described before, but also of ligase activity, transducing analyte recognition directly into the formation of a covalent bond (Hesselbert et al., 2003). The array developed with this aptazyme ligase relied on the capture and subsequent visualization of ligated aptazymes through the removal of any unligated molecules via washing steps. This enabled greater discrimination between the positive signal and the background, leading up to a lower detection limit (i.e., a low nanomolar range). In particular, radiolabeled ligase aptazymes were reacted with the biotinylated substrate (oligonucleotide) and the individual analytes or a mixture of them. The reaction solution is then spotted onto streptavidin-coated microtiter plates. In

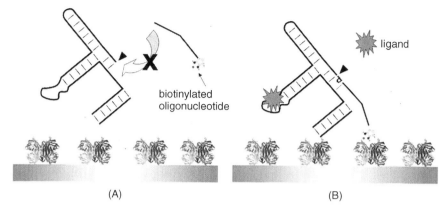

Figure 8.9   Aptazyme ligase array. [From Hesselberth et al. (2003).]

the absence of analytes, only the biotinylated aptazyme substrate binds to strep-tavidin, whereas the labeled aptazyme is washed away without any generation of radioactive signal (Figure 8.9a).

In the presence of analyte, the ligation is catalyzed and the radiolabeled aptazyme substrate is ligated to the aptazyme sequence. In this way the radiolabeled aptazyme has a biotinylated end and can then bind to streptavidin on the plate generating a radioactive signal after the washing steps (Figure 8.9b). Analyte-dependent formation of a covalent bond between the array surface and the bioreceptor/reporter molecule makes possible extremely stringent washings, leading to higher signal-to-noise ratios. The array proposed could detect different types of molecules, such as a protein (lysozyme), a peptide (HIV Rev peptide), and an organic compound (flavin mononucleotide), with detection limits of 16, 60, and 370 nM, respectively.

The exciting properties of nucleic acids have opened new scenarios in the field of nanotechnology and sensor development. In particular, the dynamic properties of nucleic acids and the thermally induced sequence-dependent separation of double-stranded DNA allow the utilization of nucleic acids with machine-like functions that can perform different motions or act as switches and motors (Beissenhirtz and Willner, 2006). The same principles have been applied for the development of aptamer-based machines, which can amplify the recognition event between the aptamer and the substrate through operation of the machine (Shlyahovsky et al., 2007). A scheme for the assay developed for cocaine is reported in Figure 8.10.

The assay is based on the use of a particular oligonucleotide (oligonucleotide A in Figure 8.10), which consists of region 1, a cocaine-specific aptamer, blocked in its unfolded structure in the absence of cocaine by a small complementary oligonucleotide (b). In the presence of cocaine, the aptamer assumes its characteristic secondary structure, displacing the blocking oligonucleotide. In the presence of a primer, polymerase, and the dNTPs, the oligonucleotide, comprising the

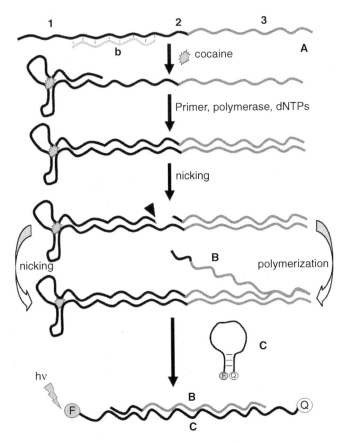

Figure 8.10   Amplified detection of cocaine by an autonomous aptamer-based machine. [From Shlyahovsky et al. (2007).]

folded aptamer and regions 2 and 3, is amplified, creating a nicking site (the arrow in Figure 8.10) for the nicking enzyme Nt. BbvC I. Scission of the strand produced results in a new replication site for polymerase and to the displacement of a new oligonucleotide (B). This nicking-polymerase activity is an autonomous operation that cannot be stopped which constitutes the "machine activity" of the aptamer/oligonucleotide and the amplification step of the assay. The displaced oligonucleotide B hybridizes with a hairpin nucleic structure (C) that is functionalized at the two ends with a quencher and a fluorophore. The fluorescence of the fluorophore is quenched by the quencher in the hairpin structure. The opening of the hairpin structure through hybridization with the oligonucleotide B produced restores the fluorescence of the dye, and the resulting fluorescence signal provides a readout signal for operation of the aptamer-based machine and the detection of cocaine. Using an assay time of 60 minutes (the optimized time for the machine to operate), a detection limit for cocaine of 5×10^{-6} M could be reached, improving the detection limit of other electrochemical or optical aptamer-based methods.

8.5 POLYMERASE CHAIN REACTION AS AN AMPLIFICATION METHOD IN APTAMER-BASED ASSAYS

Methods for the detection of specific nucleic acid sequences are much more sensitive with respect to the detection of proteins, since polymerase chain reaction (PCR) represents a very powerful amplification tool for nucleic acids. The fact that aptamers are made of DNA or RNA offers the opportunity to combine the detection of proteins with the amplification potentialities that until some years ago were restricted to the detection of nucleic acids. PCR has been used alone or combined with other molecular biology techniques for nucleic acid manipulation (proximity ligation or exonuclease protection) to develop extremely sensitive assays for the detection of a variety of proteins.

A very simple and sensitive method has recently been published based on direct amplification of an aptamer specific for the reverse transcriptase of the human immunodeficiency virus type 1 (HIV-1 RTase) (Zhang et al., 2006). After incubation of the protein with the aptamer, the complex formed is separated from the unbound aptamer by capillary electrophoresis. After collection of the aptamer–target complex, the aptamer was dissociated from the target and directly amplified by PCR. Approximately 10^2 molecules of target were detected using this method, improving the sensitivity of protein detection by several orders of magnitude.

Among the assays in which PCR has been coupled to other molecular biology techniques, Fredriksson et al. (2002) developed a very sensitive method for detection of the homodimer of the platelet-derived growth factor B-chain (PDGF-BB) based on an aptamer and termed *proximity ligation assay*. In this method (Figure 8.11), a DNA aptamer for PDGF-BB (Green et al., 1996) (A in the figure) was modified with additional sequences (B_1 and B_2 in the figure), forming a proximity probe pair; when the aptamer binds PDGF-BB, the free ends of the extensions hybridize to a connector oligonucleotide (C in the figure) added subsequently, allowing the ends to be joined (the arrow in the figure) by enzymatic DNA ligation (Cao, 2004).

The ligation products can then be amplified through the use of PCR, while unreacted probes remain silent. The amplified product was detected in real time by the use of a complementary DNA probe labeled with a fluorophore and a quencher. Homogeneous and solid-phase assays were optimized in terms of connector oligonucleotide concentration (400 nM), preincubation volume (5 μL), and concentrations of proximity probes (20 pM). With the homogeneous assay, as few as 24,000 PDGF-BB molecules were detected, about 1000-fold fewer than with an ELISA assay for the same protein. The linear range extended over a >1000-fold concentration range, and a coefficient of variation (CV) of 35% was obtained for the proximity ligation assays, which compared well with a CV value of 32% for ELISA.

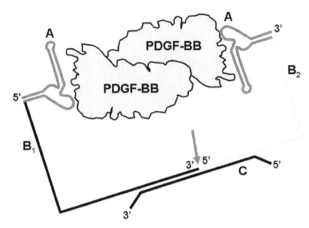

Figure 8.11   Platelet-derived growth factor B (PDGF-BB) detection using a proximity-dependent DNA ligation assay. [From Fredriksson et al. (2002).]

In solid-phase proximity ligation assay, low-femtomolar concentrations of the protein were detected using the same pair of proximity probes as those used for the homogeneous assay. This demonstrated the high sensitivity of the detection method compared with the low-picomolar detection threshold of the ELISA for the same protein.

Since this method is limited to proteins that can be bound by two aptamers or which can form a homodimer in order to bind the same aptamer in two different sites, more general methods have been developed which can also be used for proteins having only one binding site for one aptamer.

In particular, an *exonuclease protection assay* has been presented for the detection of thrombin (Wang et al., 2004). Also, in this case ligation has been used in connection with an aptamer but with the addition of a further technique, based on the enzyme exonuclease I. This enzyme can degrade DNA by releasing deoxyribonucleoside 5′-monophosphates from the 3′ termini of single-stranded DNA chains. The assay is based on the interaction between the DNA aptamer and its specific target, thrombin (Figure 8.12).

When the aptamer is bound to the protein, its DNA chain is protected from exonucleases and no degradation is present, whereas unreacted sequences are degraded by the enzyme. The intact aptamer is then hybridized to two connector oligonucleotides which are ligated at the end. The longer oligonucleotide strands are amplified and quantified by real-time PCR, giving a quantitative measurement of thrombin since the amount of product amplified corresponds to the amount of undegraded aptamer, which in turn corresponds to the amount of thrombin. The detection range of the assay can reach seven orders of magnitude, and the detection limit was on the order of 10^2 molecules of thrombin.

The same target protein has been used in an aptamer-based assay where protein detection has been observed in real time in a one-step amplification reaction

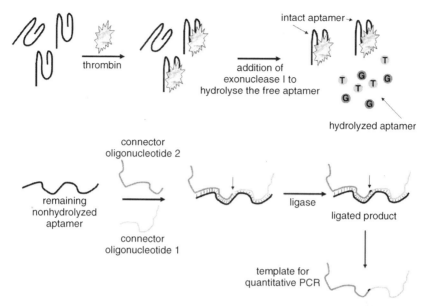

Figure 8.12   Detection of thrombin by an aptamer-based exonuclease protection assay. Thrombin is binding to its specific aptamer in solution. The free aptamer is hydrolyzed by exonuclease I, whereas the bound aptamer remains intact. The intact aptamer is then hybridized with two connector oligonucleotides and the hybridization product is ligated by DNA ligase. The ligation product is used as a template for amplification by PCR. [From (Wang et al., 2004).]

without the need for thermocycling (Di Giusto et al., 2005). The assay was based on the use of a circular DNA aptamer that consists of a thrombin-specific aptamer coupled with an unstructured template pool and a linear DNA aptamer that consists of a different thrombin-specific aptamer and a tail (Figure 8.13). When the two aptamers are binding to thrombin, they bring into proximity the tail of the linear aptamer with the loop of the circular aptamer; these have a complementary portion and can then hybridize. This hybridization forms a primer for polymerase-mediated extension and isothermal rolling circle amplification (RCA) (Baner et al., 1998). The RCA process generates a product consisting of the linear aptamer sequence and many concatenated sequences complementary to the circular aptamer template. The duplex region is designed to contain a restriction enzyme recognition site, and the presence of the corrected fragment after restriction can confirm the correct RCA product. Both fluorescent and electrochemical detection techniques were used by either adding a fluorescent dye and using a real-time PCR instrument or by using an electrochemically labeled nucleoside triphosphate during the amplification step. The high sensitivity of the circular aptamer/RCA method was demonstrated by the low picomolar detection limit achieved for the target protein, thrombin.

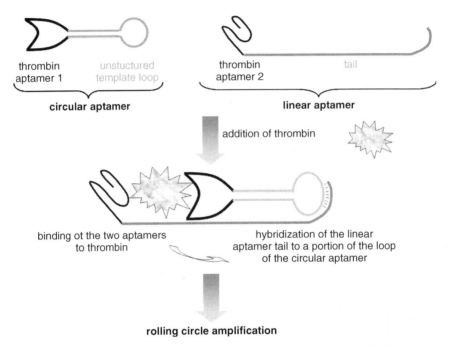

Figure 8.13 Detection of thrombin by proximity extension of circular DNA aptamers. The assay is based on the use of a thrombin-specific circular aptamer with an unstructured loop and of a thrombin-specific linear aptamer with a long tail. A region at the 3′ end of the linear aptamer tail is complementary to a portion of the circular aptamer loop. In the presence of thrombin, the two sequences are taken into proximity and the two complementary regions hybridize. This forms a primer for polymerase-mediated extension and isothermal rolling circle amplification. [From Di Giusto et al., (2005).]

8.6 CONCLUSIONS

Aptamer-based assays opened new scenarios in the field of bioanalytical chemistry. In particular, the use of aptamer as a capturing agent or receptor for the analyte represents an interesting approach to the possibility of improving analytical detection limits by coupling aptamer-based assay to various amplification strategies. In this regard it should be noted that real improvement can be assured by an interdisciplinary approach that couples instrumental analytical chemistry (based on electrochemistry, spectroscopy, etc.) with molecular biology and nanotechnology. As shown in this chapter, quantum dots together with nanoparticle-based assays, coupled to electrochemical detection, made it possible to reach attomolar detection limits. Aptazymes and aptamer-based machines, in which the enzymatic function is activated or inhibited by an effector that is the target molecule for the aptamer, is also a useful strategy for signal amplification based on the catalytic effect: Very good detection limits are assured, as in the case of ligated aptazymes, with a low nanomolar range. The latter tools are especially appealing in array systems.

Finally, ordinary enzymatic amplification methods used in nucleic acid analysis, such as polymerase chain reaction, are very powerful when applied to aptamer-based assays. Aptamers as nucleic acids can undergo an amplification step after interaction with the relative target, resulting in significant signal magnification. Actually, approximately 10^2 molecules of target were detected using this method, improving the sensitivity of protein detection by several orders of magnitude.

New combinations of aptamer-based assay with innovative ideas in molecular biology and nanotechnologies are encouraged and expected, aimed toward developing easy, sensitive, selective, and fast analytical methods.

REFERENCES

Baner, J., Nilsson, M., Mendel-Hartvig, M., Landegren, U. (1998). Signal amplification of padlock probes by rolling circle amplification. *Nucleic Acids Res* 26, 5073–5078.

Beissenhirtz, M. K., Willner, I. (2006). DNA-based machines. *Org Biomol Chem* 4, 3392–3401.

Breaker, R. R. (2002). Engineered allosteric ribozymes as biosensor components. *Curr Opin Biotechnol* 13, 31–39.

Breaker, R. R., Joyce, G. F. (1994). A DNA enzyme that cleaves RNA. *Chem Biol* 1, 223–229.

Cao, W. (2004). Recent developments in ligase-mediated amplification and detection. *Trends Biotechnol* 22, 38–44.

Centi, S., Laschi, S., Mascini, M. (2007a). Improvement of analytical performances of a disposable electrochemical immunosensor by using magnetic beads. *Talanta* 73, 394–399.

Centi, S., Tombelli, S., Minunni, M., Mascini, M. (2007b). Aptamer-based detection of plasma proteins by an electrochemical assay coupled to magnetic beads. *Anal Chem* 79, 1466–1473.

Chang, T., Tsai, C., Sun, C., Chen, C., Kuo, L., Chen, P. (2007). Ultrasensitive electrical detection of protein using nanogap electrodes and nanoparticle-based DNA amplification. *Biosens Bioelectron* 22, 3139–3145.

Chelyapov, N. (2006). Allosteric aptamers controlling a signal amplification cascade allow visual detection of molecules at picomolar concentrations. *Biochemistry* 45, 2461–2466.

Choi, J. H., Chen, K. H., Strano, M. S. (2006). Aptamer-capped nanocrystal quantum dot: a new methods for label-free protein detection. *J Am Chem Soc* 128, 15584–15585.

Clark, S. L., Remcho, V. T. (2002). Aptamers as analytical reagents. *Electrophoresis* 23, 1335–1340.

Dick, L. W., McGown, L. B. (2004). Aptamer-enhanced laser desorption/ionization for affinity mass spectrometry. *Anal Chem* 76, 3037–3041.

Di Giusto, D. A., Wlassoff W. A., Gooding, J. J., Messerie, B. A., King, G. C. (2005). Proximity extension of circular DNA aptamers with real-time protein detection. *Nucleic Acids Res* 33, e64.

Fredriksson, S., Gullberg, M., Jarvius, J., Olsson, C., Pietras, K., Gustafsdottir, S. M., Ostman, A., Landegren, U. (2002). Protein detection using proximity-dependent DNA ligation assays. *Nat Biotechnol* 20, 473–477.

Green, L. S., Jellinek, D., Jenison, R., Ostman, A., Heldin, C. H., Janjic, N. (1996). Inhibitory DNA ligands to platelet-derived growth factor B-chain. *Biochemistry* 45, 14413–14424.

Han, M., Gao, X., Su, J. Z., Nie, S. (2001). Quantum-dot-tagged microbeads for multiplexed optical coding of biomolecules. *Nat Biotechnol* 19, 631–635.

Hansen, J. A., Wang, J., Kawde, A-N., Xiang, Y., Gothelf, K. V., Collins, G. (2006). Quantum-dot/aptamer-based ultrasensitive multi-analyte electrochemical biosensor. *J Am Chem Soc* 128, 2228–2229.

Hesselberth, J. R., Robertson, M. P., Knudsen, S. M, Ellington, A. D. (2003). Simultaneous detection of diverse analytes with an aptazyme ligase array. *Anal Biochem* 312, 106–112.

Hoile, R., Yuen, M., James, G., Gilbert, G. L. (2006). Evaluation of the rapid analyte measurement platform (RAMP) for the detection of *Bacillus anthracis* at a crime scene. *Forensic Sci Int*. 171, 1–4.

Ikanovic, M., Rudzinski, W. E., Bruno, J. G., Allman, A., Carrillo, M. P., Dwarakanath, S., Bhahdigadi, S., Rao, P., Kiel, L., Andrews, C. J. (2007). Fluorescence asay based on aptamer–quantum dot binding to *Bacillus thuringiensis* spores. *J Fuoresc* 17, 193–199.

Knudsen, S., Ellington, A. D. (2006). Aptazymes: allosteric ribozymes and deoxyribozymes as biosensors. In *The Aptamer Handbook*, S. Klussmann, ed., Wiley-VCH, Weinheim, Germany, pp. 290–310.

Kotia, R. B., Li, L., McGown, L. B. (2000). Separation of nontarget compounds by DNA aptamers. *Anal Chem* 72, 827–831.

Levy, M., Cater, S. F., Ellington, A. D. (2005). Quantum-dot aptamer beacons for the detection of proteins. *ChemBioChem* 6, 2163–2166.

Li, Y., Lee, H. J., Corn, R. M. (2007). Detection of protein biomarkers using RNA aptamer microarrays and enzymatically amplified surface plasmon resonance imaging. *Anal Chem* 79, 1082–1088.

Liu, G., Wang, J., Kim, J., Jan, M., Collins, G. (2004). Electrochemical coding for multiplexed immunoassays of proteins. *Anal Chem* 76, 7126–7130.

Lu, Y., Liu, J. (2006). Functional DNA nanotechnology: emerging applications of DNAzymes and aptamers. *Curr Opin Biotechnol* 17, 580–588.

McGown, L. B., Joseph, M. J., Pittner, J. B., Vonk, G. P., Linn, C. P. (1995). The nucleic acid ligand. A new tool for molecular recognition. *Anal Chem* 67, 663A–668A.

Medintz, I. L., Uyeda, H. T., Goldman, E. R., Mattoussi, H. (2005). Quantum dot bioconjugates for imaging, labelling and sensing. *Nat Mater* 4, 435–446.

Mukhopadhyay, R. (2005). Aptamers are ready for the spotlight. *Anal Chem* 77, 115A–118A.

Nutiu, R. (2003). Structure-switching signaling aptamers. *J Am Chem Soc* 125, 4771–4778.

Pavlov, V., Xiao, Y., Shlyahosvky, B., Willner, I. (2004). Aptamer-functionalised Au nanoparticles for the amplified optical detection of thrombin. *J Am Chem Soc* 126, 11768–11769.

Polsky, R., Gill, R., Kaganovsky, L., Willner, I. (2006). Nucleic acid-functionalised Pt nanoparticles: catalytic labels for the amplified electrochemical detection of biomolecules. *Anal Chem* 78, 2268–2271.

Seetharaman, S., Zivarts, M., Sudarsan, N., Breaker, R. R. (2001). Immobilized RNA switches for the analysis of complex chemical and biological mixtures. *Nat Biotechnol* 19, 336–341.

Shlyahosvky, B., Li, D., Weizmann, Y., Nowarski, R., Kotler, M., Willner, I. (2007). Spotlighting of cocaine by an autonomous aptamer-based machine. *J Am Chem Soc* 129, 3814–3815.

Soukup, G. A., Breaker, R. R. (1999). Nucleic acid molecular switches. *Trends Biotechnol* 17, 469–476.

Tombelli, S., Minunni, M., Mascini, M. (2005). Analytical applications of aptamers. *Biosens Bioelectron* 20, 2424–2434.

Travascio, P., Witting, P. K., Mauk, A. G., Sen, D. (2001). The peroxidase activity of a hemin–DNA oligonucleotide complex: free radical damage to specific guanine bases of the DNA. *J Am Chem Soc* 123, 1337–1348.

Tuerk, C., Gold, L. (1990). Systematic evolution of ligands by exponential enrichment: RNA ligands to bacteriophage T4 DNA polymerase. *Science* 249, 505–510.

Wang, J., Liu, G., Merkoçi, A. (2003). Electrochemical coding technology for simultaneous detection of multiple DNA targets. *J Am Chem Soc* 125, 3214–3215.

Wang, X., Li, F., Su, Y., Sun, X., Li, X., Schluesener, H., Tang, F., Xu, S-Q. (2004). Ultrasensitive detection of protein using an aptamer-based exonuclease protection assay. *Anal Chem* 76, 5605–5610.

Willner, I. (2005). Nanoparticle- and nanorod-biomaterial hybrid systems for sensor, circuitry and motor applications. *Contrib Sci* 3, 79–90.

Willner, I., Baron, R., Willner, B. (2007). Integrated nanoparticle–biomolecule systems for biosensing and bioelectronics. *Biosens Bioelectron* 22 1841–1852.

Xiao, Y., Pavlov, V., Levine, S., Niazov, T., Markovitch, G., Willner, I. (2004). Catalytic growth of Au nanoparticles by NAD(P)H cofactors: optical sensors for NAD(P)$^+$-dependent biocatalyzed transformations. *Angew Chem Int Ed Engl* 43, 4519–4522.

Yao, X., Li, X., Toledo, F., Zurita-Lopez, C., Gutova, M., Momand, J., Zhou, F. (2006). Subattomole oligonucleotide and p53 cDNA determinations via a high-resolution surface plasmon resonance combined with oligonucleotide-capped gold nanoparticle signal amplification. *Anal Biochem* 354, 220–228.

Yin, Y., Alivisatos, A. P. (2005). Colloidal nanocrystal synthesis and the organic–inorganic interface. *Nature* 437, 664–670.

You, K. M., Lee, S.H., Im, A., Lee, S. B. (2003). Aptamers as functional nucleic acids: in vitro selection and biotechnological applications. *Biotechnol Bioprocess Eng* 8, 64–75.

Zhang, H., Wang, Z., Li, X-F., Le, X. C. (2006). Ultrasensitive detection of proteins by amplification of affinity aptamers. *Angew Chem Int Ed Engl* 45, 1576–1580.

PART III

APPLICATIONS

CHAPTER 9

KINETIC CAPILLARY ELECTROPHORESIS FOR SELECTION, CHARACTERIZATION, AND ANALYTICAL UTILIZATION OF APTAMERS

SERGEY N. KRYLOV

9.1 INTRODUCTION

9.1.1 Kinetic Capillary Electrophoresis

Aptamer–target binding is an example of an affinity interaction, a tight and highly selective binding of molecules (one of which is typically a biopolymer) through a number of weak noncovalent bonds. Due to the affinity nature of aptamer–target interaction, affinity methods are used for the selection, characterization, and analytical utilization of aptamers.

All affinity methods are based on noncovalent binding of a ligand (L) and a target (T), with the formation of a ligand–target complex (L · T):

$$L + T \underset{k_{off}}{\overset{k_{on}}{\rightleftharpoons}} L \cdot T \tag{9.1}$$

where k_{on} and k_{off} are rate constants of the forward and backward reactions, and $K_d = k_{off}/k_{on}$ is the equilibrium dissociation constant. All affinity methods also involve the separation of unbound L from L · T. Homogeneous affinity methods

Aptamers in Bioanalysis, Edited by Marco Mascini
Copyright © 2009 John Wiley & Sons, Inc.

do not involve the surface as a means of separation and are therefore not susceptible to nonspecific binding of L to the surface. Kinetic affinity methods are those that do not assume equilibrium in reaction (9.1) and can therefore be used for (1) selection of binding ligands with desirable k_{on} and k_{off}, (2) measuring k_{on} and k_{off}, and (3) quantitative affinity analyses involving unstable complexes (high k_{off}). The only conventional kinetic affinity method is surface plasmon resonance, a heterogeneous method with a number of inherent limitations (Wilson, 2002; Simpson and Fisher, 2005; Boozer et al., 2006).

Kinetic capillary electrophoresis (KCE) was recently introduced as a conceptual platform for *homogeneous kinetic* affinity methods (Petrov et al., 2005). KCE is defined as CE separation of species that interact with each other during their electrophoretic separation in CE. As a brief introduction to CE, the schematic of a CE instrument is depicted in Figure 9.1.

All advantages of CE over slab electrophoresis originate from the small diameter of the capillary. Due to the small diameter, the ratio between the surface area and the volume of the capillary is high, which leads to highly efficient dissipation of Joule heat. This allows for the application of very high electric fields (up to 1 kV cm^{-1}) without the consequence of overheating the separation medium. The small diameter of the capillary also allows one to avoid using gels; the run buffer can easily be retained inside the capillary, due to the capillary force. Finally, a point detector can be attached to the capillary, thus facilitating a variety of real-time HPLC-like quantitative detection schemes. CE is easily interfaced with optical (i.e., light absorbance, fluorescence, scattering), electrochemical, or mass-spectrometric detection.

KCE uses the advantages of CE to create a new analytical platform of kinetic homogeneous affinity methods. Seven KCE methods have been suggested and studied in different applications. The spectrum of proven applications includes: (1) measurements of k_{on}, k_{off}, and K_d (Berezovski and Krylov, 2002; Krylov and Berezovski, 2003; Okhonin et al., 2004a,b, 2006; Krylova et al., 2005; Petrov et al., 2005; Pang et al., 2006; Woolley et al., 2006; Larijani et al., 2007; Liu

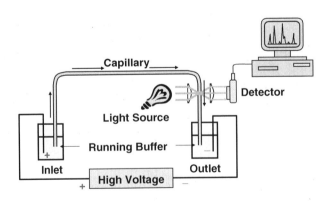

Figure 9.1   Typical CE instrument.

et al., 2007); (2) quantitative affinity analyses of proteins, DNA, and mRNA (Berezovski and Krylov, 2003; Berezovski et al., 2003; Drabovich et al., 2004, 2006, 2007; Al-Mahrouki and Krylov, 2005; Zhang et al., 2006); (3) measurement of temperature inside the capillary (Berezovski and Krylov, 2004); (4) studies on the thermochemistry of affinity interactions (Berezovski and Krylov, 2005); and (5) kinetic selection of ligands from combinatorial libraries (Berezovski et al., 2005, 2006a,b; Drabovich et al., 2005, 2006; Musheev and Krylov, 2006). This chapter is dedicated to the application of KCE methods to the selection, characterization, and analytical utilization of oligonucleotide aptamers.

KCE methods can be considered as advanced electrophoretic mobility shift assays. The quality of such assays is largely dependent on the magnitude of the shift in the mobility of the aptamer (L) upon its binding to the target. Aptamers are relatively big and highly negatively charged molecules. Accordingly, to introduce a significant shift in their mobility, the target has to be large and not as negatively charged as the aptamer. This limitation has so far restricted the application of KCE methods to proteins and peptide targets. Small molecule targets have yet to be studied in KCE-based selection, characterization, and utilization of aptamers.

Two methods have been used so far for aptamer-related applications: nonequilibrium capillary electrophoresis of equilibrium mixtures (NECEEM) and equilibrium capillary electrophoresis of equilibrium mixtures (ECEEM). An equilibrium mixture is defined as a mixture of the target and the ligand (or a mixture of ligands) which is incubated until equilibrium is reached in reaction (9.1), so that the concentrations of molecules and complexes are related as $[L][T]/[L \cdot T] = K_d$.

9.1.2 The Concept of NECEEM and ECEEM

Figure 9.2 illustrates the concept of NECEEM schematically. First, an equilibrium mixture, which contains L, T, and $L \cdot T$, is typically prepared outside the capillary. Second, a short plug of the equilibrium mixture is injected by pressure into the capillary; the capillary is prefilled with a run buffer, which is free of L, T, and $L \cdot T$. Third, a high voltage is applied to separate L, T, and $L \cdot T$ under nonequilibrium conditions. The velocity of T is assumed to be greater than that of L; the velocity of $L \cdot T$ is typically intermediate. When the zones of L, T, and $L \cdot T$ are separated, $L \cdot T$ is no longer at equilibrium with L and T; it dissociates continuously with the monomolecular rate constant k_{off}. Reassociation of L and T can be neglected if the efficiency of their electrophoretic separation is high. While the equilibrium fractions of L, T, and $L \cdot T$ migrate as short electrophoretic zones, L and T, which are produced from the dissociation of $L \cdot T$, have exponential concentration profiles. The spatial distribution of L, T, and $L \cdot T$ inside the capillary at any given time $t_1 > 0$ has three peaks and two exponential decay zones. Two peaks correspond to the equilibrium fractions of L and T. The third peak corresponds to intact $L \cdot T$ complexes that reached the detector. The exponential decay zones correspond to the L and T that dissociated from $L \cdot T$ during electrophoresis. This general concept of NECEEM is used for (1) the selection of aptamers, (2) the kinetic characterization of aptamers, and (3) the utilization of aptamers in quantitative affinity analysis of the target.

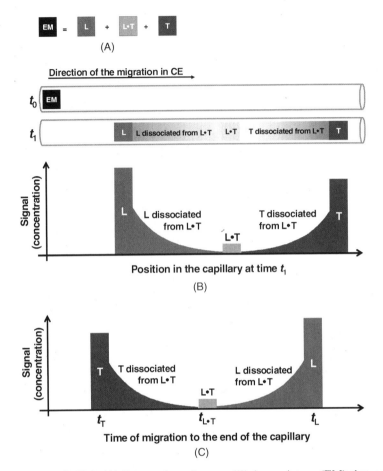

Figure 9.2   NECEEM. (A) Preparation of an equilibrium mixture (EM) that contains L, L·T, and T. (B) NECEEM-based separation of equilibrium fractions of L, L·T, and T. The top part illustrates the spatial distribution of separated components in the capillary at two different times ($t_0 = 0$ and $t_1 > t_0$) from the beginning of separation. The graph at the bottom shows concentrations of the separated components as functions of their position in the capillary at time t_1. (C) Schematic NECEEM electropherogram (signal versus migration time) assuming a point detector at the distal end of the capillary. (See insert for color representation.)

Figure 9.3 depicts the concept of ECEEM schematically. In contrast to NECEEM, in ECEEM the capillary is filled with T at a concentration identical to that in the equilibrium mixture. Therefore, quasiequilibrium will be maintained during electrophoresis if the time required for reequilibration is shorter than the characteristic time of separation. In such a case, the equilibrium mixture will be migrating as a single electrophoretic zone with an effective velocity being a function of K_d and [T]. If $K_d \gg$ [T], most of L will be free and the

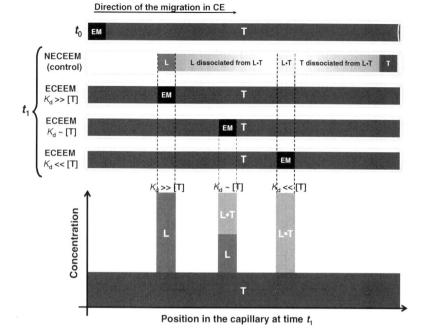

Figure 9.3   ECEEM. The equilibrium mixture (EM) contains L, L · T, and T, as depicted in Figure 9.2. (A) The spatial distribution of separated components in the capillary at two different times ($t_0 = 0$ and $t_1 > t_0$) from the beginning of separation. The distribution is also shown for NECEEM for time t_1 to demonstrate reference migration times. The spatial distribution for ECEEM is shown for three different ratios between K_d and T ($K_d \gg$ [T], $K_d \sim$ [T], and $K_d \ll$ [T]). (B) Concentrations of L, L · T, and T for three different ratios between K_d and [T] as functions of the position in the capillary at time t_1. The peaks are offset along the vertical axis for clarity. (See insert for color representation.)

zone of the equilibrium mixture will be migrating with an effective velocity equal to that of free L in NECEEM. Alternatively, if $K_d \ll$ [T], most of L will be within the complex, and the electrophoretic zone will be migrating with an effective velocity similar to that of L · T in NECEEM. If K_d is on the same order as [T], the amounts of L and L · T in the equilibrium mixture will be comparable, and the effective velocity of the electrophoretic zone of the equilibrium mixture will be intermediate to those of L and L · T in NECEEM. An ECEEM electropherogram for a single ligand is not shown, as it is trivial. It contains a plateau of T with a single peak of the equilibrium mixture whose migration time t depends on K_d and [T]:

$$\frac{1}{t} = \frac{1}{t_{L \cdot T}} \frac{[T]}{[T] + K_d} + \frac{1}{t_L} \frac{K_d}{[T] + K_d} \tag{9.2}$$

9.2 SELECTION OF APTAMERS USING KCE METHODS FOR PARTITIONING AND AFFINITY CONTROL

As mentioned above, KCE methods can be considered as advanced electrophoretic mobility shift assays. Using KCE methods for partitioning of aptamers from nonaptamers is based on two important premises. First, oligonucleotides of different sequences but identical lengths have similar electrophoretic mobilities in gel-free CE; the mobility in free solution depends on the charge-to-size ratio, which is similar for such oligonucleotides. Thus, random-sequence oligonucleotide libraries used for aptamer selection migrate as a broad but single electrophoretic zone in KCE. Second, the quality of selection depends on the magnitude of the electrophoretic shift experienced by aptamers upon binding to the target. Accordingly, the target has to be relatively large and much less negatively charged than oligonucleotides. Proteins and larger peptides are perfectly suitable targets for KCE-based partitioning. The applicability of KCE methods to small-molecule targets is still to be studied.

9.2.1 NECEEM-Based Selection of Aptamers

Foundation
Basic Principles The general concept of NECEEM described in Section 9.1.2 is used for NECEEM-based partitioning of aptamers. The only difference from the consideration in Section 9.1.2 is that the equilibrium mixture is prepared by mixing the target and the oligonucleotide library. A short plug of the equilibrium mixture is injected into the capillary, and a high voltage is used to separate the bound ligands from the unbound DNA library. The complexes with the highest k_{off} values will dissociate during the separation and form an exponential decay zone, whereas the complexes with the lowest k_{off} values will reach the end of the capillary intact. Figure 9.4 depicts two regimes of aptamer collection.

If the time window in which the fraction is collected includes both the intact complexes and the entire decay zone and cuts only the unbound library, naive aptamers with all possible k_{off} values will be collected (Figure 9.4A). If the time window collects only a fraction from the decay zone, smart aptamers with predefined k_{off} values will be collected preferentially (Figure 9.4B). The values of k_{off} of selected aptamers will be dependent on four time parameters ($t_{L \cdot T}$, t_L, t_1, and t_2) and can be calculated easily (see the inequality in Figure 9.4B). Due to the random nature of dissociation, a fraction of smart aptamers will also contain aptamers with k_{off} values outside the desirable region. Therefore, multiple rounds of SELEX have to be conducted with the same t_1-t_2 collection window in order to gradually purify the pool of smart aptamers from such random intruders.

The procedure for aptamer selection by NECEEM would typically start with measurement of the migration times of the target and the naive library (leftmost and rightmost peaks in Figure 9.4). Since the target is typically not labeled fluorescently, ultraviolet (UV) light absorption is the detection choice for the target. UV detection can also be used for the naive library. Alternatively, the

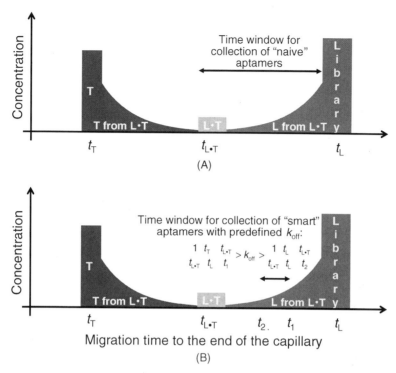

Figure 9.4   NECEEM electropherogram to illustrate (A) the selection of naive aptamers and (B) smart aptamers with predefined k_{off}. (See insert for color representation.)

naive library can be synthesized using a fluorescent tag to facilitate fluorescent detection of the library. The latter is particularly useful in studying the bulk affinity of the naive library to the target. By mixing the fluorescently labeled naive library with a target at a relatively high concentration (10^{-5} to 10^{-6} M), it is often possible to detect target–oligonucleotide complexes. Such complexes typically correspond to nonspecific binding, but they can serve as a perfect tool to determine the migration time $t_{L \cdot T}$ of the aptamer–target complexes (the middle peak in Figure 9.4). Indeed, the migration time of $L \cdot T$ is mainly a function of the sum charge and sum size ratio of the complex and is largely independent of the types of bonds involved in complex formation. Knowing $t_{L \cdot T}$ may help to determine the aptamer collection window accurately prior to the selection process. If the bulk affinity is so low that complexes of the target with the naive library are not detectable, the aptamer-collection window for the first few rounds of aptamer selection may be spanning from the peak of the target to that of the library (t_T–t_L). The collection window can be narrowed when pools of aptamers start showing detectable complexes with the progress of selection. One of the polymerase chain reaction (PCR) primers is labeled fluorescently to facilitate fluorescence detection of aptamers after the first and following rounds of selection.

Browser et al. suggested a CE-based procedure for aptamer selection, termed CE-SELEX, which is similar to NECEEM-based SELEX but does not use the kinetic capabilities of NECEEM (Mendonsa and Bowser, 2004, 2005; Mosing et al., 2005).

Background   One of the major advantages of NECEEM-based partitioning is its low background. *Background* is defined as the amount of oligonucleotide collected when a naive oligonucleotide library is sampled for partitioning without the target relative to the amount of oligonucleotide sampled. In conventional heterogeneous partitioning methods such as partitioning on nylon filters, the background is typically higher than 10^{-2}, due to the oligonucleotide sticking to the surface. NECEEM-based partitioning is a homogeneous procedure in which the influence of the surface is minimized. Moreover, if oligonucleotides adhered to the inner capillary walls, they would not contribute to the background, as aptamers are typically collected prior to oligonucleotide elution (see, e.g., Figure 9.4). The background of NECEEM-based partitioning was measured experimentally by Berezovski et al. (2005). A naive DNA library was sampled for NECEEM without any target. Fractions were collected in small time windows, and the quantity of DNA in the fractions was determined using quantitative PCR (the limit of detection of quantitative PCR was 1000 copies of the template per sample). Figure 9.5 illustrates the result of this experiment. It can be seen that the amount of DNA per 1-minute fraction in the aptamer collection window was approximately 10^6 molecules. The total amount of DNA that "licked" in the aptamer-collection window was approximately 10^7. The amount of DNA sampled was approximately 10^{12} molecules. Thus, the background of NECEEM-based partitioning was demonstrated to be 10^{-5}, which is 100 to 1000 times lower than that of conventional methods of partitioning. Although the background in NECEEM-based partitioning is exceptionally good, it is much higher than it would be if the profile of the DNA zone were Gaussian (Berezovski et al., 2005). It is not clear why the DNA profile has two pronounced wings between 8 and 19 minutes and between 25 and 30 minutes (Figure 9.5B). The nature of background in NECEEM-based partitioning is not yet understood.

Non-SELEX   The low background of NECEEM-based partitioning facilitates Non-SELEX selection of aptamers (Figure 9.6). SELEX includes multiple alternating rounds of partitioning and amplification (typically by PCR) of candidate sequences collected (Figure 9.6, left panel). The multiple rounds are required due to a high background and the poor efficiency of conventional methods of partitioning. The amplification, in turn, is required in SELEX to maintain a significant number of oligonucleotide candidates through multiple rounds of selection. Non-SELEX is faster and simpler than SELEX. Moreover, Non-SELEX seems to be out of the scope of heavily protected intellectual property, thus potentially allowing for the commercial use of aptamers without the need of acquiring a license for SELEX.

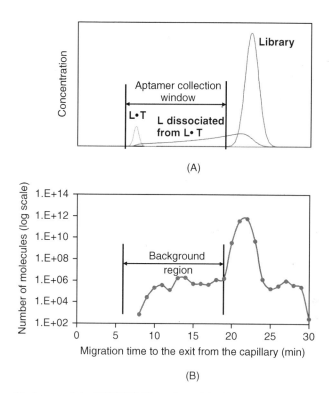

(A)

(B)

Figure 9.5 Background in NECEEM-based partitioning of aptamers. (A) Simulated NECEEM electropherogram to assist with the understanding of part (B). (B) Experimental measurement of the background. A naive DNA library was sampled for NECEEM without any target. Fractions were collected in small time windows and the quantity of DNA in the fractions was determined using quantitative PCR. The background is calculated as the amount of DNA in the aptamer-collection window divided by the total amount of DNA sampled. In this example the background was approximately 10^{-5}.

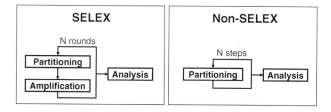

Figure 9.6 SELEX and Non-SELEX selection of aptamers.

Examples

Targets Up until now, NECEEM (and a similar method, CE-SELEX) were used to select aptamers for protein farnesyltransferase (PFTase) (Berezovski et al., 2005), MutS protein (Drabovich et al., 2005, 2006), h-Ras protein (Berezovski et al., 2006a,b), IgE (Mendonsa and Bowser, 2004), HIV-1 reverse transcriptase (Mosing et al., 2005), and Peptide Y (Mendonsa and Bowser, 2005). Aptamers for PFTase were obtained in a single round of selection. Aptamers for other targets were obtained in two to four rounds. Aptamers for MutS were developed in a regime that facilitates the selection of smart aptamers with predefined k_{off}. Aptamers for h-Ras were developed in NECEEM-based Non-SELEX.

Standard Operating Procedure A set of procedures used in the selection of aptamers for PFTase can be used as a sketch of the standard operating procedure for NECEEM-based SELEX (Berezovski et al., 2005). First, the migration time of the naive oligonucleotide library is determined using UV detection for the unlabeled library or fluorescence detection for the labeled library. In this example, the DNA library migrates as a broad but single electrophoretic zone with a migration time of 23 minutes and a peak width of approximately 3 minutes (Figure 9.7A).

Second, the migration time of the target is determined using UV detection. A sample of commercial PFTase shows four distinct peaks with migration times

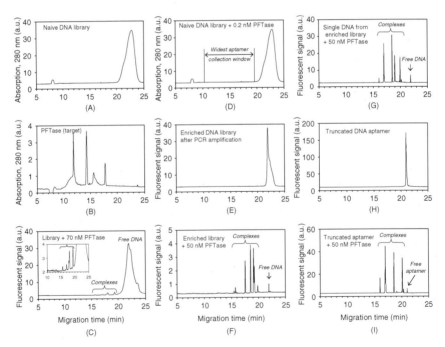

Figure 9.7 Essential steps of NECEEM-based aptamer development. See the text for details.

of 12, 14, 16, and 18 minutes (Figure 9.7B). The 12- and 16-minute peaks are broad, suggesting that they may contain more than a single component. The nature of the multiple peaks is unknown, but they can be associated with multimers of the protein. Third, the bulk affinity of the naive library to PFTase is determined. A relatively high concentration of PFTase (70 nM) is mixed with the fluorescently labeled naive library, equilibrated, and subjected to NECEEM. The resulting electropherogram contains a large peak of free DNA and several small peaks with migration times of 15 to 19 minutes, which correspond to PFTase–DNA complexes (Figure 9.7C). These small peaks can be used to define a narrow aptamer-collection window. Fourth, NECEEM-based partitioning is carried out. The naive library is mixed with PFTase at a low concentration of 0.2 nM, equilibrated, and subjected to NECEEM. Since the concentration of PFTase is 350 times lower than the one used in bulk affinity measurements, the peaks of complexes are below the limit of UV (and fluorescence) detection (Figure 9.7D).

The widest aptamer-collection window, which covers the area from the leftmost peak of PFTase to the peak of the naive DNA library, is used to collect a pool of candidate sequences. The pool is then amplified with PCR. A set of primers used to make strands complementary to candidate sequences is labeled with biotin. A second set of primers is labeled with a fluorophore. The procedure of PCR amplification of enriched pools in the first few rounds of selection, when the pool is highly heterogeneous, may require careful optimization (Musheev and Krylov, 2006). The biotin-labeled primer is used to separate the strands of the double-stranded product of PCR. The biotin-labeled complementary strands are discarded, whereas the fluorophore-labeled strands, which are amplified candidate sequences, are used for further analysis or a second round of selection. The amplified enriched pool can be analyzed by CE for purity. The width of the peak for the enriched pool is much narrower than that of the naive library, as the enriched pool has much lower heterogeneity (Figure 9.7E). The enriched pool is then analyzed for bulk affinity to the target. It is mixed with PFTase at 50 nM concentration, equilibrated, and sampled for NECEEM. The resulting NECEEM electropherogram contains a small peak of free DNA at 22 minutes and a series of peaks of PFTase–DNA complexes in the window between 15 and 20 minutes (Figure 9.7F).

Comparison of Figure 9.7C and F demonstrates a dramatic improvement in the affinity of the enriched library with respect to that of the naive library; the naive library bound only a small fraction of PFTase, whereas the enriched library bound almost all PFTase. Multiple peaks of complexes correspond to multiple components of the PFTase sample. A question arises as to whether or not a single aptamer can bind all the components. To answer this question, candidate sequences are cloned into bacteria using a standard procedure (www.emdbiosciences.com/docs/docs/PROT/TB183.pdf), and the binding of a single sequence to the PFTase sample is studied using NECEEM. A single sequence exhibits a binding pattern similar to that of the heterogeneous pool (Figure 9.7G), suggesting that a single aptamer can bind to all components of

the PFTase sample. Such a binding pattern indicates that every PFTase component contains the same epitope. The latter is an indirect indication that multiple
components in the PFTase sample may be multimers of PFTase.

Finally, it is interesting to know whether or not the constant regions are
involved in aptamer formation. To answer this question, an aptamer sequence
without the constant regions but with a fluorophore tag is synthesized. When it is
sampled for CE without PFTase, it shows a sharp peak, confirming the ultimate
homogeneity of the sample (Figure 9.7H). To test the affinity of the truncated
aptamer to PFTase, the protein is mixed with the aptamer at a concentration of
50 nM, equilibrated, and sampled for NECEEM. The resulting NECEEM electropherogram is similar to that of the aptamer with constant regions (Figure 9.7I).
Thus, the constant regions in this particular case are not involved in formation
of the secondary structure of the aptamer. In the selection described, one round
is sufficient to obtain the enriched DNA library and isolate individual aptamers
with nanomolar affinity. If one round is not enough to generate aptamers of
desirable affinity, multiple rounds of selection should be used. Due to the very
high efficiency of NECEEM-based partitioning, the number of rounds required is
typically below five. Moreover, there are published and unpublished observations
that additional rounds lead to decreasing affinity (Mendonsa and Bowser, 2004,
2005; Mosing et al., 2005). The nature of decreasing affinity is still not known,
but the phenomenon emphasizes the need to monitor the affinity of enriched
pools after every round of selection. NECEEM provides a highly efficient and
accurate tool for measuring the bulk affinity of pools and thus facilitates a great
level of control over the progress of SELEX.

Smart Aptamers with Predefined k_{off}   Smart aptamers with predefined ranges
of k_{off} values were selected for MutS protein. Fractions of candidate oligonucleotides were collected in two rounds of selection within two aptamer-collection
windows (Figure 9.8). Every round of selection consisted of NECEEM partitioning, collection of an enriched pool of oligonucleotide candidates, PCR amplification, separation of strands, and measuring the bulk affinity of the enriched library
with NECEEM. In region I, the theoretically predicted range of k_{off} was between
0 and 1.0×10^{-3} s^{-1}; in region II it was between 1.7×10^{-3} and 2.5×10^{-3} s^{-1}
(Figure 9.8A). After only two rounds of NECEEM-based SELEX in regions I
and II, the pools of DNA had experimental bulk k_{off} values of 0.4×10^{-3} and
1.7×10^{-3} s^{-1}, respectively (Figure 9.8B). This experiment proved the theoretical
consideration presented above.

Non-SELEX   Aptamers for h-Ras protein were selected through
NECEEM-based Non-SELEX. The essence of the procedure and the
results of a binding assay are shown in Figure 9.9. The equilibrium mixture
of the naive DNA library and the target is prepared and subjected to the first
step of NECEEM-based partitioning. The first enriched library is collected
in a vial with several microliters of the target solution. The mixture is
incubated to establish the first aptamer-enriched equilibrium mixture. The

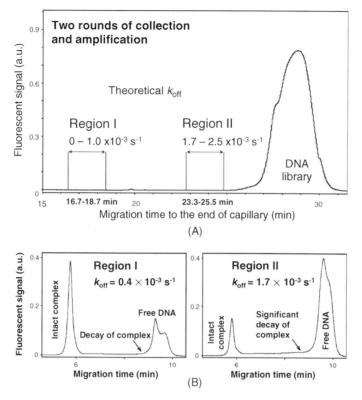

Figure 9.8 NECEEM-based selection of smart aptamers with predefined k_{off}. (A) Two aptamer-collection windows with respect to the migration time of the naive DNA library. The intact complexes (not shown) elute in region I. Region II corresponds to aptamers that dissociated from the complexes during electrophoresis. (B) NECEEM-based affinity analysis of enriched libraries after two rounds of selection in regions I (left) and II (right). Experiments depicted in part (B) were performed with a shorter capillary than experiments shown in part (A); accordingly, migrations times are shorter in (B).

enriched equilibrium mixture is then subjected to the second step of NECEEM partitioning. Aptamers are again collected into the target solution to establish the second aptamer-enriched equilibrium mixture. The procedure is repeated one more time to collect the third aptamer-enriched fraction. Since Non-SELEX selection does not involve PCR amplification between the partitioning steps, every DNA molecule in every fraction has a unique sequence in the randomized region.

In the example described, every fraction collected was analyzed by quantitative PCR to determine the number of DNA molecules and bulk K_d for the three aptamer-enriched libraries. The analytical procedure and its results are shown in Figure 9.9B. The three enriched libraries contained 3×10^7, 2×10^5, and 3×10^3

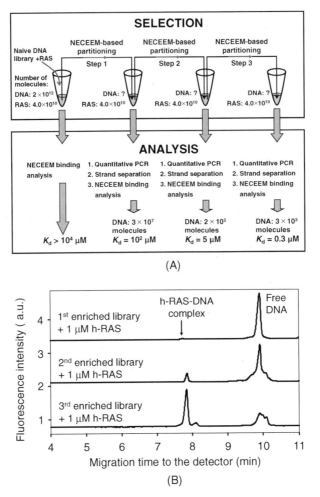

Figure 9.9 Non-SELEX selection of aptamers for h-Ras protein: (A) procedure for selection and accompanying analyses of pools; (B) results of NECEEM binding analyses of pools after each of the three steps of Non-SELEX.

DNA molecules, respectively, and had K_d values of 100, 5, and 0.3 μM, respectively. The bulk K_d of the naive library to h-Ras was greater than 10^4 μM, suggesting that three steps of Non-SELEX selection improved bulk affinity by more than four orders of magnitude.

Less than a 10% fraction of the aptamer-enriched equilibrium mixtures can be sampled for steps 2 and 3, due to limitations of currently available commercial CE instrumentation. The partial sampling makes it statistically improbable to select aptamers with an abundance lower than $(F^{m-1}N)^{-1}$, where F is the fraction of ligands sampled, m is the number of steps of selection, and N is the number of different DNA sequences in the naive library sampled for the first step. For the parameters used in the proof-of-principle work ($F = 0.03$, $m = 3$, and $N = 2 \times 10^{12}$), the minimum abundance is 5×10^{-10}. Designing instrumentation and methodology that will facilitate the complete sampling of collected ligands for all steps on Non-SELEX is an essential task in making Non-SELEX a highly practical approach.

9.2.2   ECEEM-Based Selection of Aptamers

Foundation   Whereas NECEEM can be used to select smart aptamers with predefined k_{off} values, ECEEM can facilitate the selection of smart aptamers with predefined K_d values (Drabovich et al., 2005, 2006). The ECEEM procedure used for aptamer selection follows the basic concept of ECEEM described above. The capillary is filled with a solution of the target and a short plug of the equilibrium mixture of the target and the oligonucleotide library is injected into the capillary. A high voltage is applied and the target–ligand complexes migrate under the conditions of quasiequilibrium with the target present in the capillary. This means that ligands spend some time within the complex and some time as free molecules. It is assumed that k_{off} and k_{on} are high enough to maintain the quasiequilibrium between complexes and free molecules. In such a case, the effective migration time t of the ligand will depend on K_d and the concentration of free target [T] as described in equation (9.2). The effective migration time of the ligand can change between $t_{L \cdot T}$ and t_L depending on K_d and [T], which is assumed to be equal to the overall concentration of the target [T] because of the constant supply of target from the run buffer. As a result, the interaction of the library with the constant flow of the target distributes ligands along the capillary according to their K_d values:

$$K_d(t) = [T]\frac{t_L}{t_{L \cdot T}}\frac{t - t_{L \cdot T}}{t_L - t} \tag{9.3}$$

Ligands with the same K_d values migrate as a single peak with a small width (Figure 9.3); thus, according to equation (9.3), the theoretical range of K_d values in a fraction collected between times t_1 and t_2 will be

$$[T]\frac{t_L}{t_{L \cdot T}}\frac{t_2 - t_{L \cdot T}}{t_L - t_1} > K_d > [T]\frac{t_L}{t_{L \cdot T}}\frac{t_1 - t_{L \cdot T}}{t_L - t_1} \tag{9.4}$$

In an experiment, peak broadening and nonspecific interactions with other components in the system may introduce a bias of unbound ligands into the fraction collected. That is why the selection procedure may require a few rounds of SELEX to approach the theoretically predicted K_d values.

It is necessary to know t_L and $t_{L \cdot T}$ for ECEEM-based selection of smart aptamers. These two parameters can be obtained from a single NECEEM experiment. If the naive library has a bulk affinity that allows for detecting the peak of the complex, obtaining t_L and $t_{L \cdot T}$ is trivial. If bulk affinity of the naive library is too low to detect the peak of the complex, NECEEM-based partitioning can be used to obtain an enriched library for which the complexes are detectable, and such complexes can be used to find $t_{L \cdot T}$. In this case, NECEEM improves the capabilities of ECEEM. In general, different KCE methods can augment each other's analytical capabilities.

The unique feature of ECEEM for the selection of aptamers arises from its simple mathematical description. The range of the affinity distribution along the capillary is determined by [T], $t_{L \cdot T}$, and t_L. Changing these parameters can theoretically facilitate the selection of aptamers with affinities ranging from picomolar to millimolar values of K_d. It should be noted that the simple mathematical description is applicable only under an assumption of quasiequilibrium in the system. The reequilibration time in reaction (9.1) should be much shorter than the characteristic time of electrophoretic separation.

Example   So far, experimental evaluation of ECEEM has been conducted only for the selection of DNA aptamers for MutS protein (Drabovich et al., 2005, 2006). The migration times of DNA, t_L, and of the complexes, $t_{L \cdot T}$, were found from a NECEEM experiment with the naive DNA library (Figure 9.10A). Aptamer fractions were then collected within three ligand-collection windows (Figure 9.10B) in three rounds of ECEEM-based SELEX. Each round consisted of ECEEM separation, fraction collection, PCR amplification, separation of strands, and measuring the bulk affinity of the enriched library with NECEEM. The progression of K_d during three consecutive rounds of ECEEM-based SELEX is shown in Figure 9.10B. The solid curve illustrates the theoretical dependence of K_d on migration time, calculated using equation (9.3). The naive DNA library had a bulk affinity to MutS of approximately 2 μM. In only three rounds of SELEX, the affinity of the enriched aptamer pool approached the theoretically predicted value. The difference in the affinity of the pools was over two orders of magnitude. These three pools provided rich material for obtaining a large panel of aptamers with significantly varying K_d values for multiaptamer calibration-free affinity analysis of MutS with an ultrawide dynamic range (see Section 9.4.2).

9.2.3   Optimization of PCR

NECEEM- and ECEEM-based methods of aptamer partitioning have exceptionally high efficiency. For the overall aptamer-selection procedure to be efficient, however, the high efficiency of the new partitioning methods has to be matched by a high efficiency of PCR. PCR amplification of random DNA libraries used in aptamer selection has been studied (Musheev and Krylov, 2006). Using CE as an analytical tool, fundamental differences were found between PCR amplification of homogeneous DNA templates and that of large libraries of random DNA sequences. For a homogeneous DNA template, product formation proceeds until primers are exhausted, whereas for a random DNA library as a template, product

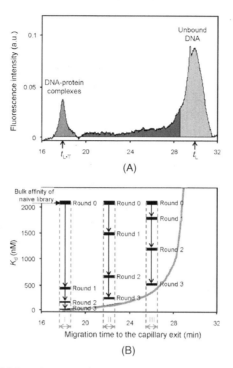

Figure 9.10   ECEEM-based selection of smart aptamers with predefined K_d for MutS protein: (A) a NECEEM electropherogram for the determination of t_L and $t_{L \cdot T}$, and (B) the progression of K_d of pools collected in three consecutive rounds of SELEX within time windows I, II, and III. The curve shows the theoretical dependence of K_d on migration time defined by equation (9.3).

accumulation stops when PCR primers are still in excess of the products. The products then convert rapidly to by-products and virtually disappear after only five additional cycles of PCR. Interestingly, the yield of the products decreases with the increasing length of DNA molecules in the library. It was also proven that the initial number of DNA molecules in the PCR mixture has no effect on the rate of by-product formation. The increase in the *Taq* DNA polymerase concentration in the PCR mixture increases the yield of PCR products with respect to that of by-products. These findings indicate that standard procedures of PCR amplification of homogeneous DNA samples cannot be transferred directly to the PCR amplification of random DNA libraries. To ensure highly efficient selection, PCR has to be optimized for the amplification of random DNA libraries.

Asymmetric PCR with the excess of the primer that is extended to aptamer candidates over the primer that is extended to complementary aptamer sequences has been used successfully in quick testing of affinity between rounds of SELEX. Asymmetric PCR is slower, but it allows us to avoid the strand separation procedure, as complementary strands are generated in small amounts and affect measurements only slightly. Asymmetric PCR has also been tested for the preparative preparation of enriched libraries in SELEX and can further speed up the procedure and reduce its cost.

9.2.4 Future of KCE Methods for Aptamer Selection

So far, only NECEEM and ECEEM have been evaluated for aptamer selection and only in application to the selection of DNA aptamers. Despite the demonstration of outstanding results, the application of KCE methods to aptamer selection is still in its infancy. The view of aptamers as useful analytical and therapeutic tools guarantees that work on KCE-based selection of aptamers will continue and other KCE methods will be adapted for this application.

Non-SELEX is an attractive alternative to SELEX, due to increased speed and suitability for easy automation. Non-SELEX leads to a final pool with all sequences being unique. The analysis of such sequences would be much more efficient if single-molecule sequencing were used. In this respect, the current effort in the development of single-molecule sequencing technologies is quite encouraging (Bayley, 2006).

If Non-SELEX proves to be applicable to the selection of smart aptamers, the technology will eventually be adapted to the selection of smart ligands from nonamplifiable libraries of different natures. The challenges in such an application lie in the need (1) to separate the library from the target without separating the components of the library themselves and (2) to arrange for sensitive detection when fluorescence detection cannot be used (it is difficult to label small molecules fluorescently without affecting their binding ability to the target). In this respect, NECEEM and ECEEM can be applied immediately to the selection of ligands from small DNA- or RNA-encoded libraries in which the oligonucleotide sequence barcodes the structure of the small molecule to which it is attached (Garner et al., 2004; Kanna et al., 2004). Oligonucleotide tags (rather than small molecules) define the electrophoretic properties of such libraries; therefore, the electrophoretic properties of the encoded libraries are similar to those of pure oligonucleotide libraries. For the same reason, NECEEM and ECEEM can also be applied directly to RNA-display libraries. KCE methods promise great potential in the development of smart drug candidates and smart affinity ligands of different natures.

9.3 MEASUREMENTS OF BINDING PARAMETERS OF TARGET–APTAMER INTERACTION BY KCE METHODS

9.3.1 Foundation

Finding k_{on}, k_{off}, K_d, ΔS, and ΔH in a fast and accurate fashion is important for assessing the progression of selection as well as for final characterization of aptamers. A single NECEEM electropherogram contains sufficient data for finding both K_d and k_{off}. To increase the accuracy of measurements, multiple NECEEM experiments can be performed with varying concentrations of target and aptamer.

NECEEM starts with the equilibrium mixture in which the concentrations of L, T, and L · T are controlled by K_d: $K_d = [L][T]/[L \cdot T]$. Therefore, NECEEM has a memory of the initial equilibrium that is necessary for finding K_d. The complex dissociates during NECEEM and the kinetics of complex dissociation is recorded in the decay parts of the electropherogram, allowing for the calculation of k_{off}. The areas of peaks and decay areas in a NECEEM electropherogram are proportional to the amounts of corresponding species. A single NECEEM electropherogram can be used to find the four measurable parameters (i.e., A_1, A_2, A_3, and $t_{L \cdot T}$) needed for the determination of K_d and k_{off} (Figure 9.11). A_1 is the area of the peak corresponding to L, which was unbound in the equilibrium mixture. A_2 is the area of the exponential decay zone left by L that had dissociated from L · T during electrophoresis. A_3 is the area of the peak corresponding to L · T that was still intact at the time of passing the detector. Finally, $t_{L \bullet T}$ is the migration time of the complex. The values of K_d and k_{off} can be calculated using the following algebraic formulas or their variations (Berezovski and Krylov, 2002; Krylov and Berezovski, 2003):

$$K_d = \frac{[T]_0(1 + A_1/(A_2 + A_3)) - [L]_0}{1 + (A_2 + A_3)/A_1} \tag{9.5}$$

and

$$k_{off} = \ln \frac{A_2 + A_3}{A_2} \bigg/ t_{L \bullet T} \tag{9.6}$$

where $[T]_0$ and $[L]_0$ are total concentrations of T and L in the equilibrium mixture: $[T]_0 = [T] + [L \cdot T]$ and $[L]_0 = [L] + [L \cdot T]$. Advantageously, only the areas and the migration time associated with a single species are required. This simplifies the use of fluorescence detection since fluorescent labeling of oligonucleotide aptamers is relatively easy (either by organic synthesis or by incorporating a fluorescently labeled primer in PCR). Labeling a target (e.g., protein) is not as trivial, but luckily, it is not required for NECEEM-based measurements of K_d and k_{off}. In general, a major step in the method development for NECEEM involves finding conditions for good-quality separation of L from L · T. In the instance of oligonucleotide aptamers and protein targets, such separation is achieved easily in simple buffers, due to a dramatic difference in charge/size ratios between DNA and the majority of proteins.

If fluorescence detection is used, the potential change of the quantum yield of fluorescence upon binding of a fluorescently labeled aptamer to the target should be taken into consideration. Relative quantum yields of fluorescence can be measured in NECEEM in a simple procedure and included in equations (9.5) and (9.6) to correct for its change through dividing the areas by the relative quantum yield (Krylov and Berezovski, 2003). When on-column detection is

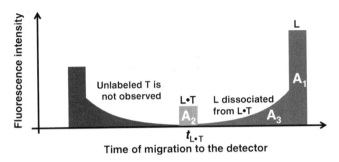

Figure 9.11 NECEEM electropherogram illustrating finding the four parameters needed for quantitative measurements: A_1 is the peak area of L that was free in the equilibrium mixture, A_2 is the peak area of intact $L \cdot T$ at the time of its passing the detector, A_3 is the peak area of L dissociated from $L \cdot T$ during the separation, and $t_{L \cdot T}$, is the migration time of $L \cdot T$ from the capillary inlet to the detector.

used, the areas must be divided by the migration times of corresponding species. These rules originate from the basic CE principles and are common for all KCE methods.

The equilibrium mixture is typically prepared in an incubation buffer, whereas dissociation occurs in an electrophoresis run buffer. The values of K_d and k_{off} are thus measured for the incubation and run buffers, respectively. If the incubation buffer and the electrophoresis run buffer are identical, K_d and k_{off} are determined under the same set of conditions and k_{on} can be calculated as $k_{on} = k_{off}/K_d$. It is typically possible to match the incubation and run buffers for aptamer selection, as both the formation of protein–oligonucleotide complexes and electrophoresis are well suited to TRIS-based buffers. An example of difficulty in such matching is when T is a protein requiring the use of a high salt concentration to preserve its native structure. CE cannot tolerate high salt concentrations in the run buffer, due to the high Joule heating, which can lead to increased temperature inside the capillary. The increased temperature can decrease the quality of separation and lead to incorrectly measured binding parameters. Overly high temperatures can also denature the protein target. This emphasizes the need for temperature control inside the capillary.

9.3.2 Temperature Control inside the Capillary

The ability to control the temperature inside the capillary depends on the quality of heat exchange between the buffer inside the capillary and the environment outside the capillary. The best heat exchange efficiency is achieved through washing the capillary with a liquid heat exchanger. This approach has been implemented in some commercially available CE instruments. The amount of heat generated in CE depends on the buffer, the diameter of the capillary, and the electric field used. Depending on the amount of heat generated, heat exchange can be more or less efficient, and the temperature inside the capillary can differ to a certain degree from that of the heat exchanger. This emphasizes the need for accurate measurement of the temperature inside the capillary. Typically, spectroscopic methods are used for these measurements. Spectroscopic methods rely on the dependence of spectroscopic properties of molecular probes (such as fluorophores) on temperature. Although easy to use, these methods have two serious limitations. First, they measure temperature only at the detection point, whereas in most instruments, thermostabilization of the capillary at the detection point is less efficient than in the rest of the capillary. This makes spectroscopic measurements inaccurate. Second, a calibration curve for spectroscopic methods needs to be built with the same instrument, which is to be calibrated. On the other hand, a calibration curve can be built only with a verified instrument. This contradiction is like "the chicken and the egg" problem: To calibrate a nonverified instrument, one has to assume that it is verified. The two limitations make spectroscopic methods of temperature determination largely impractical.

KCE on its own offers an alternative to spectroscopic methods; KCE-based measurements are based on the dependence of one of the binding parameters (k_{on}, k_{off}, or K_d) on temperature. First, a calibration curve of binding parameter versus temperature is built for an affinity pair of choice, which can, for example, be an aptamer and its protein target. Advantageously, the calibration curve does not need to be built with a CE instrument in question. It can be built with another (verified) CE instrument or with a non-CE technique such as SPR. Second, the same binding parameter is measured in the CE instrument in question and the calibration curve is used to find the temperature inside the capillary. In addition to solving the chicken and the egg problem, the KCE approach also measures an effective temperature in the total volume of the capillary, which is a more relevant value than that measured at the detection point.

9.3.3 Examples

Measurement of K_d and k_{off} NECEEM has been used so far to measure K_d and/or k_{off} for the interaction of (1) DNA aptamers with their protein targets (Berezovski et al., 2005, 2006a,b; Drabovich et al., 2005, 2006); (2) proteins with nonaptamer DNA (Berezovski and Krylov, 2002; Krylov and Berezovski, 2003; Okhonin et al., 2004a,b 2006; Krylova et al., 2005; Pang et al., 2006; Woolley et al., 2006; Larijani et al., 2007); and (3) protein–peptide interaction (Yang et al., 2005). Figure 9.12 shows a representative experimental NECEEM electropherogram for the interaction between fluorescently labeled ssDNA and a protein. When defining the areas, it is important to define boundaries between them accurately. The boundary between A_1 and A_2 is typically more difficult to define visually, but it can be found by comparing the peaks of free aptamer in the presence and absence of the target. Our estimates show that the uncertainty in defining the boundaries between the areas leads to experimental errors in the range of 10%, which is acceptable for most applications. Mathematical modeling of a NECEEM electropherogram can be used to find both K_d and k_{off} from

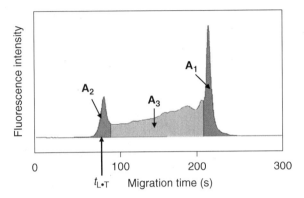

Figure 9.12   Example of an experimental NECEEM electropherogram with four parameters needed for quantitative measurements (see the Figure 9.10 legend for details).

nonlinear regression analysis without a need to define the areas (Okhonin et al., 2004a; Petrov et al., 2005). Most practitioners of NECEEM use the area method, as it is simple, fast, and reasonably accurate.

Determination of K_d and k_{off} based on NECEEM is not only fast and accurate, but is also characterized by a wide and adjustable dynamic range. The upper limit of measurable K_d values depends on the highest concentration of T available and can be as high as millimolar. Such a high upper limit allows for the measurement of K_d values for very low affinities: for example, bulk affinities of naive oligonucleotide libraries (Berezovski et al., 2005, 2006a,b; Drabovich et al., 2005, 2006). The lower limit of measurable K_d depends on the concentration limit of detection. For fluorescence detection, it can be as low as picomolar. The dynamic range of measurable k_{off} values is defined by the migration time of the aptamer–target complex, which can easily be regulated by the length of the capillary, electric field, or electroosmotic velocity. The practically proven dynamic range of k_{off} spans from 10^{-4} to 1 s^{-1} (Berezovski and Krylov, 2002; Krylov and Berezovski, 2003; Yang et al., 2005).

Measurement of Temperature Inside the Capillary   The use of NECEEM for measuring temperature inside the capillary has been demonstrated experimentally by measuring k_{off} as a function of temperature for the interaction between SSB protein and fluorescently labeled ssDNA (Berezovski and Krylov, 2004). The calibration curve was built with a verified Beckman MDQ CE instrument, which uses a liquid heat exchanger to remove excess heat efficiently from the capillary. This calibration curve was then used to measure the temperature inside the capillary in a custom-built instrument with the capillary exposed to the ambient atmosphere at 20°C. It was found that the temperature inside the capillary was 15°C higher than the ambient temperature. As affinity interactions are very sensitive to temperature, the unjustified assumption that the capillary temperature is equal to the ambient temperature could potentially lead to dramatic misinterpretation of experimental results. KCE with a well-controlled temperature inside the capillary can be used to study thermochemistry and measure the ΔH and ΔS values of affinity interactions.

Thermochemistry of Taq DNA Polymerase Binding Its Aptamer
NECEEM with reliable temperature control was used for detailed study of the thermochemistry of interaction between *Taq* DNA polymerase and its DNA aptamers. *Taq* DNA polymerase is widely used in PCR and an aptamer for the enzyme was previously selected for hot-start PCR. It binds the enzyme and inhibits it at low temperatures, thus slowing down the amplification of nonspecific DNA hybrids. It dissociates from the enzyme, however, at higher temperatures, making the enzyme functional. Figure 9.13 depicts the temperature dependencies of three binding parameters, k_{on}, k_{off}, and K_d, and van't Hoff plots for the calculation of ΔH and ΔS (Berezovski and Krylov, 2005). The value of K_d is constant below 36°C but grows rapidly as the temperature surpases 36°C. This indicates that the equilibrium stability of the complex decreases when the temperature exceeds 36°C. Two reasons may account for this decrease in stability: first, the decreased rate of complex formation, k_{on}, and second, the increased rate of complex dissociation, k_{off}. In the example presented, k_{off} changes gradually following the general Arrhenius trend. In contrast, k_{on} changes abruptly at 36°C, indicating that the equilibrium stability decreases due to the decreased rate of complex formation. The most likely explanation of this phenomenon is that the secondary structure of the aptamer starts "melting" at 36°C. On the other hand, if the aptamer is already bound to the protein, the protein stabilizes its secondary structure and the temperature increase does not lead to structural changes. This example proves that knowledge of temperature dependencies of k_{on} and k_{off}, which can be obtained using KCE methods, helps us to understand mechanisms of temperature dependencies of affinity complex stability.

9.4 QUANTITATIVE AFFINITY ANALYSIS OF A TARGET USING APTAMER AS AN AFFINITY PROBE

9.4.1 Foundation

An unknown concentration of a spectroscopically invisible target can be measured through its affinity reaction with a spectroscopically visible aptamer; the aptamer is used here as an affinity probe. Typically, in such analyses, the kinetic and equilibrium binding parameters of interaction between the target and its aptamer are unknown; instead, a calibration curve is built for the dependence of a signal (e.g., fluorescence, absorbance, radioactivity, etc. of the aptamer) on the concentration of target. Classical affinity methods thus have a general disadvantage: A new calibration curve has to be built for every new concentration of aptamer.

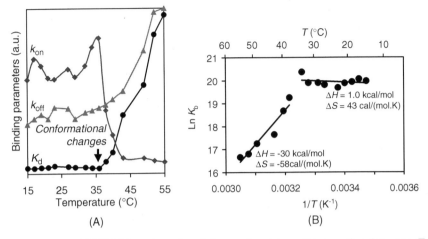

Figure 9.13 NECEEM-based study of thermochemistry of interaction between *Taq* DNA polymerase and its DNA aptamer. (A) Temperature dependencies of k_{on}, k_{off}, and K_d; (B) a van't Hoff plot used for the determination of ΔS and ΔH for this interaction.

KCE methods provide a calibration-free alternative to classical affinity analyses utilizing aptamers. The general theory of such measurements has not yet been developed, but their feasibility is quite evident, as concentrations of T and L in the starting solution (or in the equilibrium mixture) are parameters in equations used for the determination of k_{on}, k_{off}, and K_d [see, e.g., equation (9.5) and equations in Okhonin et al., 2004a and 2006]. When determining the binding parameters, the concentrations of L and T are assumed to be known. This consideration can be reversed: If the binding parameters and the concentration of L are known, the unknown concentration of T can be found. KCE methods using an apparatus for simple mathematics [NECEEM, SweepCE; Okhonin et al., 2004b and ppKCE (Okhonin et al., 2006)] are immediately applicable to calibration-free affinity measurements of the unknown concentration of T. It is important to emphasize that every KCE method can be used for quantitative measurements of concentrations if the kinetic nature of the method is appreciated and the experimental results are analyzed accordingly. Here, only affinity measurements with NECEEM are described.

NECEEM appears to be the simplest KCE method for affinity-based measurements of the concentration of the target using the aptamer as an affinity probe. If the K_d value is known, the unknown concentration of the target can be found from an algebraic equation obtained by rearranging equation (9.5):

$$[T]_0 = K_d \frac{A_2 + A_3}{A_1} + [L]_0 \frac{1}{1 + A_1/(A_2 + A_3)} \qquad (9.7)$$

where areas A_1, A_2, and A_3 refer to those in Figure 9.11. As is seen from equation (9.7), NECEEM does not require a typical calibration procedure because the K_d value serves as a "calibration" parameter. This equation includes the total concentration of the aptamer, $[L]_0 = [L] + [L \cdot T]$, and requires no recalibration if it changes (the change of $[L]_0$ is often required to change the dynamic range of the method). Moreover, the method can be used even if the $L \cdot T$ complex completely dissociates during separation. In this case, $A_2 \approx 0$ and equation (9.7) reduces to (Berezovski et al., 2003)

$$[T]_0 = K_d \frac{A_3}{A_1} + [L]_0 \frac{1}{1 + A_1/A_3} \tag{9.8}$$

Due to this important feature, aptamers with high k_{off} values can still be used for quantitative affinity analyses by NECEEM. This also makes the method applicable to systems in which $L \cdot T$ migrates so slowly that $L \cdot T$ dissociates to an undetectable level by the time it reaches the detector.

When the k_{off} value is much lower than the reciprocal migration time of $L \cdot T$, no detectable dissociation of $L \cdot T$ occurs and only the peaks of L and $L \cdot T$ are observed (no decay zones are detected). Thus, A_3 can be assumed to equal zero and equation (9.7) can be reduced to:

$$[T]_0 = K_d \frac{A_2}{A_1} + [L]_0 \frac{A_2}{A_1 + A_2} \tag{9.9}$$

If a mixture of several aptamers with different K_d values is used in an aptamer-based multiprobe affinity analysis (see below), the concentration of the target can be found by solving the following algebraic equation:

$$\sum_{i=1}^{n} \frac{[L]_{0_i}}{K_{d_i} + [T]_0 - f \sum_{j=1}^{n} [L]_{0_j}} = \frac{f \sum_{i=1}^{n} [L]_{0_i}}{[T]_0 - f \sum_{i=1}^{n} [L]_{0_i}} \tag{9.10}$$

where K_{d_i} and $[L]_{0_i}$ are the parameters of the ith aptamer ($i = 1, \ldots, n$) and f is a fraction of bound aptamers:

$$f = \frac{\sum_{i=1}^{n} [L_i \cdot T]}{\sum_{i=1}^{n} [L_i]} \tag{9.11}$$

The value of f can be found experimentally from a NECEEM electropherogram using the formula

$$f = \frac{A_1 + A_2}{A_1 + A_2 + A_3} \tag{9.12}$$

NECEEM-based quantitative affinity analysis is simple, fast, and accurate. The limit of detection is defined by the sensitivity of the detection system used. For best systems utilizing laser-induced fluorescence detection, the mass limit of detection can be as low as 1000 molecules and the concentration limit of detection can reach picomolar.

DNA aptamers as affinity probes considerably enrich the capabilities of KCE in analyses of a variety of targets. Aptamers have a number of advantages over antibodies. First, DNA aptamers are highly negatively charged and thus do not adhere to the bare silica of uncoated capillaries. Second, aptamers are relatively small molecules; this simplifies the separation of free DNA (L) from the DNA–target complex (L·T). Third, aptamers can be synthesized chemically; moreover, this synthesis is now available commercially at very low cost. Fourth, aptamers can easily be end-labeled with fluorophores without affecting the aptamer–target interaction, which facilitates their fluorescence detection. If constant regions (used for PCR amplification) are not truncated, quantitative PCR can be used to detect aptamers quantitatively, thus improving the limit of detection to as few as 100 molecules (Zhang et al., 2006). Advantageously, aptamers can be selected and characterized using KCE methods (see above); thus a single instrumental and conceptual platform can be used both for obtaining aptamers and for using them analytically.

9.4.2   Example

An important example of NECEEM-based quantitative affinity assay is the analysis of a protein with an ultrawide dynamic range (Drabovich et al., 2007). Protein concentration can vary over several orders of magnitude in many physiological and pathological processes; therefore, studies of these processes require affinity analysis of proteins with a very wide dynamic range of accurately measured concentrations. Such an analysis can be realized with multiple affinity probes that bind the target with significantly different values of K_d (Marvin et al., 1997). Every probe in a multiprobe affinity analysis is responsible for detection of the target in the range of concentrations of one to two orders of magnitude around its K_d value. A multiprobe affinity analysis has been demonstrated only for small-molecule targets (Marvin et al., 1997; Ohmura et al., 2003). Such an analysis of proteins was out of reach due to the lack of high-selectivity affinity probes with a wide range of K_d values. Obtaining antibodies with a K_d range of two orders of magnitude or more is virtually impossible without compromising their selectivity. Aptamers provide a vital alternative. Smart DNA aptamers for MutS protein with predefined K_d values in a range of over two orders of magnitude were selected using the ECEEM methods (Drabovich et al., 2005, 2006) (see Section 9.2.2). Three aptamers with K_d values of 7.6, 46, and 810 nM were used successfully to develop the first aptamer-based multiprobe affinity analysis (Drabovich et al., 2007).

Using NECEEM, binding curves f versus $[T]_0$ for the three individual aptamers (Figure 9.14A) and for their equimolar mixture (Figure 9.14B) were

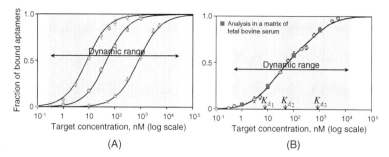

Figure 9.14 Concentration dynamic range for the affinity analysis of MutS protein using: (A) 1 nM of three individual aptamers with K_d values of 7.6, 46, and 810 nM, and (B) a mixture of 1 nM of these aptamers. Solid lines were calculated with equation (9.10) using the values of K_d and the concentrations shown. All points except for the squares in part (B) correspond to analyses of MutS in the bare TRIS–acetate buffer. The squares in (A) correspond to analyses of MutS in the presence of fetal bovine serum in the buffer.

built. The theoretical curves fit the experimental data perfectly, suggesting that aptamers bind MutS with 1:1 stoichiometry and that no interference occurs between the aptamers in the mixture. The perfect fit of the experimental data into the theoretical curve also suggests that building calibration curves is not necessary in multiaptamer affinity analysis; equation (9.10) can be used directly to find unknown $[T]_0$ using known K_d, f, and $[L]_0$. As shown in Figure 9.14, the dynamic range for single-probe analysis is approximately two orders of magnitude. The three individual probes cover the overall dynamic range of five orders of magnitude. Analysis based on a mixture of the three probes has a dynamic range of more than four orders of magnitude. These results prove that smart aptamers with predefined K_d values can facilitate multiprobe affinity analysis of proteins with an ultrawide dynamic range. Moreover, multiaptamer analysis can be performed in biological fluids such as serum (see Figure 9.14B), which suggests its potential utility in applications ranging from basic research to clinical analysis.

9.5 CONCLUSIONS

KCE methods represent a new generation of affinity methods for selection, characterization, and analytical utilization of oligonucleotide aptamers. This development can be considered still in its infancy, with only the surface of capabilities being scratched. Much more needs to be done to understand how methods other than NECEEM and ECEEM can be used for the selection of aptamers and what their properties are in other applications. KCE itself requires a great deal of work to establish a complete space of KCE methods and classify them. A new nomenclature is required for KCE methods to become attractive to a wide community of molecular scientists. Joint work of teams that use electrophoresis for affinity studies as well as those that select aptamers and apply them in analysis and therapy will be needed.

REFERENCES

Al-Mahrouki, A. A., Krylov, S. N. (2005). Calibration-free quantitative analysis of mRNA. *Anal Chem* 77, 8027–8030.

Bayley, H. (2006). Sequencing single molecules of DNA. *Curr Opin Chem Biol* 10, 628–637.

Berezovski, M., Krylov, S. N. (2002). Nonequilibrium capillary electrophoresis of equilibrium mixtures: A single experiment reveals equilibrium and kinetic parameters of protein–DNA interactions. *J Am Chem Soc* 124, 13674–13675.

Berezovski, M., Krylov, S. N. (2003). Using DNA-binding proteins as an analytical tool. *J Am Chem Soc* 125, 13451–13454.

Berezovski, M., Krylov, S. N. (2004). Using nonequilibrium capillary electrophoresis of equilibrium mixtures for the determination of temperature in capillary electrophoresis. *Anal Chem* 76, 7114–7117.

Berezovski, M., Krylov, S. N. (2005). Thermochemistry of protein–DNA interaction studied with temperature-controlled nonequilibrium capillary electrophoresis of equilibrium mixtures. *Anal Chem* 77, 1526–1529.

Berezovski. M., Nutiu, R., Li, Y., Krylov, S. N. (2003). Affinity analysis of a protein–aptamer complex using nonequilibrium capillary electrophoresis of equilibrium mixtures. *Anal Chem* 75, 1382–1386.

Berezovski, M., Drabovich, A., Krylova, S. M., Musheev, M., Okhonin, V., Petrov, A., Krylov, S. N. (2005). Nonequilibrium capillary electrophoresis of equilibrium mixtures: a universal tool for development of aptamers. *J Am Chem Soc* 127, 3165–3171.

Berezovski, M. V., Musheev, M. U., Drabovich, A. P., Jitkova, J. V., Krylov, S. N. (2006a). Non-SELEX: selection of aptamers without intermediate amplification of candidate oligonucleotides. *Nat Protoc* 1, 1359–1369.

Berezovski, M., Musheev, M., Drabovich, A., Krylov, S. N. (2006b). Non-SELEX selection of aptamers. *J Am Chem Soc* 128, 1410–1411.

Boozer, C., Kim, G., Cong, S., Guan, H., Londergan, T. (2006). Looking towards label-free biomolecular interaction analysis in a highthroughput format: a review of new surface plasmon resonance technologies. *Curr Opin Biotechnol* 17, 400–405.

Drabovich, A., Krylov, S. N. (2004). Single-stranded DNA-binding protein facilitates gel-free analysis of polymerase chain reaction products in capillary electrophoresis. *J Chromatogr A* 1051, 171–175.

Drabovich, A., Krylov, S. N. (2006). Identification of base pairs in single nucleotide polymorphisms by Muts protein-mediated capillary electrophoresis. *Anal Chem* 78, 2035–2038.

Drabovich, A., Berezovski, M., Krylov, S. N. (2005). Selection of smart aptamers by equilibrium capillary electrophoresis of equilibrium mixtures (ECEEM). *J Am Chem Soc* 127, 11224–11225.

Drabovich, A., Berezovski, M., Krylov, S. N. (2006). Selection of smart aptamers by methods of kinetic capillary electrophoresis. *Anal Chem* 78, 3171–3178.

Drabovich, A. P., Okhonin, V., Berezovski, M., Krylov, S. N. (2007). Smart aptamers facilitate multi-probe affinity analysis of proteins with ultra-wide dynamic range. *J Am Chem Soc* 129.

Gartner, Z. J., Tse, B. N., Grubina, R., Doyon, J. B., Snyder, T., Liu, D. R. (2004). DNA-templated organic synthesis and selection of a library of macrocycles. *Science* 305, 1601–1605.

Kanan, M. W., Rozenman, M. M., Sakurai, K., Snyder, T. M., Liu, D. R. (2004). Reaction discovery enabled by DNA-templated synthesis and in vitro selection. *Nature* 431, 545–549.

Krylov, S. N., Berezovski, M. (2003). Non-equilibrium capillary electrophoresis of equilibrium mixtures: appreciation of kinetics in capillary electrophoresis. *Analyst* 128, 571–575.

Krylova, S. M., Musheev, M., Nutiu, R., Li, Y., Lee, G., Krylov, S. N. (2005). Tau protein binds single-stranded DNA sequence specifically: the proof obtained in vitro with nonequilibrium capillary electrophoresis of equilibrium mixtures. *FEBS Lett* 579, 1371–1375.

Larijani, M., Petrov, A. P., Krylov, S. N., Martin, A. A. (2007). AID is a sequence-independent single-stranded DNA binding protein: preferential mutation of hot-spot motifs is imposed during catalysis. *J Biol Chem* 27, 20–30.

Liu, Z., Drabovich, A. P., Krylov, S. N., Pawliszyn, J. (2007). Dynamic kinetic capillary isoelectric focusing: new insights into protein–DNA interactions. *Anal Chem* 79, 1097–1100.

Marvin, J. S., Corcoran, E. E., Hattangadi, N. A., Zhang, J. V., Gere, S. A., Hellinga, H. W. (1997). The rational design of allosteric interactions in a monomeric protein and its applications to the construction of biosensors. *Proc Natl Acad Sci U S A* 94, 4366–4371.

Mendonsa, S. D., Bowser, M. T. (2004). In vitro evolution of functional DNA using capillary electrophoresis. *J Am Chem Soc* 126, 20–21.

Mendonsa, S. D., Bowser, M. T. (2005). In vitro selection of aptamers with affinity for neuropeptide Y using capillary electrophoresis. *J Am Chem Soc* 127, 9382–9383.

Mosing, R. K., Mendonsa, S. D., Bowser, M. T. (2005). Capillary electrophoresis-SELEX selection of aptamers with affinity for HIV-1 reverse transcriptase. *Anal Chem* 77, 6107–6112.

Musheev, M. U., Krylov, S. N. (2006). Selection of aptamers by systematic evolution of ligands by exponential enrichment: addressing the polymerase chain reaction issue. *Anal Chim Acta* 564, 91–96.

Ohmura, N., Tsukidate, Y., Shinozaki, H., Lackie, S. J., Saiki, H. (2003). Combinational use of antibody affinities in an immunoassay for extension of dynamic range and detection of multiple analytes. *Anal Chem* 75, 104–110.

Okhonin, V., Berezovski, M., Krylov, S. N. (2004a). Capillary electrophoresis: a non-stopped-flow method for measuring bimolecular rate constant of complex formation between protein and DNA. *J Am Chem Soc* 126, 7166–7167.

Okhonin, V., Krylova, S. M., Krylov, S. N. (2004b). Nonequilibrium capillary electrophoresis of equilibrium mixtures, mathematical model. *Anal Chem* 76, 1507–1512.

Okhonin, V., Petrov, A., Berezovski, M., Krylov, S. N. (2006). Plug-plug kinetic capillary electrophoresis: a method for direct determination of rate constants of complex formation and dissociation. *Anal Chem* 78, 4803–4810.

Pang, Z., Liu, X., Al-Mahrouki, A., Berezovski, M., Krylov, S. N. (2006). Selection of surfactants for cell lysis in chemical cytometry to study protein-DNA interactions. *Electrophoresis* 27, 1489–1494.

Petrov, A., Okhonin, V., Berezovski, M., Krylov, S. N. (2005). Kinetic capillary electrophoresis (KCE): a conceptual platform for kinetic homogeneous affinity methods. *J Am Chem Soc* 127, 17104–17110.

Simpson, J. T., Fisher, R. J. (2005). Measurement of protein–protein interactions using surface plasmon resonance spectroscopy. In *Protein–Protein Interactions*, 2nd ed., pp. 355–375.

Wilson, W. D. (2002). Tech sight: analyzing biomolecular interactions. *Science* 295, 2103–2105.

Woolley, G. A., Jaikaran, A., Berezovski, M., Calarco, J. D., Krylov, S. N., Smart, O. S., Kumita, J. R. (2006). Reversible photo-control of DNA binding by a designed GCN4-bZIP protein. *Biochemistry* 45, 6075–6084.

Yang, P., Whelan, R. J., Jameson, E. E., Kurzer, J. H., Argetsinger, L. S., Carter-Su, C., Kabir, A., Malik, A., Kennedy, R. T. (2005). Capillary electrophoresis and fluorescence anisotropy for quantitative analysis of peptide–protein interactions using JAK2 and SH2-Bβ as a model system. *Anal Chem* 77, 2482–2489.

Zhang, H., Wang, Z., Li, X-F., Le, X. C. (2006). Ultrasensitive detection of proteins by amplification of affinity aptamers. *Angew Chem Int Ed Engl* 45, 1576–1580.

CHAPTER 10

APTAMERS FOR SEPARATION OF ENANTIOMERS

CORINNE RAVELET and ERIC PEYRIN

10.1 INTRODUCTION

Drug enantiomers often differ in their pharmacokinetics (especially metabolism) and in their pharmacodynamics, resulting in stereoselective clinical responses. This has attracted considerable regulatory, academic, and commercial interest toward improving the clinical benefit/risk ratio of the drugs concerned. Increasingly, drug development involves pure enantiomers either de novo or of previously marketed racemic drugs ("chiral switch"). In addition, although living organisms are composed predominantly of L-amino acids, it is now well established that D-amino acids, found in higher-order organisms in the form of free amino acids, peptides, and proteins, play a preponderant role in biochemistry and physiopathology. For example, free D-serine is known to be an important endogenous synaptic regulator, while free D-asparate seems to play a role as a messenger in the maturation and differentiation of Leyding cells (Fuchs et al., 2005). The detection of D-amino acids is also of importance in estimating chronological age in forensic sciences, in assessing food quality, and in analyzing materials of extraterrestrial origin. Therefore, it is of great interest to develop efficient methodologies for the analysis and purification of enantiomers. Notably, various chiral selectors (CSs) have been introduced in liquid chromatography (LC) and capillary electrophoresis (CE) or capillary electrochromatography (CEC) for the separation and quantification of enantiomers.

Three general strategies can be described for enantioselective methods using CSs (Francotte, 2001). The first is to adapt the solute to a particular CS: for

Aptamers in Bioanalysis, Edited by Marco Mascini
Copyright © 2009 John Wiley & Sons, Inc.

example, via the precolumn derivatization of the racemate. The second, the approach used most, involves classically the random screening of CSs or a selection based on experience or from a database for the resolution of a given racemate. These "conventional" CSs include, notably, cyclodextrins, ligand-exchange selectors, cellulose and amylose derivatives, macrocyclic antibiotics, synthetic polymers, alkaloids, or proteins (Francotte, 2001). However, the enantioselectivity is not predetermined for this type of CS. Therefore, a third strategy has been reported, and concerns the development of CSs specifically designed against the racemate to resolve target-specific CSs. To date, the CSs have been designed using three different approaches: molecular imprinting technology, the production of antibodies, or a combination strategy.

At the present time, two categories of combinatorial methodologies have been described for the generation and development of CSs designed specifically against the racemate to resolve. A combination strategy from a small library of low-molecular-mass selectors has been the first combinatorial method described. More recently, a very efficient approach from a very large library of single-stranded oligonucleotides has allowed the creation of a new class of CSs based on nucleic DNA and RNA aptamers. Since the first reports of the systematic evolution of ligands by exponential enrichment (SELEX) procedure by three independent laboratories (Ellington and Szostak, 1990; Robertson and Joyce, 1990; Tuerk and Gold, 1990), the development of the in vitro selection has allowed the discovery of aptamers against various targets, including amino acids, nucleosides, drugs, peptides, proteins, and cells (Jayasena, 1999). The use of aptamers as tools in analytical chemistry is a very promising and exciting field of research, due to their ability to bind target molecules with an affinity equal or superior to those of antibodies. Various analytical aptamer-based formats have been exploited, including ELONA assays (Ito et al., 1998), biosensors (aptasensors) (Potyrailo et al., 1998), aptazymes (Famulok, 2005), flow cytometry (Davis et al., 1996), and separation techniques (German et al., 1998; Deng et al., 2001). Particular applications of aptamers as specific ligands for enantiomeric separation in liquid chromatography and capillary electrophoresis are addressed in this chapter.

10.2 GENERATION AND PROPERTIES OF ENANTIOSELECTIVE APTAMERS

Although in most cases the aptamers were not selected for enantioselective binding, efficient monitoring of the selection procedure has often permitted very high specificity, exemplified by the ability of the aptamer to bind the target stereospecifically. As hypothesized by Famulok and co-workers (Geiger et al., 1996), it seems likely that a SELEX procedure designed for high-affinity binding would result in the isolation of enantiospecific oligonucleotides. Of course, a necessary condition is that the target immobilization to the matrix (e.g., chromatographic support, magnetic beads) must allow adequate exposure of the key enantiomer functional groups to the oligonucleotide pool during in vitro selection. If necessary, and in order to isolate nucleic acid sequences with higher stereospecificity, a SELEX methodology involving counterselection with the non-target enantiomer can be carried out so that nonenantioselective sequences can be removed successfully from the nucleic acid pool (Klussmann et al., 1996). Classically, the in vitro selection of aptamers has been carried out using an enantiopure target immobilized on a matrix. Alternatively, it has been shown recently by Sawai and co-workers that enantiospecific aptamers can also be isolated using a racemic mixture (Shoji et al., 2007). It was shown that although the selection was carried out using a racemic thalidomide derivative, the aptamer clone selected showed a high binding affinity for the R-form of thalidomide but not for the S-form. This result is of great interest, as it therefore appears possible to generate enantioselective aptamers even using a nonresolved racemic mixture of compounds (Shoji et al., 2007).

Various enantioselective DNA and RNA aptamers have been isolated against chiral compounds. They are commonly characterized by dissociation constants in the millimolar to submicromolar range, and these binding affinities are frequently associated with extreme stereoselectivity. For example, enantioselectivity values ranging from about 100 to more than 10,000 have been reported for amino acids and derivatives (histidine, tryptophane, arginine, tyrosinamide), nucleosides (adenosine), drugs (thalidomide), or oligopeptides (vasopressin) (Geiger et al., 1996; Klussmann et al., 1996; Williams et al., 1997; Michaud et al., 2004; Majerfeld et al., 2005; Majerfeld and Yarus, 2005; Shoji et al., 2007). It is very likely that such very impressive stereoselectivity originates from the adaptive conformational transition that is classically associated with aptamer–target complex formation (Patel et al., 1997). This allows the aptamer-binding pocket to adjust its recognition surface with the target enantiomer and to maximize complementarity through tightly packed contacts (Patel et al., 1997; Michaud et al., 2003). In contrast, the nontarget enantiomer is assumed to engage fewer stabilizing contacts with the oligonucleotidic pocket, resulting in a lower binding affinity. From a kinetic point of view, it is also important to point out that target–aptamer complex formation is generally characterized by small rate constants (Gebhardt et al., 2000), as expected for an association process that includes rate-limiting slow structural changes mediating the binding surface complementarity.

10.3   IMMOBILIZED APTAMERS FOR ENANTIOMERIC SEPARATION BY LIQUID CHROMATOGRAPHY

Nucleic acid aptamers were first exploited by our group for use in target-specific chiral separations by LC. Immobilized DNA, RNA and L-RNA (spiegelmer) oligonucleotides have been used successfully for the separation of enantiomers of various compounds. Table 10.1 summarizes the aptamer-based CSPs reported to date.

TABLE 10.1 Aptamers Used as Target-Specific Chiral Stationary Phases for LC Enantiomeric Separation

Oligonucleotide	Target	Separation System	α^a/h^b	Stability
DNA	Oligopeptide (arginine-vasopressin)	Narrowbore LC	nd^c/35–40	≥5 months
	Adenosine	Micro-LC	3,6/nd (20°C)	≥2 months
DNA	Tyrosinamide	Micro-LC	80/nd (20°C)	1 month
RNA	Arginine	Micro-LC	nd/35–60	few days
L-RNA	Arginine	Micro-LC	nd/nd	≥2 months
	Aromatic amino acids and derivatives	Micro-LC	1,3–30/25–120 (10°C)	≥3 months
	Histidine	Micro-LC	20/nd (10°C)	≥2 months

$^a\alpha$, apparent enantioselectivity.
^bh, reduced theoretical plate height (efficiency decreases as h increases).
^{c}nd, not determined. The too-low retention factor of the nontarget enantiomer cannot permit accurate calculation of the apparent enantioselectivity.

10.3.1 Stationary-Phase Preparation and Column Packing

In all cases, aptamer immobilization on chromatographic support has been achieved using the well-known biotin–streptavidin bridge. This type of ligand immobilization was chosen because the procedure appears to be easy and rapid to perform and the biotin–streptavidin bridge is known to maintain the receptor (aptamer)-binding ability efficiently (Deng et al., 2001). Prior to immobilization, 5′-biotinylated aptamers were renaturated by heating oligonucleotides in an aqueous buffer, similar to that of the selection medium, and left to stand at room temperature for 30 minutes. Immobilization of aptamers was commonly attained by mixing streptavidin POROS chromatographic particles (20 μm) with the aptamer solution for several hours. Subsequently, the aptamer-modified particles were packed in-house, in most cases into microbore columns (internal diameters in the range 300 to 800 μm). The amount of biotinylated aptamer immobilized was found to be about 15 to 20 nmol per 100 μL of column volume. Nonmodified streptavidin particles were also packed into the microbore columns as control stationary phases.

10.3.2 DNA Aptamer-Based CSPs

The first example of aptamer-based target-specific chromatographic chiral separation has been reported using an antioligopeptide DNA aptamer. The enantiomers of arginine-vasopressin were separated using an immobilized 55-base DNA aptamer (Michaud et al., 2003) known to bind the D-enantiomer of the oligopeptide stereopecifically (Williams et al., 1997). The influence of various chromatographic parameters (such as column temperature, eluent pH, and salt concentration) on L- and D-peptide retention was investigated to provide information about the binding mechanism and then to define the utilization conditions of the aptamer column. Very important apparent enantioselectivity was observed, the nontarget ennatiomer being roughly not retained by the column. More, it was demonstrated by a complete thermodynamic analysis that both dehydration at the binding interface, charge–charge interactions, and adaptive conformational transitions contributed to the specific D-peptide–aptamer complex formation. Furthermore, it was established that the aptamer column was stable during an extended period of time (more than five months). Due in part to the slow association–dissociation kinetics mediated by the aptamer conformational change, low efficiency was observed for the D-enantiomer, as exemplified by its high reduced plate height (Table 10.1). However, it was also shown that the efficiency as well as the peak shape can be improved significantly by decreasing the flow rate (Michaud et al., 2003).

From a practical point of view, it was fundamental to extend the applicability of such target-specific aptamer CSP to the resolution of small bioactive molecule enantiomers. So, in a subsequent work, such an approach has been applied to the chiral resolution of small molecules of biological interest (Michaud et al., 2004). The DNA aptamers used as test CSs have been selected against D-adenosine (Huizenga and Szostak, 1995) and L-tyrosinamide (Vianini et al., 2001). An apparent enantioseparation factor of about 3.5 (at 20°C) was observed for the anti-D-adenosine aptamer chiral stationary phase, while a very high enantioselectivity was obtained with the immobilized anti-L-tyrosinamide aptamer (ca. 80 at 20°C). This made possible baseline resolution even at a relatively high column temperature (about 30°C). Representative chromatograms are presented in Figure 10.1 for the two aptamer-based CSPs. The anti-D-adenosine aptameric stationary phase can be used for two months without loss of selectivity, while some performance degradation was observed for the anti-L-tyrosinamide column over this period (Table 10.1) (Michaud et al., 2004).

10.3.3 RNA Aptamer-Based CSPs and the Mirror-Image Strategy

Most of the aptamers reported in the literature are related to RNA sequences (70% of aptamers are RNA oligonucleotides). Thus, the enantioselective properties and the stability of an anti-L-arginine D-RNA aptamer target-specific chiral stationary phase have been tested (Brumbt et al., 2005). It was found that such an immobilized ligand was very quickly degraded by RNases under usual chromatographic utilization and storage. To overcome this severe limitation for practical use, it appeared fundamental to develop an RNA molecule intrinsically resistant to the classically cleaving RNases. A very interesting strategy involving the mirror-image approach has been developed to design biostable L-RNA ligands (also called *spiegelmers*) for potential therapeutic or diagnostic applications (Klussmann et al., 1996). As the structure of nucleases is inherently chiral, the RNases accept only a substrate in the correct chiral configuration (i.e., the "natural" D-oligonucleotide) (Figure 10.2).

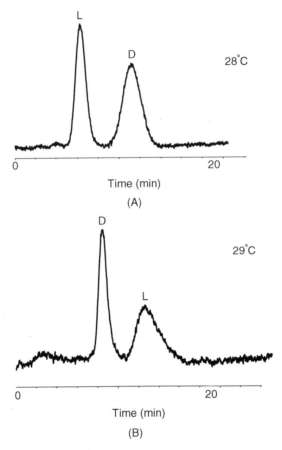

Figure 10.1 (A) Chromatographic resolution of adenosine enantiomers using a D-adenosine target-specific DNA aptamer as CSP: column, 370×0.76 i.d. mm; mobile phase, phosphate buffer 20 mM, KCl 25 mM, MgCl$_2$ 1.5 mM (pH 6.0); amount of D-,L-adenosine injected, 70 pmol; injection volume, 100 nL; flow rate, 50 μL min^{-1}; detection at 260 nm. (B) Chromatographic resolution of tyrosinamide enantiomers using an L-tyrosinamide target-specific DNA aptamer as CSP: column, 250×0.76 i.d. mm; mobile phase, phosphate buffer 20 mM, KCl 25 mM, MgCl$_2$ 1.5 mM (pH 6.0); amount of D-,L-tyrosinamide injected, 70 pmol; injection volume, 100 nL; flow rate, 20 μL min^{-1}; detection at 224 nm.

So L-oligonucleotides are expected to be unsusceptible to naturally occurring enzymes. This concept has been applied successfully to create a biostable RNA CSP. It was demonstrated that a chiral stationary phase based on L-RNA (i.e., the mirror image of the natural D-RNA aptamer) was stable for an extended period of time (about 1600 column volumes of mobile phase) under the usual chromatographic conditions of storage and experiments. In addition, as expected

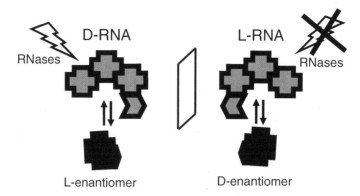

Figure 10.2   Mirror-image strategy for the generation of biostable L-RNA aptamers (spiegelmers).

from the principle of chiral inversion (i.e., the mirror image of the natural aptamer recognizes with the same affinity and specificity the mirror image of the target), D-arginine interacted with the L-RNA stationary phase, whereas L-arginine was not retained significantly by the column. This was responsible for the reversed elution order of enantiomers relative to that obtained using the various D-RNA columns (Figure 10.3) (Brumbt et al., 2005).

Subsequently, an immobilized anti-D-histidine L-RNA aptamer microcolumn has been described, and the retention and enantioselective properties of the CSP have been evaluated (Ruta et al., 2007a). The minimal sequence of the 40-mer RNA aptamer used was obtained via a SELEX procedure directed against L-histidine. This oligonucleotide was known to discriminate strongly between L- and D-histidine ($K_d \sim 10$ μM for the L-histidine/aptamer association and about 10,000 μM for the D-histidine–aptamer association). As designed previously, the mirror-image approach was used to create a biostable L-RNA CSP in order to use such an aptamer in a routine chromatographic context. The effects on solute retention of the variation in various operating parameters, including the mobile-phase pH and the $MgCl_2$ concentration as well as the column temperature, were assessed. The results suggested that (1) the protonated form of histidine was involved in stereospecific RNA binding, and (2) Mg^{2+} was essential for target enantiomer binding to the specific aptamer sites. From a practical point of view, it appeared that the baseline resolution in a minimum analysis time can be achieved at a column temperature of 35°C for an eluent containing 10 mM of $MgCl_2$, pH 5.5 (Ruta et al., 2007a).

10.3.4   Class-Specific Aptamer-Based CSPs

Another major drawback of aptamers is related to the too-high target specificity obtained when using the SELEX procedure. For a broad application in the chiral separation field, such a high specificity is assumed to be of poor practical interest,

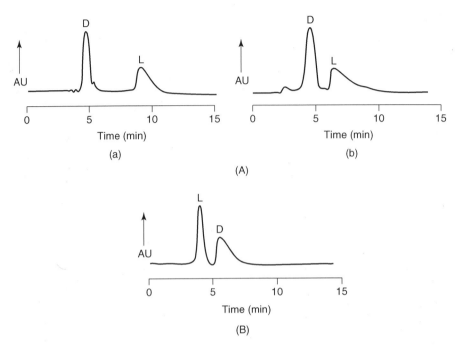

Figure 10.3   (A) Chromatograms for the resolution of arginine using an anti-L-arginine D-RNA chiral stationary phase. Amount of D-,L-arginine injected: (a) 10 ng and (b) 100 ng. Column: 370 × 0.76 i.d. mm; mobile phase, phosphate buffer 25 mM, NaCl 25 mM, MgCl$_2$ 5 mM (pH 7.3); column temperature, 4°C; injection volume, 100 nL; flow rate, 50 μL min^{-1}; detection at 208 nm. (B) Chromatogram for the resolution of arginine using an anti-D-arginine L-RNA chiral stationary phase. Column: 370 × 0.76 i.d. mm; mobile phase, phosphate buffer 25 mM, NaCl 25 mM, MgCl$_2$ 5 mM (pH 7.3); column temperature, 4°C; amount of D-,L-arginine injected, 50 ng; injection volume, 100 nL; flow rate, 50 μL min^{-1}; detection at 208 nm. [From Brumbt et al. (2005), with permission.]

as each aptamer CSP is expected to resolve only one racemate. However, we have recently reported an aptamer-based chiral stationary phase that was able to resolve racemates not only of the target but also of various related compounds (Ravelet et al., 2005). An anti-L-tyrosine RNA aptamer was chosen because the sequence was selected from a degenerate pool derived from a dopamine aptamer selected previously. The tyrosine-binding site of the aptamer would have some remembrance of the modalities of recognition of dopamine, which would explain its ability to bind with relative high affinity some structurally related analogs. The enantiomers of tyrosine and analogs (11 enantiomeric pairs) were separated using an immobilized anti-tyrosine-specific L-RNA aptamer. It was found also that immobilized RNA aptamer could be used under hydroorganic mobile-phase conditions without alteration of the stationary-phase stability (about three months of experiments) (Ravelet et al., 2005).

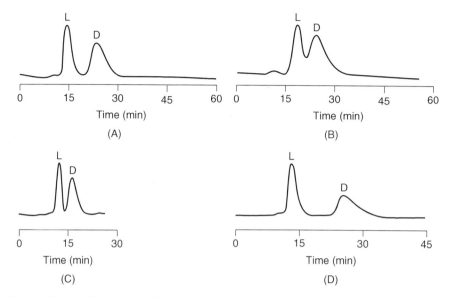

Figure 10.4   Chromatographic resolution of (a) tryptophan, (b) 2-quinolylalanine, (c) *N*-acetyltryptophan, and (d) 1-methyltryptophan using an anti-D-tyrosine L-RNA aptamer chiral stationary phase. Column, 350×0.76 i.d. mm; mobile phase, TRIS–HCl buffer 8 mM, NaCl 25 mM, MgCl$_2$ 5 mM (pH 7.4); column temperature, $10°$C; injected concentration, 0.50 mM; injection volume, 100 nL; flow rate, 15 μL min^{-1}; detection at 220 nm. [From Ravelet et al. (2005), with permission.]

10.4   APTAMERS FOR ANALYSIS OF ENANTIOMERS BY CAPILLARY ELECTROPHORESIS

10.4.1   Aptamers as Chiral Additives in the Background Electrolyte for CE Enantiomeric Separation

Affinity CE can be defined as electrophoretic separations where the separation patterns are influenced by molecular binding interactions taking place during the separation process. Whereas many of the affinity techniques have been adapted to capillary electrophoresis, one area of great importance is related to enantiomeric separation using chiral selectors added to the background electrolyte. An interest in using aptamers as target-specific chiral additives for CE enantiomeric separation has recently been demonstrated (Ruta et al., 2006) using a truncated sequence (53 bases) of an anti-L-arginine RNA aptamer (Geiger et al., 1996). The aptamer–target association has high affinity, characterized by a dissociation constant K_d of 330 nM at 25°C. In addition, this aptamer is known to be very strongly enantioselective (12,000) (Geiger et al., 1996). An L-RNA aptamer was used as a biostable target-specific chiral additive, and the separation of arginine

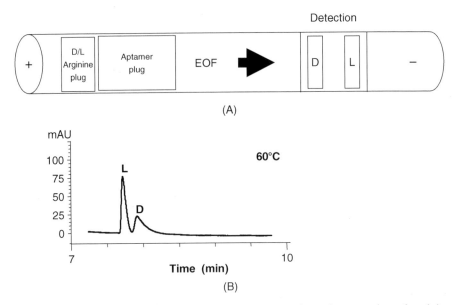

Figure 10.5 (a) Partial-filling CE method for the enantiomeric separation of arginine using an anti-D-arginine L-RNA aptamer as a chiral additive (EOF, electroosmotic flow). (b) Enantiomeric separation of arginine for a capillary temperature of 60°C, an enantiomer concentration of 0.25 mg mL^{-1}, and a 500s aptamer plug. Running buffer, 25 mM TRIS-HCl, 25 mM KCl, 10 mM Mg^{2+} (pH 7.0); capillary, 50 µm i.d. uncoated fused-silica capillary with an extended light path (total and effective lengths of 64.5 and 56 cm, respectively). Separation zone, 80 µM anti-D-arginine L-RNA aptamer applied at 50 mbar; applied voltage, 15 kV; sample injection, 50 mbar for 4 s; detection wavelength, 195 nm.

enantiomers was carried out using the partial filling mode (Figure 10.5a) (Ruta et al., 2006).

In this technique, the running buffer partially filled in the capillary containing the ultraviolet (UV)-absorbing components (here, the chiral selector) and the remaining portion is filled with the neat running buffer without such components. The experiments have to be performed under conditions where the UV-absorbing components do not migrate toward the detection end of the capillary or migrate at a velocity slower than the analytes. Thus, the analyte can be detected in the running buffer without the interference of UV-absorbing components. The arginine enantiomers were positively charged, whereas the aptameric chiral selector was negatively charged. So the RNA species migrated more slowly than the sample using an uncoated fused-silica capillary, in cationic mode. The sample migrated through the aptamer zone when voltage was applied, allowing the separation of enantiomers in relation to their respective interaction with the chiral selector: A stronger interaction between the solute and the chiral selector resulted in a lower mobility of the solute toward the cathodic end of the capillary. An additional advantage is that the chiral selector was introduced into the capillary hydrodynamically. Therefore, the aptamer working solution was not electrolyzed and can be reused for several times. Finally, the partial-filling mode allowed reducing the consumption of the chiral selector since less than 60 pmol of aptamer was used per run (Ruta et al., 2006). The analyses were performed over a wide temperature range (15 to 60°C). Whereas the L-arginine peak was clearly observed, the target enantiomer could not be detected in the separation window for the temperature range 15 to 50°C. The D-arginine peak was visualized only when the capillary temperature increased to very high temperatures (55 to 60°C). A representative electropherogramm is presented in Figure 10.5b for a capillary temperature of 60°C. As the target enantiomer was not detected over the low-to-moderate temperature range, this was consistent with the fact that D-arginine formed a very stable complex with the chiral selector. Such an observation was in agreement with the very high enantioselective properties observed previously by Famulok and co-workers at ambient temperature (Geiger et al., 1996). Further experiments have been performed based on a study of the influence of the target sample load as well as aptamer plug length on the target CE migration, efficiency, and band shape. The results demonstrated that binding site heterogeneity, slow mass transfer kinetics, and nonlinear effects governed the target enantiomer behavior. It was suggested that the three RNA conformations, identified by additional thermal unfolding experiments over the 50 to 60°C temperature range, could be responsible, at least in part, for the heterogeneous distribution of the target-binding sites of the aptamer chiral selector (Ruta et al., 2006).

10.4.2   Aptamers for the Design of an Affinity CE-Based Enantioselective Competitive Assay

As an alternative to the classical solid-phase techniques, affinity CE-based assays also represent a powerful technology for the detection and quantification of analytes. For example, ACE immunoassays offer several clear advantages over the commonly used immunoassays: They consume less sample and reagent, eliminate washing steps, are compatible with automation and on-line analysis, do not require antibody or analyte immobilization on a solid support, avoid nonspecific binding of antibody or analyte to the surface, and present wide analyte applicability (German et al., 1998). Taken into account all these attractive background features, an ACE-based assay was designed using the anti-D-arginine L-RNA aptamer used previously (see above) as a target-specific receptor (Geiger et al., 1996) for detection of the minor enantiomer in a nonracemic mixture at a very low level.

As arginine is a low-molecular-mass species bearing only one net positive charge, whereas RNA is much heavier and highly negatively charged, the charge/mass ratio of the analyte–aptamer complex is close to that of the free aptamer. So a direct ACE assay based on separation of the free aptamer from the complex cannot be easily achieved, so that a competitive binding format was developed. Furthermore, an on-capillary mixing approach was carried out in which the reaction and the separation occurred simultaneously in the same capillary by introducing solutions of sample and reagents in the form of individual zones (plugs). The general scheme used is illustrated in Figure 10.6a (Ruta et al., 2007b).

The electrophoretic experiments were conducted using a poly(vinyl alcohol) (PVA)-coated capillary. This coating shields the silanol groups of the fused silica and eliminates the electroosmotic flow (EOF). The capillary was first prefilled with the background electrolyte and three different plugs were introduced hydro-dynamically using the short-end injection method: (1) a plug containing a known amount of the L-RNA aptamer, (2) a sample plug containing an unknown amount of target (D-arginine), and (3) a plug containing a known amount of the labeled target (referred to as *label*). For use as label, D-arginine was tagged with a dansyl group. The linker, its attachment position, and the chemistry used for the labeling were chosen to mimic the structure of the immobilized arginine that was used as target in the SELEX methodology. This allowed avoiding possible alteration of the binding properties of the labeled target. More, the label remained positively charged under the running buffer conditions used. When an electric field was applied, both the target and the label moved toward the cathodic end, whereas the negatively charged aptamer migrated in the opposite direction (Figure 10.6a). The target was first in contact with the aptamer zone, causing formation of the D-arginine–aptamer complex. The slow complex dissociation kinetics (Geiger et al., 1996; Ruta et al., 2006) allowed separation of free from complexed target. The unbound D-arginine migrated toward the cathode while the target–aptamer complex and the free aptamer moved with similar mobility toward the anodic end. Then, labeled D-arginine passed through the aptamer zone and bound to the free aptamer-binding sites. As evoked above for the target, the unbound fraction of label was separated from the label–aptamer complex zone. The free labeled D-arginine band, moving toward the cathodic end, was detected at the ultraviolet detector window (Figure 10.6A) (Ruta et al., 2007b).

D- and L-arginine were injected as a sample plug over the concentration range 0.5 to 200 μM. The effect of the arginine concentration on the assay response is presented in Figure 10.6b. The label-free fraction was enhanced with the increasing sample target concentration because fewer aptamer-binding sites were available. The calibration curve presents the sigmoidal shape typical of competitive assays. In contrast, no binding of the nontarget enantiomer to the aptamer was obtained, demonstrating that the cross-reactivity of the aptameric receptor with the opposite enantiomer was negligible. These enantioselective properties of the anti-D-arginine L-RNA aptamer were accounted to analyze enantiomer mixtures of arginine, and as little as 2 μM of D-arginine in the presence of 20 mM of L-arginine can be detected (Ruta et al., 2007b). The assay allows detection of 0.01% of the minor enantiomer in a nonracemic mixture, one order of magnitude lower than the 0.1 to 1% detection limits, which are typically attainable with currently available stereoselective analysis methods.

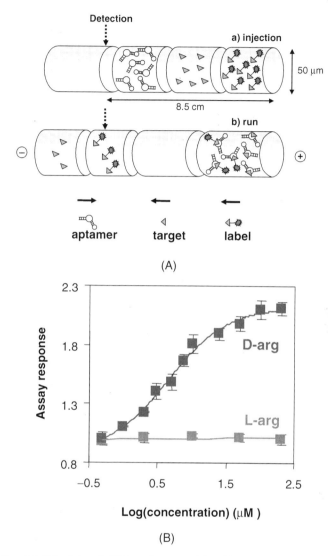

Figure 10.6   (a) Principle of CE-based competitive binding assay using on-capillary mixing of the various species. Arrows indicate the migration direction of the interacting species when an electric field is applied. (b) Standard curves obtained for D- (square) and L- (diamond) arginine enantiomers over the concentration range 0.5 to 200 μM. The relative peak height corresponds to the ratio of free-label peak height for an enantiomeric concentration in the sample plug to the free-label peak height for the blank (no enantiomer in the sample plug). Experimental conditions: running buffer, 25 mM TRIS-HCl, 25 mM KCl, 10 mM $MgCl_2$, (pH 7.0); coated PVA capillary, 50 μm i.d. with an extended light path (total length 64.5 cm); short-end injection method (effective length, 8.5 cm; negative polarity, cathode at the inlet and anode at the outlet; applied voltage, −25 kV); hydrodynamic injection (−50 mbar): 20-s aptamer (100 μM) plug (∼24 nL), 32-s sample (∼41 nL) plug, 4-s label (400 μM) plug (∼5 nL); temperature, 12°C; UV detection at 250 nm. (See insert for color representation.)

10.5 CONCLUSIONS

The SELEX technology has proven to be particularly valuable to create aptamer-based CSs specifically designed to resolve the racemate. Such new target-specific CSs present some very interesting features. First, aptameric CSs are characterized by a predetermined enantioselectivity and commonly exhibit very high separation factors (see Table 10.1). Moreover, as reported very recently, such high chiral recognition properties can account efficiently for the development of powerful enantioselective CE assays for the detection of very low enantiomeric impurity (Ruta et al., 2007b). Another potentially interesting application area may be in the selective removal of enantiomeric impurities. The strategy for such "enantiopolishing" would be to run the sample over a chiral stationary phase prepared against the wanted enantiomer, leading to faster elution of any optical impurity and collection of enantiomer of higher enantiomeric purity.

Second, it is easy to reverse the enantiomeric elution order through the change in configuration of the selector or the enantiomer used as target. This appears of very great interest for the determination of enantiomeric excess.

More overall, these chiral selectors of high or very high affinity and selectivity could find applications in various other analytical fields, such as the design of enantioselective sensors. However, such CSs also present some major drawbacks. Obviously, one important general limitation of selectors especially designed for a single analyte is related to an application range that is restricted in most cases to a narrow set of compounds. This renders these CSPs inefficient in generic approaches. In comparison with the most commonly used CSs, including cellulose derivatives, cyclodextrins, and antibiotic macrocyclics, the aptamer CSs are commonly characterized by slower mass transfer kinetics, leading to poorer performance efficiency (see Table 10.1). In addition, due to the relatively high cost of aptamers, the applications have been limited to miniaturized systems, including capillary electrophoresis and micro-LC (see Table 10.1). This constraint probably precludes, at least at the present time, any applications at the preparative level. Moreover, the stability of the aptamer CSPs over much more than several months has to be evaluated for the application to preparative-scale separation. Another major problem is related to the SELEX methodology. Although highly efficient, this requires sophisticated equipment and expensive reagents, and can be relatively time consuming. Finally, the number of aptamers directed against small chiral molecules of importance, such as drugs, is relatively restrained at this time. However, improved methods of selection have recently been developed (Murphy et al., 2003, Mendonsa and Bowser, 2005) that can lead to rapid generation of new aptamers.

REFERENCES

Brumbt, A., Ravelet, C., Grosset, C., Ravel, A., Villet, A., Peyrin, E. (2005). Chiral stationary phase based on a biostable L-RNA aptamer. *Anal Chem* 77, 1993–1998.

Davis, K. A., Abrams, B., Lin, Y., Jayasena, S. D. (1996). Use of a high affinity DNA ligand in flow cytometry. *Nucleic Acids Res* 24, 702–706.

Deng, Q., German, I., Buchanan, D., Kennedy, R. T. (2001). Retention and separation of adenosine and analogues by affinity chromatography with an aptamer stationary phase. *Anal Chem* 73, 5415–5421.

Ellington, A. D., Szostak, J. W. (1990). In vitro selection of RNA molecules that bind specific ligands. *Nature* 346, 818–822.

Famulok, M. (2005). Allosteric aptamers and aptazymes as probes for screening approaches. *Curr Opin Mol Ther* 7, 137–143.

Fuchs, S. A., Berger, R., Klomp, L. W. J., de Koning, T. J. (2005). D-Amino acids in the central nervous system in health and disease. *Mol Gen Met* 85, 168–180.

Francotte, E. R. (2001). Enantioselective chromatography as a powerful alternative for the preparation of drug enantiomers. *J Chromatogr A* 906, 379–397.

Gebhardt, K., Shokraei, A., Babaie, E., Lindqvist, B. H. (2000). RNA aptamers to S-adenosylhomocysteine: kinetic properties, divalent cation dependency, and comparison with anti-S-adenosylhomocysteine antibody. *Biochemistry* 39, 7255–7265.

Geiger, A., Burgstaller, P., von der Eltz, H., Roeder, A., Famulok, M. (1996). RNA aptamers that bind L-arginine with sub-micromolar dissociation constants and high enantioselectivity. *Nucleic Acids Res* 24, 1029–1036.

German, I. D., Buchanan, D., Kennedy, R. T. (1998). Aptamers as ligands in affinity probe capillary electrophoresis. *Anal Chem* 70, 4540–4545.

Huizenga, D. E., Szostak, J. E. (1995). A DNA aptamer that binds adenosine and ATP. *Biochemistry* 34, 656–665.

Ito, Y., Fujita, S., Kawazoe, N., Imanishi, Y. (1998). Competitive binding assay for thyroxine using in vitro selected oligonucleotides. *Anal Chem* 70, 3510–3512.

Jayasena, S. D. (1999). Aptamers: an emerging class of molecules that rival antibodies in diagnostics. *Clin Chem* 45, 1628–1650.

Klussmann, S., Nolte, A., Bald, R., Erdmann, A., Furste, J. P. (1996). Mirror-image RNA that binds D-adenosine. *Nat Biotechnol* 14, 1112–1115.

Majerfeld, I., Yarus, M. (2005). A diminutive and specific RNA binding site for L-tryptophan. *Nucleic Acids Res* 33, 5482–5493.

Majerfeld, I., Puthenvedu, D., Yarus, M. (2005). RNA affinity for molecular L-histidine: genetic code origins. *J Mol Evol* 61, 226–235.

Mendonsa, S. D., Bowser, M. T. (2005). In vitro selection of aptamers with affinity for neuropeptide Y using capillary electrophoresis. *J Am Chem Soc* 127, 9382–9383.

Michaud, E., Jourdan, E., Villet, A., Ravel, A., Grosset, C., Peyrin, E. (2003). A DNA aptamer as a new target-specific chiral selector for HPLC. *J Am Chem Soc* 125, 8672–8679.

Michaud, M., Jourdan, E., Ravelet, C., Villet, A., Ravel, A., Grosset, C., Peyrin, E. (2004). Immobilized DNA aptamers as target-specific chiral stationary phases for resolution of nucleoside and amino acid derivative enantiomers. *Anal Chem* 76, 1015–1020.

Murphy, M. B., Fuller, S. T., Richardson, P. M., Doyle, S. A. (2003). An improved method for in vitro evolution of aptamers and applications in protein detection and purification. *Nucleic Acids Res* 31, e110.

Patel, D. J., Suri, A. K., Jiang, F., Jiang, L., Fan, P., Kumar, R. A., Nonin, S. (1997). Structure, recognition and adaptive binding in RNA aptamer complexes. *J Mol Biol* 272, 645–664.

Potyrailo, R. A., Conrad, R. C., Ellington, A. D., Hieftje, G. M. (1998). Adapting selected nucleic acid ligands (aptamers) to biosensors. *Anal Chem* 70, 3419–3425.

Ravelet, C., Boulkedid, R., Ravel, A., Grosset, C., Villet, A., Fize, J, Peyrin, E. (2005). A L-RNA aptamer chiral stationary phase for resolution of target and related compounds. *J Chromatogr A* 1076, 62–70.

Robertson, D. L., Joyce, G. F. (1990). Selection in vitro of an RNA enzyme that specifically cleaves single-stranded DNA. *Nature* 344, 467–468.

Ruta, J., Ravelet, C., Grosset, C., Fize, J., Ravel, A., Villet, A., Peyrin, E. (2006). Enantiomeric separation using an l-RNA aptamer as chiral additive in partial-filling capillary electrophoresis. *Anal Chem* 78, 3032–3039.

Ruta, J., Grosset, C., Ravelet, C., Fize, J., Villet, A., Ravel, A., Peyrin, E. (2007a). Chiral resolution of histidine using an anti-D-histidine L-RNA aptamer microbore column. *J Chromatogr B* 845, 186–190.

Ruta, J., Ravelet, C., Baussanne, I., Decout, J. L., Peyrin, E. (2007b). An aptamer-based enantioselective competitive binding assay for the trace enantiomer detection. *Anal Chem.* 79, 4716–4719.

Shoji, A., Kuwahara, M., Ozaki, H., Sawai, H. (2007). Modified DNA aptamer that binds the (R)-isomer of a thalidomide derivative with high enantioselectivity. *J Am Chem Soc* 129, 1456–1464.

Tuerk, C., Gold, L. (1990). Systematic evolution of ligands by exponential enrichment: RNA ligands to bacteriophage T4 DNA polymerase. *Science* 249, 505–510.

Vianini, E., Palumbo, M., Gatto, B. (2001). In vitro selection of DNA aptamers that bind L-tyrosinamide. *Bioorg Med Chem* 9, 2543–2548.

Williams, K. P., Liu, X. H., Schumacher, T. N., Lin, H. Y., Ausiello, D. A., Kim, P. S., Bartel, D. P. (1997). Bioactive and nuclease-resistant L-DNA ligand of vasopressin. *Proc Natl Acad Sci U S A* 94, 11285–11290.

CHAPTER 11

APTAMER-MODIFIED SURFACES FOR AFFINITY CAPTURE AND DETECTION OF PROTEINS IN CAPILLARY ELECTROPHORESIS AND MALDI–MASS SPECTROMETRY

LINDA B. McGOWN

11.1 INTRODUCTION

The establishment of proteomic approaches to biomarker discovery and disease profiling in recent years has tested the limits of existing tools for the capture and detection of low-abundance proteins in biological samples. Affinity-binding reagents have played a crucial role in the translation of proteomic discoveries to clinical diagnostics, due to their ability to isolate target proteins from complex protein mixtures. Antibodies have been unrivaled as affinity reagents for proteins, due to their strong and selective binding; however, drawbacks associated with their production, stability, and manipulation have prompted researchers to seek alternatives. Foremost among alternatives are aptamers (Tuerk and Gold, 1990; McGown et al., 1995), which offer affinity on a par with that of monoclonal antibodies, but with important advantages: (1) once an aptamer to a target protein has been identified, it can be synthesized, chemically modified, and manipulated with ease; (2) aptamers are chemically stable and can be folded and unfolded reversibly for capture and release of the target protein, allowing aptamer-modified surfaces to be reused indefinitely (McGown et al., 1995; Jayasena, 1999; Brody and Gold, 2000; Tombelli et al., 2005).

Aptamers have been employed successfully over the past decade in chromatography, capillary electrophoresis, sensing, imaging, and protein isolation and purification (Jayasena, 1999; Brody and Gold, 2000; Tombelli et al., 2005; Schou and Heegaard, 2006; Ravelet et al., 2006). In this chapter we describe the use of aptamer-modified surfaces for affinity protein capture and detection in capillary electrophoresis (CE) (Connor and McGown, 2006) and in matrix-assisted laser desorption/ionization mass spectroscopy (MALDI-MS) (Dick and McGown, 2004; Cole et al., 2007). These recent entries in protein analysis offer simple, robust, reusable platforms for isolation, preconcentration, and specific detection of low-abundance proteins in biological samples. Further, in this chapter we extend the family of DNA-binding ligands beyond combinatorially selected aptamers to include genomic-inspired oligonucleotides that bind with high affinity to protein targets through interactions that may, in some cases, hold biological as well as analytical significance (Connor et al., (2006)).

11.2 APTAMER-MODIFIED CAPILLARIES IN AFFINITY CAPILLARY ELECTROPHORESIS

We first demonstrated the use of aptamer-coated capillaries for affinity capture in CE using the thrombin-binding DNA aptamer (TBA) as a model system (Connor and McGown, 2006). The TBA is a 15-mer of sequence 5′–GGTT-GGTGTGGTTGG–3′ that forms a G-quadruplex structure (Figure 11.1) (Bock et al., 1992; Kelly et al., 1996). It is one of the most widely studied, well-characterized aptamers, and therefore is commonly used in initial investigations of analytical applications. In our affinity CE approach, the aptamer is attached through a covalent linker to the inner surface of a bare fused-silica glass capillary. The protein sample is loaded onto the modified capillary via applied pressure and allowed to incubate. The capillary is then washed to remove unbound and nonspecifically associated proteins. Finally, the bound protein is released and eluted under appropriate conditions and collected for analysis.

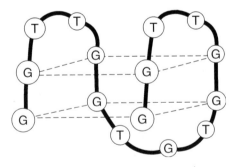

Figure 11.1   G-quadruplex structure of the thrombin-binding aptamer. [From Bock et al. (1992) and Kelly et al. (1996).]

In order to quantitate protein capture and binding capacity in the TBA coated capillary, eluate was collected after each step (load, wash, and elute), and total amounts of protein in each were determined using fluorescence spectroscopy. The experiment was also performed using a bare (unmodified) capillary and a capillary coated with an oligonucleotide that has the same base composition as the TBA but in a scrambled sequence that does not form a G-quadruplex structure and does not bind with thrombin (Bock et al., 1992). The results (Table 11.1) showed that the aptamer-coated capillary captures approximately three times as much thrombin as the bare and scrambled oligonucleotide-coated capillaries. The scrambled oligonucleotide yielded no more capture than the bare surface, indicating that thrombin capture at the TBA surface is due to specific affinity interactions. Since thrombin capture occurs only if the TBA is in the G-quadruplex conformation, we could further conclude that the immobilized TBA at the surface is able to form the G-quadruplex.

Protein-binding selectivity of the TBA-coated capillary was investigated using human serum albumin (HSA), which does not bind to TBA. Since the fluorescence spectra of thrombin and HSA overlap, MALDI-time-of-flight mass spectromtery (MALDI-TOF-MS) was used to distinguish between the two proteins in the fractions collected. The results in Table 11.1 show that HSA was not retained on the TBA-coated capillary, nor did it affect the capture of thrombin from equimolar mixtures of the two proteins.

The results of the capture experiments in Table 11.1 were used to determine the affinity-binding capacity of the TBA-coated capillaries. Using the total protein recovered on the TBA-coated capillary after subtraction of the nonspecific recovery from the bare or scrambled oligonucleotide-coated capillary, we were

TABLE 11.1   Affinity Capture of Thrombin in TBA-Coated Capillaries and Binding Capacity Determinations

Type	Column Length (cm)	I.D. (mm)	Vol. (μL)	Inner Surface Area (cm^2)	Protein Sample	Protein Loaded (pmol)	Protein Captured (pmol)	Coverage (mol nm^{-2})
TBA	47	100	3.7	1.5	Thrombin	570	64	2.2×10^{-25}
TBA	27	100	2.1	0.85	Thrombin	290	29	1.7×10^{-25}
TBA	27	25	0.13	0.21	Thrombin	290	10	2.3×10^{-25}
TBA	47	100	3.7	1.5	Thrombin + HSA	570 each	59	2.0×10^{-25}
TBA	47	100	3.7	1.5	HSA	570	0	0
Scrambled	47	100	3.7	1.5	Thrombin	570	20	6.8×10^{-26}
Bare	47	100	3.7	1.5	Thrombin	570	21	7.1×10^{-26}

Source: Connor and McGown (2006).

able to determine that the binding capacity is approximately 0.08 molecule of thrombin per square nanometer.

11.3 APTAMER-MODIFIED SURFACES FOR AFFINITY MALDI-MS

11.3.1 Overview

Recent advances in MALDI-MS have led to protein capture array technology based on the use of MALDI probe surfaces modified with various binding agents, such as antibodies, metals, hydrophobic phases, lectins, enzymes, and protein receptors (Hutchens and Yip, 1993; Merchant and Weinberger, 2000; Isaaq et al., 2002; Tang et al., 2004). We recently added aptamers to the protein capture toolbox for affinity MALDI-MS (Dick and McGown, 2004; Cole et al., 2007). In our approach, the aptamers are attached to fused-silica MALDI probe surfaces in spots approximately 1 cm in diameter using the same covalent attachment method that we used to modify the fused-silica capillary surfaces in the affinity CE studies described above. Samples are incubated on the aptamer spots, followed by rinsing to remove unbound and loosely associated proteins as well as salts and other sample concomitants. The low-pH MALDI matrix is then applied, causing the aptamers to unfold and release the captured proteins. The spot is dried to crystallize the sample–matrix mixture and analyzed using MALDI-TOF-MS. Following analysis, the spot is rinsed to remove residual matrix and protein, and buffer is applied to refold the aptamer. The reconstituted surface can then be reused. The stability of the DNA oligonucleotides and the reversibility of folding allow the spots to be reused indefinitely.

Preconcentration of protein at the aptamer spot can be achieved by performing multiple-sample incubation/rinse cycles before applying MALDI matrix. Analogous to protein capture using antibodies, we have observed that capture of proteins at aptamer-coated surfaces is essentially irreversible under the conditions used for protein capture. The detectability of a protein at an aptamer-coated spot of a given size will therefore be limited by surface capacity, since repeated applications of sample (each followed by incubation and rinsing to remove unbound proteins and other sample concomitants) will eventually saturate the surface binding sites with target protein. Using the binding capacity of 0.08 protein molecule per square centimeter that was determined for thrombin using the similarly modified fused-silica capillaries (see above), a MALDI spot that is 1 cm in diameter should have a capacity of approximately 2×10^{-11} mol proteins, or 10^{13} protein molecules. Since the incident laser beam in the MALDI-MS experiment probes an area much smaller than $1 \, cm^2$, the spot size could be reduced to increase the concentration of protein per unit surface area for a given applied sample volume, thereby reducing the number of applications needed to saturate the surface binding sites.

In the following sections we describe our results for affinity MALDI-MS of two proteins, thrombin and immunoglobulin E (IgE). IgE provides a good contrast to thrombin in terms of both size (thrombin is 36 kDa and IgE is

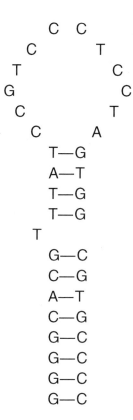

Figure 11.2   Proposed stem–loop structure of the IgE-binding aptamer. [From Wiegand et al. (1996).]

200 kDa) and aptamer structure [the TBA forms a Hoogsteen hydrogen-bonded G-quadruplex (Figure 11.1), whereas the IgE-binding aptamer (5′-GGGGCACG-TTTATCCGTCCCTCCTAGTGGCGTGCCCC–3′) is thought to form a secondary stem–loop structure through intramolecular Watson–Crick hydrogen bonding between complementary bases (Figure 11.2)] (Wiegand et al., 1996).

11.3.2   Affinity MALDI-MS of Thrombin

Our first demonstration of aptamer-modified surfaces for affinity MALDI-MS used the thrombin–TBA system described for affinity CE studies in the preceding section (Dick and McGown, 2004). In the MALDI-MS experiments, 3 µL of sample was applied to the aptamer-coated and scrambled oligonucleotide-coated MALDI probe surfaces and incubated for 30 minutes. The surfaces were then rinsed with buffer, followed by application of the MALDI matrix and MALDI-TOF-MS analysis.

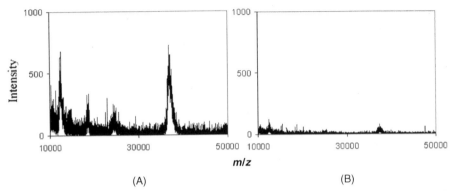

Figure 11.3 MALDI-TOF-MS results for 50 pmol of thrombin incubated for 30 minutes at room temperature on (A) a thrombin-binding DNA aptamer-coated spot and on (B) a scrambled oligonucleotide-coated spot, followed by rinsing to remove unbound or loosely bound protein. [From Dick et al. (2004).]

As typified by the results shown in Figure 11.3, the aptamer-coated spots yielded strong signals for thrombin, while the scrambled oligonucleotide-coated spots yielded little or no signal. The very low signals that were occasionally observed in some of the scrambled oligonucleotide spots, such as in the example in Figure 11.3, are attributed to incomplete removal of nonspecifically adsorbed thrombin during the rinse step. Subsequent to the thrombin study, we upgraded to a new MALDI-TOF-MS system that provides much greater detectability, allowing us to reduce sample loading by several orders of magnitude (see Section 11.3.3). The use of highly diluted samples dramatically reduced nonspecific adsorption and facilitated removal of nontarget proteins and other interfering sample concomitants.

Selective capture of thrombin from an equimolar mixture of thrombin and human serum albumin (HSA) is shown in Figure 11.4. There is a very small peak around 67,000 m/z, corresponding to the molecular weight of HSA, in addition to the large peak from thrombin. Repeated trials using the same amount of HSA alone gave results ranging from no signal to peaks of the same small magnitude as shown for the protein mixture in Figure 11.4. The HSA peak is probably due to residual protein that is weakly associated with the aptamer or adsorbed to exposed fused-silica surfaces. The results demonstrate the high specificity of protein capture by the aptamer.

Experiments were also conducted using pooled human plasma. The predominant, stable form of the protein in plasma is prothrombin, a zymogen that is cleaved to yield the activated thrombin upon appropriate signaling (Blomback and Hanson, 1979). The presence of any detectable thrombin in its active (i.e., cleaved) form is probably due to stabilizers in the pooled plasma preparation. The plasma concentration of prothrombin is about 90 mg/L (Blomback and Hanson, 1979), which corresponds to about 2.5 μM, or 7.5 pmol on the aptamer spot.

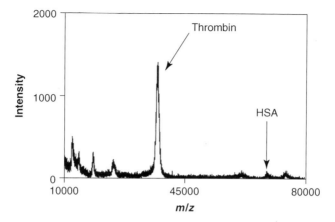

Figure 11.4   MALDI-TOF-MS results for 25 pmol each of thrombin and HSA incubated for 4 hours on a thrombin-binding DNA aptamer-coated spot, followed by rinsing to remove unbound or loosely bound protein. [From Dick et al. (2004).]

Figure 11.5 shows results for plasma incubated at an aptamer spot (A) and a scrambled spot (B). At the aptamer spot, the appearance of a peak at the molecular weight corresponding to thrombin demonstrates the selective capture of thrombin from the complex plasma matrix by the aptamer-modified surface. The mass spectrum shows an additional peak near 58,000 m/z which corresponds to the mass of prothrombin. There are also peaks corresponding to the m/z values of HSA^{+1} and HSA^{+2}. This is not surprising, since albumin is present at millimolar concentrations (i.e., a 1000-fold excess over prothrombin) in plasma. The peaks corresponding to thrombin and prothrombin that were observed at the aptamer spot are not seen at the scrambled spot. Peaks from singly and doubly charged albumin are present in both spectra, which again indicates nonspecific adsorption of the highly abundant HSA.

11.3.3   Affinity MALDI-MS of IgE

Following the thrombin experiments, we investigated the affinity MALDI-MS platform for capture and detection of IgE in simple solution and in pooled human serum using the DNA aptamer to IgE (Cole et al., 2007). IgE is the least abundant of the immunoglobulins in serum, normally occurring at a level of approximately 800 pM (Gould et al., 2003). This is 10^5 lower than the most abundant immunoglobulin, immunoglobulin G (IgG), which is normally present at approximately 100 μM in human serum (Gould et al., 2003), giving us the opportunity to study the effectiveness of the aptamer platform for low-abundance proteins in biological samples. The experiments were performed using a new MALDI-TOF-MS system that provides higher sensitivity than the system used for the thrombin work, allowing us to probe state-of-the-art limits of detection.

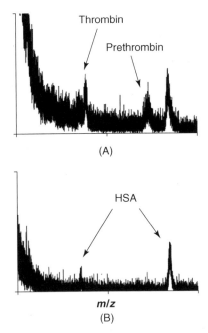

Figure 11.5   Results for thrombin capture from pooled human plasma: (A) aptamer spot incubated with 5 μL of plasma and rinsed; (B) scrambled spot incubated with 5 μL plasmaand rinsed. [From Dick et al. (2004).]

Figure 11.6 shows results for MALDI-MS of 1 μL of 500 nM IgE solution (500 fmol IgE) on an unmodified stainless steel probe surface (Figure 11.6A) and at fused-silica probe surfaces modified with the IgE aptamer (Figure 11.6B) and scrambled IgE control (Figure 11.6C). The mass spectrum at the aptamer surface exhibits the numerous, multiply charged ions of the 200-kDa protein that are consistent with the spectrum of IgE at a stainless steel MALDI probe. There is a small amount of nonspecific capture at the scrambled control surface. Figure 11.6D shows the results for a 10,000-fold dilution of the IgE solution (50 amol applied) at an aptamer spot. Interestingly, the spectrum at the lower concentration is dominated by peaks at approximately 67,000, 33,500, and 22,000 m/z, which correspond to the +3, +6, and +9 ions of IgE. This is attributed to dominance of these higher charged ions at the high matrix/protein ratios for these samples as well as the apparent favorability of ions with multiples of +3 charge. Alternative explanations based on fragmentation of IgE were not borne out by experiments in which IgE was treated with dithiothreitol (DTT) to break the disulfide bonds between the various chains or subjected to high laser powers to aid degradation (results not shown).

Figure 11.7 shows results for 1 amol IgE, including a first application of 1 μL of 1 pM IgE to an aptamer spot (Figure 11.7A), a blank run with matrix only

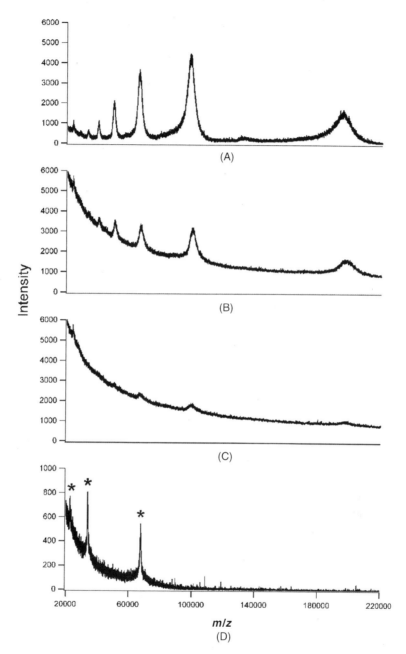

Figure 11.6   Mass spectra of IgE at various surfaces: (A) 500 fmol of IgE on an unmodified stainless steel probe surface; (B) 500 fmol of IgE on an aptamer spot; (C) 500 fmol of IgE on a scrambled spot; (D) 50 amol of IgE on an aptamer spot; asterisks denote peaks at 22,200, 33,500, and 67,000 m/z, corresponding to +9, +6, and +3 ions of IgE. Note that the intensity scale of (A) cannot be compared with that of (B) to (D). [From Cole et al. (2007).]

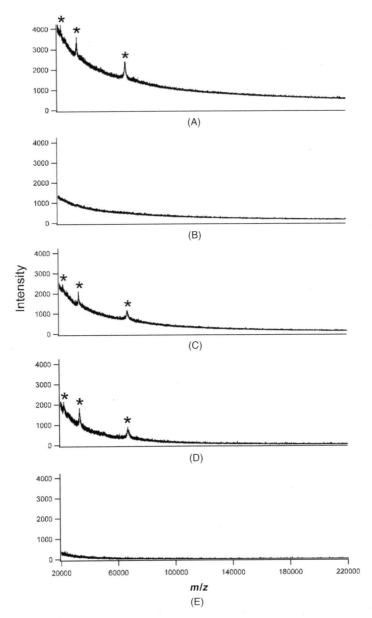

Figure 11.7 Capture and preconcentration of IgE: (A) first trial of 1 amol of IgE (1 μL of 1 pM IgE) on an aptamer spot (run 1); (B) blank run with matrix only at same spot after rinsing to remove matrix and protein from run 1; (C) second trial of 1 amol IgE at same spot (run 2); (D) 1 amol IgE on aptamer spot, applied in 10 successive incubation/rinse cycles, each of 0.1 amol IgE (1 μL of 100 fM IgE) prior to application of MALDI matrix; (E) 1 amol of IgE (1 μL of 1 pM IgE) on scrambled spot. Asterisks denote peaks at 22,200, 33,500, and 67,000 *m/z*, corresponding to +9, +6, and +3 ions of IgE. [From Cole et al. (2007).]

following cleaning and reconstitution of the spot (Figure 11.7B), a second run of 1 µL of 1 pM IgE on the same spot (Figure 11.7C), and results for 1 amol IgE that was applied in 10 successive incubation/rinse cycles of 0.1 amol IgE (1 µL of 100 fM IgE) prior to application of MALDI matrix and analysis at an aptamer spot and a scrambled spot (Figure 11.7D and E, respectively). The results illustrate the detection of 1 amol IgE, the reusability of the aptamer spot, the ability to preconcentrate IgE at an aptamer spot, and the absence of IgE at the scrambled spot even after multiple incubation and rinse cycles.

Figure 11.8 shows the results for protein capture from commercial human serum. Undiluted serum on an aptamer spot (Figure 11.8A) shows several peaks, including a series at 150,000, 75,000, 50,000, and 37,500 m/z that is attributed to IgG (150 kDa) and a series at 67,000, 33,500, and 22,200 m/z that may be due to HSA or to the +3, +6, and +9 ions of IgE (see below). Analysis of serum that was spiked with IgE by mixing equal volumes of serum and 50 pM IgE solution prior to application to the aptamer spot (Figure 11.8B) increased the latter peaks, suggesting that the peaks in the spectrum of the unspiked sample are due to IgE. Upon 1000-fold dilution of the serum, the IgG peaks dominate the spectrum at both the aptamer and scrambled spots, showing only small contributions from the IgE peaks at the aptamer spot (Figure 11.8C) and none at the scrambled spot (Figure 11.8D). Like the undiluted serum, spiking the diluted sample with IgE suppresses the IgG peaks and enhances the IgE peaks at the aptamer spot (Figure 11.8E), indicating effective competition by IgE for the aptamer sites. This is demonstrated further in the spectrum of a 10,000-fold dilution of the serum (Figure 11.8F), which resembles that of low concentrations of IgE with only minor peaks from other proteins.

To determine the effect of incubation time on IgE capture and further evaluate potential interference from albumin, aliquots containing 50 fmol each of IgE and albumin were analyzed at two aptamer spots, one after 3 minutes of incubation and the other after 30 minutes of incubation, as in the other experiments in this work. After 3 minutes (Figure 11.9A) the peaks at approximately 67,000, 33,500, and 22,000 m/z dominate the spectrum, with minor peaks at 100,000 and 50,000 m/z that correspond to the +2 and +4 ions of IgE that are expected at this relatively high IgE concentration. These peaks become more intense after a 30-minute incubation (Figure 11.9B), but there is no sign of the albumin peaks at 133,000 and 44,000 m/z, confirming that only IgE is captured and/or detected at the aptamer surface. The expanded regions in Figure 11.9A and B show development of the IgE parent peak at 200,000 m/z in going from 3 minutes to 30 minutes incubation, further confirming identification of the captured protein as IgE.

Since IgE normally occurs at levels of 800 pM in human serum (Gould et al., 2003), its detection in 10^3- to 10^4-fold dilutions of serum is consistent with the detection of 1 pM to 100 fM IgE in standard IgE solutions. This detectability compares favorably with other published results, including label-free (Liss et al., 2002; Xu et al., 2005) and fluorescent-labeled (Gokulrangan et al., 2005;

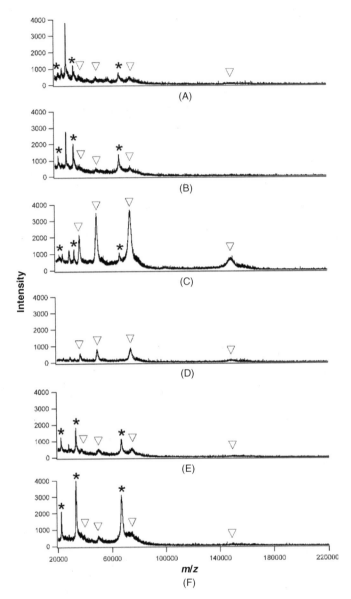

Figure 11.8   Capture of IgE from human serum: (A) undiluted serum (1 μL) on an aptamer spot; (B) 50 : 50 dilution of serum with 50 pM IgE on an aptamer spot (1 μL of 50% serum with 25 amol of IgE added); (C) 1000-fold dilution of serum (1 μL) on an aptamer spot; (D) same as (C) on a scrambled spot; (E) 1000-fold dilution of serum diluted 50:50 with 50 pM IgE on an aptamer spot (1 μL of 2000-fold diluted serum with 25 amol of IgE added); (F) 10,000-fold dilution of serum diluted 50 : 50 with 50 pM IgE on an aptamer spot (1 μL of 20,000-fold serum diluted with 25 amol of IgE added). Asterisks denote peaks at 22,200, 33,500, and 67,000 m/z, corresponding to +9, +6, and +3 ions of IgE. Inverted triangles denote peaks at 37,500, 50,000, 75,000, and 150,000 m/z, corresponding to +4, +3, +2, and +1 ions of IgG. [From Cole et al. (2007).]

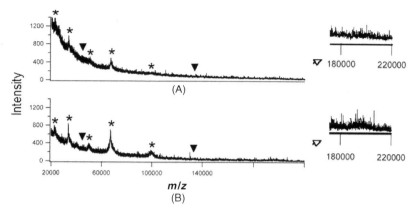

Figure 11.9   Capture of human serum albumin (HSA) and IgE: (A) 50 fmol each of IgE and albumin at an aptamer spot after 3 minutes of incubation; (B) same as (A) after 30 minutes of incubation. Asterisks denote peaks at 22,200, 33,500, 50,000, 67,000, and 100,000 m/z, corresponding to +9, +6, +4, +3, and +2 ions of IgE. Inverted solid triangles denote peaks at 22,000, 33,250, and 66,700 m/z, corresponding to +3, +2, and +1 ions of albumin, and at 44,000 and 133,000 m/z, corresponding to +3 and +1 ions of albumin dimer. Arrows point to expanded views of the area around the IgE parent peak at 200,000 m/z. [From Cole et al. (2007).]

Stadtherr et al., 2005) detection of IgE in simple solution that provided detectability down to 100 pM IgE [corresponding to 5 fmol using a 50-μL aliquot in the case of one immobilized aptamer sensor (Xu et al., 2005)], and use of fluorescent-labeled IgE aptamer in the run buffer in affinity CE that gave a detection limit of 46 pM IgE in simple solution but yielded detectable signals only for serum that was spiked with 5 nM IgE and not for native IgE in the serum (German et al., 1998). In the present work we achieved capture and detection of native IgE in human serum and found that dilution of the serum by at least 10^3-fold allowed detection of native IgE with little interference from other serum proteins. This is comparable to detectability of a commercial antibody-based ELISA kit that offers 75 pM detection (Stadtherr et al., 2005).

It is interesting that dilution of the serum sample actually improves the detectability of IgE, even though all sample components, including IgE, are equally diluted. This is presumably due to decreased rates of competition for binding sites between IgE and lower-affinity binders that occur in vastly greater abundance in the sample. Although the eventual equilibrium should be the same, the time it takes to reach equilibrium at the surface in the absence of stirring or other agitation increases when the total number of protein molecules greatly exceeds the number of binding sites. Under such conditions, nonspecific proteins may build up at the surface and block access of the few target proteins to the binding sites. Even for the highly diluted serum samples, Figure 11.9 shows that IgE capture at the surface slowly evolves over the course of 30 minutes.

The capacity of a typical aptamer-coated spot that is $1\,cm^2$ in diameter is approximately 10^{13} protein molecules. If $1\,\mu L$ of serum contains approximately 10^{13} total protein molecules, dilution of the sample by 10^4 reduces the total protein molecules in the 1-μL application volume to 10^9, which is much less than the binding capacity of the spot. Dilution of the sample should therefore increase the availability of aptamer sites to IgE and accelerate equilibration.

11.3.4 Summary

The studies described in this section demonstrate the power and simplicity of affinity MALDI-MS using aptamer-modified surfaces to capture and detect low-abundance proteins in biological fluids, to preconcentrate proteins directly at the MALDI probe surface, and to confirm the identity of the affinity-captured protein through examination of the mass spectrum. The reusable surfaces could easily be adapted to microarray formats, and the procedure is amenable to automation.

11.4 BEYOND APTAMERS: GENOME-INSPIRED DNA BINDING LIGANDS

In a variation on the aptamer concept, we recently introduced the use of genomic DNA sequences for affinity protein capture. This began with our unexpected discovery that insulin exhibits affinity binding to the G-quadruplex DNA structure formed by a two-repeat sequence of the insulin-linked polymorphic region (ILPR) of the human insulin gene promoter region (Gokulrangan et al., 2005). The ILPR is a noncoding minisatellite in the human insulin gene promoter region located 363 base pairs upstream of the insulin gene (Bell et al., 1982; Hammond-Kosack et al., 1992; Catasti et al., 1996; Lew et al., 2000). The ILPR is one of several genetic loci associated with genetic susceptibility to insulin-dependent diabetes mellitus (Bell et al., 1982). It contains tandem repeats of a G-rich repeat unit of the consensus sequence $5'-ACAG_4TGTG_4-3'$. Oligonucleotides containing at least two tandem repeats can form intramolecular G-quadruplex structures in vitro (Hammond-Kosack et al., 1992; Catasti et al., 1996; Lew et al., 2000). A proposed structure for the G-quadruplex formed by four tandem repeats is shown in Figure 11.10 (Hammond-Kosack et al., 1992).

Insulin binding to the G-quadruplex formed by a sequence of two of the tandem ILPR repeats, hereinafter referred to as $ILPR_2$, was studied using both the affinity CE and affinity MALDI-MS techniques described above. It is important to note that the surface-immobilized DNA that is intrinsic to both capture schemes is a more realistic model for G-quadruplexes in genomic DNA than would be provided by free DNA in solutions, since G-quadruplexes in genomic DNA would be anchored in the chromosomal strand.

In the first set of experiments, MALDI-MS was used to compare insulin capture at surfaces coated with $ILPR_2$ to capture at surfaces coated with a single

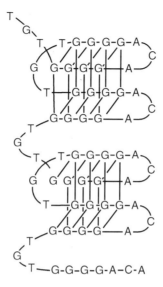

Figure 11.10 G-quadruplex structure proposed for four repeats of the G-strand of the ILPR. [From Hammond-Kosack et al. (1992).]

ILPR sequence (ILPR$_1$) that does not form an intramolecular G-quadruplex, and the G-quadruplex-forming TBA and non-G-quadruplex-forming scrambled TBA sequences that were described above in the thrombin studies. Figure 11.11A shows the spectrum of insulin on a bare (uncoated) spot that was not rinsed prior to application of MALDI matrix. This spectrum exhibits the expected insulin peak at 5808 *m/z* and serves as a standard for comparison with spectra of the other surfaces. Figure 11.11B shows the spectrum for insulin on a bare spot that was rinsed prior to the application of matrix. The absence of any significant peaks confirms that protein is effectively removed from the bare surface in the rinsing step. The spectrum of insulin captured at the ILPR$_2$ (Figure 11.11C) clearly indicates insulin capture by the presence of peaks at 5808 and 2900 *m/z*, corresponding to singly and doubly charged insulin ion, respectively. By comparison, very little insulin capture is indicated at the ILPR$_1$ surface (Figure 11.11D) or the scrambled TBA surface (Figure 11.11E), neither of which can form an intramolecular G-quadruplex structure. The specificity of insulin for the ILPR G-quadruplex over the TBA G-quadruplex is indicated by the much smaller degree of capture at the TBA surface (Figure 11.11F).

Table 11.2 shows quantitative results for insulin capture experiments from 10- and 1-μM insulin standard solutions. At both concentrations, insulin capture was at least 10-fold greater for ILPR$_2$ than any of the other surfaces (discounting the bare unrinsed surface). The standard deviations in all cases are much lower for the ILPR$_2$ surface than for the other surfaces, indicating that

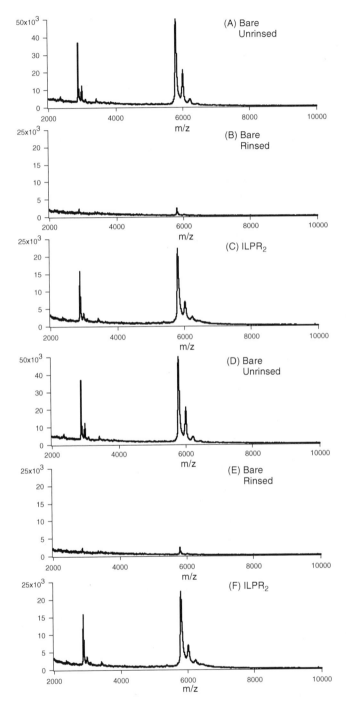

Figure 11.11 MALDI-MS of 10 μM insulin on various surfaces: (A) bare fused silica, unrinsed; (B) bare fused silica, rinsed; (C) $ILPR_2$; (D) $ILPR_1$; (E) scrambled; (F) TBA. [From Connor et al. (2006).]

TABLE 11.2   Results of Insulin Capture by Affinity MALDI-MS at Various Surfaces[a]

Sample	Surface	No. Spots	Mean Peak Intensity	%RSD	Mean Peak Area	%RSD
Insulin (10 μM)	Bare unrinsed	3	110	50	110	70
	ILPR$_2$	**3**	**100**	**20**	**100**	**40**
	ILPR$_1$	6	8	80	4	90
	TBA	4	2	80	1	130
	Scrambled	3	10	100	8	120
	Bare rinsed	3	5	60	2	100
Insulin (1 μM)	**ILPR$_2$**	**3**	**100**	**9**	**100**	**20**
	ILPR$_1$	6	7	80	4	80
	Scrambled	3	5	30	3	70

[a] Averaged over several spots and normalized to ILPR$_2$ (in bold), with percent relative standard deviation (%RSD).

retention of insulin due to affinity capture at the ILPR$_2$ surface is much more reproducible than retention due to nonspecific adsorption events at the other surfaces.

Affinity CE was then used to study cross-reactivity between the G-quadruplex-binding proteins (insulin and thrombin) and the G-quadruplex-forming DNA (ILPR$_2$ and TBA) using ILPR$_2$- and TBA-coated capillaries. Protein identities were confirmed using MALDI-TOF-MS. Fluorescence spectra of the elute collections, shown in Figure 11.12, were used to compare protein capture. Based on the fluorescence spectral peak areas, we found that five times more insulin is captured on the ILPR$_2$ capillary as on the TBA capillary, and at least three times more thrombin is captured on the TBA capillary as on the ILPR$_2$ capillary. These results confirm that there is selectivity among the various G-quadruplex structures toward their respective protein targets.

The ability of ILPR$_2$-coated surfaces to capture insulin from biological samples was investigated using commercial human pancreatic nuclear extract (HPNE) in both affinity MALDI-MS and affinity CE. Representative affinity MALDI-MS results are shown in Figure 11.13 for HPNE at a bare surface that was not rinsed prior to MALDI-MS analysis, and for capture at ILPR$_2$ and ILPR$_1$ surfaces. The spectrum of HPNE at the bare, unrinsed surface (Figure 11.13A) shows a number of peaks resulting from multiple proteins, as expected for a nuclear extract. The spectrum for ILPR$_2$ (Figure 11.13B) shows a large peak near 3400 m/z, assigned tentatively to the beta chain of insulin. Human insulin consists of a 21-amino acid alpha chain (2382 Da) and a 30-amino acid beta chain (3430 Da) that are linked by two disulfide bonds. The presence of the individual chain is attributed to the presence in the commercial HPNE preparation of 500 μM dithiothreitol (DTT), which can break the disulfide bonds between the alpha and beta chains in the insulin molecule. This assignment was confirmed using both affinity CE

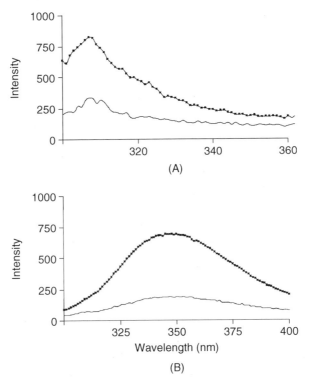

Figure 11.12   Fluorescence emission spectra of collections of captured proteins: (A) insulin captured from 1 capillary volume of 200 μM insulin and eluted from ILPR$_2$- and TBA-coated capillaries; (B) thrombin captured from 1 capillary volume of 156 μM thrombin and eluted from TBA- and ILPR$_2$-coated capillaries. [From Connor et al. (2006).]

and affinity MALDI-MS to analyze capture from standard insulin samples that were treated with DTT prior to analysis. No peaks are evident in the spectrum for the ILPR$_1$ surface (Figure 11.13C), indicating no protein capture.

This work demonstrates the ability of the immobilized ILPR$_2$ oligonucleotide to capture human insulin from standard solutions and from nuclear extracts of pancreatic cells with high selectivity in both affinity MALDI-mass spectrometry and affinity CEC. The availability of a DNA-binding ligand to human insulin offers an alternative to antibodies for in vitro and in vivo detection and sensing of insulin as well as its isolation and purification from biological samples. Although the genome-inspired ILPR$_2$ binding ligand to insulin was not derived combinatorially and is not therefore an "aptamer," it offers the same advantages, such as ease of production and manipulation, stability, relatively small size, and reusability. Future studies of other genomic-inspired DNA oligonucleotides may lead to the identification of new affinity-binding ligands to other biological targets. In addition, this pathway to discovery of DNA-binding ligands may offer

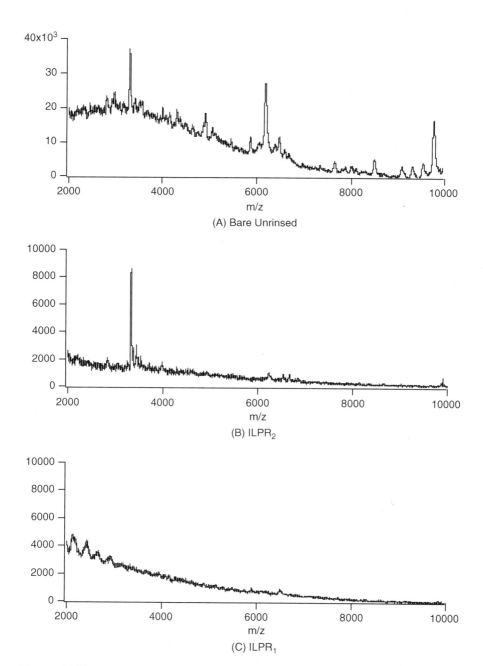

Figure 11.13   Affinity MALDI-MS of HPNE on (A) bare fused silica, unrinsed;
(B) ILPR$_2$; and (C) ILPR$_1$ surfaces. [From Connor et al. (2006).]

new insight into the role of three-dimensional DNA conformations, such as the G-quadruplex in nuclear processes and cellular transformations.

Acknowledgment

Portions of this work were supported by the U.S. National Institutes of Health.

REFERENCES

Bell, G. I., Selby, M. J., Rutter, W. J. (1982). The highly polymorphic region near the human insulin gene is composed of simple tandemly repeating sequences. *Nature* 295, 31–35.

Blomback, B., Hanson, L. A. (1979). *Plasma Proteins*, Wiley, New York, 1979.

Bock, L. C., Griffin, L. C., Latham, J. A., Vermaas, E. H., Toole, J. J. (1992). Selection of single-stranded DNA molecules that bind and inhibit human thrombin. *Nature* 355, 564–566.

Brody, E., Gold, L. (2000). Aptamers as therapeutic and diagnostic agents. *Rev Mol Biotechnol* 74, 5–13.

Catasti, P., Chen, X., Moyzis, R. K., Bradbury, E. M., Gupta, G. (1996). Structure–function correlations of the insulin-linked polymorphic region. *J Mol Biol* 264, 534–545.

Cole, J. R., Dick, L. W., Jr., Morgan, E. J., McGown, L. B. (2007). Affinity capture and detection of immunoglobulin E in human serum using an aptamer-modified surface in matrix assisted laser desorption/ionization mass spectrometry. *Anal Chem* 79, 273–279.

Connor, A. C., McGown, L. B. (2006). Aptamer stationary phases for protein capture in affinity capillary chromatography. *J Chromatogr A* 1111, 115–119.

Connor, A. C., Frederick, K. A., Morgan, E. J., McGown, L. B. (2006). Insulin capture by an insulin-linked polymorphic region G-quadruplex DNA oligonucleotide. *J Am Chem Soc* 128, 4986–4991.

Dick, L. W., Jr., McGown, L. B. (2004). Aptamer-enhanced laser desorption/ionization for affinity mass spectrometry. *Anal Chem* 76, 3037–3041.

German, I., Buchanan, D. D., Kennedy, R. T. (1998). Aptamers as ligands in affinity probe capillary electrophoresis. *Anal Chem* 70, 4540–4545.

Gokulrangan, G., Unruh, J. R., Holub, D. F., Ingram, B., Johnson, C. K., Wilson, G. S. (2005). DNA aptamer-based bioanalysis of IgE by fluorescence anisotropy. *Anal Chem* 77, 1963–1970.

Gould, H. J., Sutton, B. J., Beavil, A. J., Beavil, R. L., McCloskey, N., Coker, H. A., Fear, D., Smurthwait, L. (2003). The biology of IgE and the basis of allergic disease. *Annu Rev Immunol* 21, 579–628.

Hammond-Kosack, M. C., Dobrinski, B., Lurz, R., Docherty, K., Kilpatrick, M. W. (1992). The human insulin gene linked polymorphic region exhibits an altered DNA structure. *Nucleic Acids Res* 20, 231–236.

Hutchens, T. W., Yip, T. T. (1993). New desorption strategies for the mass spectrometric analysis of macromolecules. *Rapid Commun Mass Spectrom* 7, 576–580.

Isaaq, H. J., Veenstra, T. D., Conrads, T. P., Felschow, D. (2002). The SELDI-TOF MS approach to proteomics: protein profiling and biomarker identification. *Biochem Biophys Res Commun* 292, 587–592.

Jayasena, S. D. (1999). Aptamers: an emerging class of molecules that rival antibodies in diagnostics. *Clin Chem* 45, 1628–1650.

Kelly, J. A., Feigon, J., Yeates, T. O. (1996). Reconciliation of the x-ray and NMR structures of the thrombin-binding aptamer d(GGTTGGTGTGGTTGG). *J Mol Biol* 256, 417–422.

Lew, A., Rutter, W. J., Kennedy, G. C. (2000). Unusual DNA structure of the diabetes susceptibility locus IDDM2 and its effect on transcription by the insulin promoter factor Pur-1/MAZ. *Proc Natl Acad Sci U S A* 97, 12508–12512.

Liss, M., Petersen, B., Wolf, H., Prohaska, E. (2002). An aptamer-based quartz crystal protein biosensor. *Anal Chem* 74, 4488–4495.

McGown, L. B., Joseph, M. J., Pittner, J. B., Vonk, G. P., Linn, C. P. (1995). The nucleic acid ligand: a new tool for molecular recognition. *Anal Chem* 67, 663A–668A.

Merchant, M., Weinberger, S. R. (2000). Recent advancements in surface-enhanced laser desorption/ionization-time of flight-mass spectrometry. *Electrophoresis* 21, 1164–1177.

Ravelet, C., Grosset, C., Peyrin, E. J. (2006). Liquid chromatography, electrochromatography and capillary electrophoresis applications of DNA and RNA aptamers. *J Chromatogr A* 1117, 1–10.

Schou, C., Heegaard, N. H. H. (2006). Recent applications of affinity interactions in capillary electrophoresis. *Electrophoresis* 27, 44–59.

Stadtherr, K., Wolf, H., Lindner, P. (2005). An aptamer-based protein biochip. *Anal Chem* 77, 3437–3443.

Tang, N., Tornatore, P., Weinberger, S. R. (2004). Current developments in SELDI affinity technology. *Mass Spectrom Rev* 23, 34–44.

Tombelli, S., Minunni, M., Mascini, M. (2005). Analytical applications of aptamers. *Biosens Bioelectron* 20, 2424–2434.

Tuerk, C., Gold, L. (1990). Systematic evolution of ligands by exponential enrichment: RNA ligands to bacteriophage T4 DNA polymerase. *Science* 249, 505–510.

Wiegand, T. W., Williams, P. B., Dreskin, S. C., Jouvin, M. H., Kinet, J. P., Tasset, D. (1996). High-affinity oligonucleotide ligands to human IgE inhibit binding to Fc epsilon receptor I. *J Immunol* 157, 221–230.

Xu, D., Xu, D., Yu, X., Liu, Z., He, W., Ma, Z. (2005). Label-free electrochemical detection for aptamer-based array electrodes. *Anal Chem* 77, 5107–5113.

CHAPTER 12

STRATEGY FOR USE OF SMART ROUTES TO PREPARE LABEL-FREE APTASENSORS FOR BIOASSAY USING DIFFERENT TECHNIQUES

BINGLING LI, HUI WEI, and SHAOJUN DONG

12.1 INTRODUCTION

Biosensors, which usually need biological molecules as recognition elements, are playing a growing role in various fields, due to their effective selectivity, high sensitivity, and easy performance. Thus, even for only one useful target, it is worth large numbers of designs and improvements to develop detection that is effective enough for a variety of applications. In this way, more and more techniques and novel recognition elements are being brought into bioassays.

Aptamers are screened through the systematic evolution of ligands by exponential enrichment (SELEX) process as functional oligonucleotides (Ellington and Szostak, 1990; Robertson and Joyce, 1990; Tuerk and Gold, 1990). Because of their generally impressive selectivity and affinity, they have already attracted much analyst attention and were quick to be used in analytical assays as novel recognition elements. Aptamers exhibit multifarious advantages over traditional recognition elements (Osborne and Ellington, 1997; Jayasena, 1999; Famulok et al., 2000; Tombelli et al., 2005a). For example, they are easy to synthesize in vitro, easy to modify, and flexible to design. These inherent advantages as oligonucleotides not only make target diversification but also improve detection efficiency. They can also help analysts simplify the analytical process and fabricate a sensing platform smartly and conveniently.

Aptamers in Bioanalysis, Edited by Marco Mascini
Copyright © 2009 John Wiley & Sons, Inc.

Until now, a large number of aptasensors developed in the past were rationally designed. Several of them have been proven feasible in practical biosamples. Various techniques to get a detectable signal have been used or combined, such as optical transduction (Lee and Walt, 2000), quartz crystal microbalance (QCM) (Liss at al., 2002), surface plasmon resonance (SPR) (Tombelli et al., 2005b), fluorescence (Nutiu et al., 2004; Pavlov et al., 2005; Yang et al., 2005; Rupcich et al., 2006), colorimetry (Lu and Liu, 2006, 2007), chromatography (Ravelet et al., 2006), and electrochemistry (Willner and Zayats, 2007). On the whole, among all the convenience-aimed aptasensors, label-free seems to be one of the most widely used methods with whatever detection techniques are adopted.

Label-free was at first employed to describe a process of avoiding a label (or rather, a radioactive label) on analytes [the analyte label-free (ALF) mode] and has attracted more and more attention since it appeared. The term has become a synonym for techniques that can release any one operation from labeling and is now considered to be one of the most promising areas in analytical science. In addition, such a label-free method is useful not only in basic research in laboratories but also exhibits significant market potential and promise in practical applications.

To some degree, the advantages of label-free detection are remarkable and are summarized in Table 12.1. As shown in the table, label-free is one of the most effective methods to make detection simpler and more convenient. However, its maturation requires greater effort and appears likely to escalate as a popular process in several traditional and new detection techniques. Undoubtedly, the development of aptamers endows this technique with more potential in future bioassays, due to the unique properties of aptamers.

As a matter of fact, most developed aptasensors have been designed in an ALF mode. Generally, the aptamers need to be assigned the responsibility to both recognize and produce signals. This means the detectable signals are produced mainly depending on the labeled probes on aptamers or other changes that aptamers undergo (e.g., conformational change) in the presence of targets. First, like some label-free immunoassays, this achieves detection while avoiding target destruction after labeling and holds the affinity at a higher degree. Second,

TABLE 12.1 Advantages of Label-Free Detection

Studying molecular interactions without the relatively complex step of modifying molecules and thus simplifying the experimental process

Experiencing the influence of label on physicochemical or binding properties and thus retaining higher levels of activity and affinity

Avoiding problems caused by secondary detection or auxiliary reagents and thus increasing sensitivity and specificity

Providing the potential for assays to be cheaper (no labeling or reagent costs)

Shortening assay development times, thus reducing the time required to complete detection

TABLE 12.2 Submodes in the Analyte Label-Free Mode

	SOALF Mode		SFALF Mode	
	POSOALF Mode	PFSOALF Mode	POSFALF Mode	PFSFALF Mode
Substrate to immobilize[a]	On	On	Off	Off
Signal probe labeling[b]	On	Off	On	Off
Smart degree[c]	*	**	**	***

[a]The aptamers need to be fixed onto certain types of substrates for recognizing, separating, and enriching targets or providing a collection platform for signals.
[b]Aptamers will be labeled with certain probes to produce detectable signals.
[c]The more asterisks there are, the smarter the mode is.

complicated steps of chemical modification on targets are not necessary, leading to relatively easier operations.

As a step further, ALF could be divided into the following modes, which are listed in Table 12.2.

1. *Substrate-on ALF (SOALF)*. Concretely, despite the term *analyte label-free*, the aptamers need to be fixed onto certain types of substrates to recognize, separate, and enrich targets or to provide a collection platform for the signals. Then, depending on the methods producing the signals, SOALF will be separated into two modes:

    a. *Probe-on SOALF (POSOALF)*. For this type of sensor, aptamers will not only be fixed but will also be labeled with probes to produce detectable signals.

    b. *Probe-off SOALF (PFSOALF)*. For this mode, signal probes will also be cut off. The signals are produced primarily based on changes in the sensing surface in the presence of targets (e.g., conformational changes, charge changes).

2. *Substrate-off ALF (SFALF)*. As a further simplified mode, substrates for immobility are not excluded. In this type of sensor, sensing systems are more smartly designed, and detectable signals are delivered only through disturbances resulting from the aptamer–target binding process. Obviously, it is a more convenient mode to use as an apsensor. Again, two discrete modes are included.

    a. *Probe-on SFALF (POSFALF)*. Similar to POSOALF, probes need to be labeled onto the aptamers to produce signals.

    b. *Probe-off SFALF (PFSFALF)*. This is the simplest mode of all: no fixation, no separation, no label, and no probe.

Now, in conjunction with Table 12.2, we introduce these easy label-free modes in detail to help us to understand how useful aptamers are in making detection simple and convenient. In terms of their degree of smartness, most attention will be focused on POSFALF and PFSFALF, especially PFSFALF.

It should be noted that there is another type of in vitro functional oligonucleotides selected: deoxyribozymes/ribozymes (DNAzymes/RNAzymes) (Breaker and Joyce, 1994, 1995; Cuenoud and Szostak, 1995; Li and Sen, 1996; Breaker, 1997; Sen and Geyer, 1998). Different from the aptamers, they are catalytic nucleic acids capable of catalyzing a broad range of reactions, including cleaving nucleic acid substrate (Breaker and Joyce, 1994, 1995), ligation (Cuenoud and Szostak, 1995), phosphorylation (Li and Breaker, 1999), and porphyrin metallation (Li and Sen, 1996). Because they have also been in common use for bioassays in recent years, sensors adopting these DNAzymes/RNAzymes are included. In the following section we use the word *aptasensors* to represent sensors that use either aptamers or DNAzymes/RNAzymes.

12.2 ELECTROCHEMICAL APTASENSORS

Electrochemical aptasensors have been developed widely for some period of time, especially in the most recent few years (Willner and Zayats, 2007). Due to the properties of the technique itself, most electrochemical aptasensors depend on electrodes for electron transfer. Thus, the SOALF mode could make sense. Here we introduce mainly easy routes in this mode rather than in the SFALF mode.

12.2.1 POSOALF Mode

Most electrochemical aptasensors reported belong to this mode, for it requires redox probes to produce electrochemical signals on the electrode surface. Wang's group has employed an indirect amplified electrochemical route using aptamer–protein complexes binding with inorganic nanocrystals as labels and has accomplished the multianalysis of various proteins by their specific aptamers (Hansen et al., 2006). A gold electrode is functionalized with aptamers specific to α-thrombin and lysozyme. The α-thrombin and lysozyme are modified with CdS and PbS quantum dots (QDs), respectively, and both are bound to the corresponding aptamers associated with the surface. The QD-functionalized proteins acted as tracer labels for analysis of the proteins. In the presence of nonfunctionalized α-thrombin or lysozyme, displacement of the corresponding labeled proteins followed by dissolution of the remaining QDs captured on the electrode surface, and then detection by electrochemical stripping of the ions released, enables quantitative detection of the two proteins. As described, QDs here are not taken as stable and strong fluorescence probes but endowed with a novel electrochemical function. Dong's group uses a similar sensing process but for the first time uses the QDs as an amplified electrochemiluminesence (ECL) co-reactant in aptasensors (as shown in Figure 12.1A) (Guo et al., 2008). Using these semiconductor nanocrystals, both lysozyme and α-thrombin are sensitively detected.

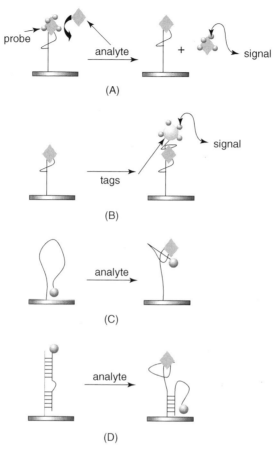

Figure 12.1   Principles of different types of POSOALF. (A) Probe-modified targets will be replaced by free targets and then collected in an electrochemical cell to produce a detectable signal (Guo et al., 2008). (B) Sandwich-type sensors in which aptamer-modified tags are generally used as signal elements (Ikebukuro et al., 2005; Mir et al., 2006; Polsky et al., 2006; Centi et al., 2007; Zheng et al., 2007; Zhou et al., 2007). (C) In the presence of analytes, conformational changes in aptamers will change the distance between the probe and the electrode. Then detectable electrochemical signals can be produced (Xiao et al., 2005). (D) TREAS mode. After adding analytes, the duplex will be destroyed and the modified probe will approach the electrode. Then an increased electrochemical signal can be detected (Xiao et al., 2005; Zuo et al., 2007).

Sandwich-type sensing platforms are also used widely in electrochemical aptasensors (Willner and Zayats, 2007), especially for common model molecules such as α-thrombin (Mir et al., 2006; Polsky et al., 2006; Centi et al., 2007) and platelet-derived growth factor (PDGF) (Zhou et al., 2007), which possess two active aptamer-binding sites. This type of sensor usually contains an aptamer-modified electrode as a detection substrate and another aptamer-modified electrochemical active material as a signal-producing probe (Figure 12.1B).

Until now, a variety of probes, such as horseradish peroxidase (HRP) (Mir et al., 2006) pyroquininoline quinone glucose dehydrogenase (PQQ-GDH) (Ike-bukuro et al., 2005), and nanoparticles (Polsky et al., 2006; Zheng et al., 2007), have been used in sandwich-type sensors. For example, a sensing platform is constructed by immobilizing one α-thrombin aptamer (15-mer) onto a gold electrode for capturing the target and the other aptamer (29-mer) modified with PQQ-GDH for detection (Zhou et al., 2007). In the presence of α-thrombin, the PQQ-GDH-modified aptamer will interact with the protein and an electrochemical signal is produced after the addition of glucose. In this way, as little as 10 nM α-thrombin can be detected selectively and a linear response is obtained between 40 and 100 nM.

Obviously, the sensors above are fabricated relatively complicatedly, because most sensors need multiple steps of modification, label, or separation. These steps might display their advantages, such as high selectivity and sensitivity, but still not avoid the defaults that label usually faces. As a matter of fact, most electrochemical aptasensors have been designed for speed and simplicity, which exemplify the inherent advantages of electrochemistry. Hence, most electrochemical aptasensors reported are generally simpler.

An original label-free electrochemical aptasensor is derived from the electronic aptamer-based (E-AB) sensors of Heeger and Plaxco (Xiao et al., 2005a). As shown in Figure 12.1C, this sensor is designed for α-thrombin, and the signal generation depends only on obvious, target binding–induced aptamer conformational change on a redox-aptamer-modified electrode. The aptasensor developed is sensitive and reusable. More important, it does not require modification of the analyses (Xiao et al., 2005a). Although this type of E-AB does not avoid the probe-label process completely, it does release analytes from modification or relatively complicated experiment steps. In successive works, a series of electrochemical aptasensors have been developed based on this principle, with the targets extended to other proteins (Lai et al., 2007), small molecules (Baker et al., 2006), and even metal ions (Xiao et al., 2007). The probe was extended to ferrocene as well by O'Sullivan's group (Radi et al., 2006). The signal-produced mode is also changed to improve the sensing efficiency according to the different targets.

As a further step, a target-responsive electrochemical aptamer switch (TREAS) was explored. TREAS is, in fact, a duplex-to-complex mode usually containing a duplex or partial duplex hybridized between an aptamer-containing strand and its complementary or partly complementary strand (Xiao et al., 2005b; Zuo et al., 2007). Under given conditions, these duplexes could be split in the presence of targets and induce a detectable signal change. The advantages of TREAS rest on the fact that they can be designed smartly to improve the sensors' performance and are widely adopted in various electrochemical methods. For example, the initial E-AB is improved from a signal-off to a signal-on sensor; thus, a one-order increase in sensitivity can been obtained (Figure 12.1D) (Xiao et al., 2005). Fan and co-workers have realized small-molecule ATP detection using a TREAS aptasensor (Zuo et al., 2007).

12.2.2 PFSOALF Mode

In this mode redox probes to produce detectable signals are no longer covalently labeled onto the aptamers in this mode. Probes dissolved in electrolyte solution can interact with aptamers immobilized on an electrode via (1) electrostatic repellence (for negatively charged probes), (2) electrostatic adsorption (for positively charged probes), and (3) intercalation (for DNA intercalators).

12.2.3 Electrochemical Impedimetric Aptasensors

Electrochemical impedance spectroscopy (EIS) is a rapidly developing electrochemical technique that has been incorporated in the design of biosensing systems to utilize its advantages, which include sensitivity, low cost, and convenience, as well as being label-free (Katz and Willner, 2003; K'Owino and Sadik, 2005; Pejcic and De Marco, 2006; Daniels and Pourmand, 2007). As noted by Radi et al. (2005), "EIS is an electrochemical technique for the investigation of bulk and interfacial electrochemical properties of any type of solid or liquid material connected to or part of an appropriate electrochemical transducer. Any intrinsic property of a material or a specific process that could affect the interfacial properties of an electrochemical system can potentially be studied by EIS." That is why EIS is popular in label-free sensor fabrication. This technique is well suited to monitoring the various stages necessary for characterizing and detecting the recognition event when an immobilized molecule interacts with its ligand (the analyte).

Aptamer-based impedimetric bioassays are developing rapidly. In 2005, two papers in this field published in the same issue of *Analytical Chemistry* (Radi et al., 2005; Xu et al., 2005). They brought EIS into an aptasensing process and gave simple PFSOALF examples of electrochemical detection. These aptamer-based impedimetric bioassays are designed to focus on protein detection in a general electrode–aptamer–target mode (the principle is illustrated in Figure 12.2A).

In Radi's work, a gold electrode is functionalized with thiolated α-thrombin aptamer (TBA) and blocked with neutral 2-mercapoethanol (MCE) to act as a sensing interface for α-thrombin (Radi et al., 2005). This negatively charged interface repels the anionic redox couple of $[Fe(CN)_6]^{3-/4-}$ and therefore repels the electron-transfer process. The electron-transfer resistance (Ret) produced is greater than that of bare electrodes. For α-thrombin the pH is nearly neutral, about 7.5, with a molecular mass of about 35,000 Da. When it binds its aptamer on the electrode here, the Ret increases due to an integrated function of both its bulk size and the resistive hydrophobic layer insulating the conductive support. The protein is detected simply by this further increase in electron transfer impedance (Ret), with a sensitive detection limit of about 5 nM. Similarly, in Xu's work,

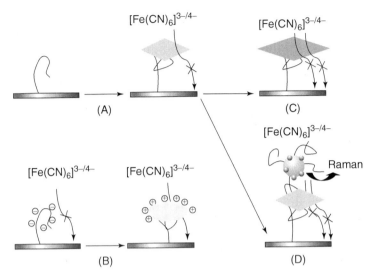

Figure 12.2   Principles of different types of impedimetric aptsensors: (A, B) the simplest modes for protein detection; (C) denaturalization by guanidine hydrochloride to amplify the signal; (2006); (D) scheme of a sandwich-type sensor. Aptamer/R6G-modified AuNPs are used as amplifying elements. [(A) from Radi et al. (2005); (B) from Rodriguez et al. (2005); (C) from Xu et al. (2006); (D) from Wang et al. (2007); Li and Dong (2008).]

using the same principle, immunoglobulin E (IgE) is taken as the target (Xu et al., 2005). The difference lies in the fact that the sensing surface is constructed on an array configuration of electrodes, which provides the ability to examine the effect of different base mutations in the aptamer sequence on the affinity for IgE–aptamer binding.

At almost the same time, another impedimetric aptasensor was developed using lysozyme as an analyte (Figure 12.2B) (Rodriguez et al., 2005). Despite the same electrode–aptamer–target method being used, the result is different. A negative sensing surface to repel a $[Fe(CN)_6]^{3-/4-}$ probe is generated through a biotinylated aptamer linked to a streptavidin-functionalized indium tin oxide electrode. Then the association of protein leads to a Ret decrease rather than an increase. That is attributed to the positive charge exhibited by lysozyme at about pH 7. The +8 net charges are high enough to provide an excess positive charge, which facilitates access of the probes and the resulting redox reaction. The impedimetric aptasensors described above are the simplest to use for protein detection. Recently, the design has been improved by a series of amplified methods to realize higher sensitivity (Xu et al., 2006; Li and Dong, 2008). Taking α-thrombin as a model molecule, Fang's group added guanidine hydrochloride disposal after target binding (Figure 12.2C) (Xu et al., 2006). It was found that protein denatured by guanidine hydrochloride can remain on the electrode and lead to a further enhanced Ret signal. Due to this smart design, about a 10-fold higher sensitivity than that obtained without amplification is achieved.

Another example comes from Dong's group (Li and Dong, 2008). For α-thrombin detection, this work is based on the fact that one α-thrombin molecule has two active sites for its 15-mer aptamer. As shown in Figure 12.2D, gold nanoparticles (AuNPs) are first functionalized by aptamers for recognizing protein and rhodamine 6G (R6G) molecules for blocking the surface. During the sensing process, thiolated α-thrombin-binding aptamer is first immobilized on the gold electrode to capture the target, and once in the presence of the α-thrombin, the bifunctionalized AuNPs could bind further to α-thrombin, forming a sandwich-type sensing system on the electrode. For negatively charged bifunctionalized AuNPs, the Ret signal is obviously amplified (Figure 12.3) (Li and Dong, 2008).

Through such an amplified method, a detection limit of 0.02 nM is realized. In fact, a disadvantage for impedimetric sensing is that the detection is easily interfered with by the nonadsorption of other materials on the electrode. So here, R6G molecules modified on AuNPs can provide a very selective surface-enhanced Raman scattering (SERS) method to realize qualitative recognition using gold nanoparticle–induced "hotspots" (as shown in Figure 12.4) (Wang et al., 2007). As shown, although it belongs to the POSOALF mode, the foregoing advantages still provide the potential to carry out mixed-technique analysis more easily.

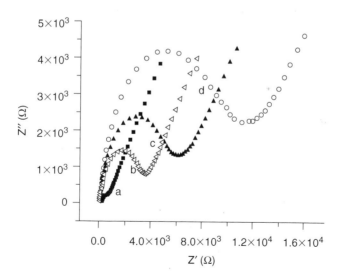

Figure 12.3   Niquist plots of (a) a bare Au electrode, (b) an Au/TBA/MCE system, (c) an Au/TBA/MCE/α-thrombin system, and (d) an Au/TBA/MCE/α-thrombin/AuNPs system. The concentration of α-thrombin was 45 nM. TBA, α-thrombin-binding aptamer; MCE, blockers modified after the TBA, 2-mecaptoethanol). [From Li and Dong (2008), with permission. Copyright © 2007 Elsevier.]

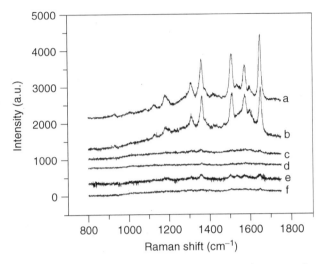

Figure 12.4 SERS spectra of Raman reporters with various proteins. (a) 100 nM α-thrombin. (b) 200 nM β-thrombin, 200 nM γ-thrombin, and 100 nM α-thrombin. (c) 200 nM β-thrombin and 200 nM γ-thrombin. (d) 400 nM β-thrombin. (e) 1 μM BSA and (f) without protein. [From Wang et al. (2007), with permission. Copyright © 2007 Royal Society of Chemistry.]

Undoubtedly, bulk proteins are very suitable targets for impedimetric aptasensors. However, the conditions are not the same for small molecules because most small molecules have a low molecular mass or low net charge that is not strong enough to affect conditions on the electrode surface.

Recently, strategies to overcome the shortcomings noted above have been developed to make EIS detection available for small molecules. Usually, the TREAS mode is needed. In one of the strategies, TREAS is designed with a part-duplex DNA consisting of an amine-functionalized aptamer–containing strand and its partially complementary strand (Figure 12.5A) (Zayats et al., 2006). Linked by dithiobis (succinimidylpopionate), the part duplex is immobilized onto a gold electrode through the amine group on the apatmer-containing strand. In the presence of analytes, the part duplex is separated due to aptamer–target interaction, leading to loss from the electrode of the partially complementary strand. The disassociation results in removal of the negative charge from the electrode surface and thus decreased interfacial electron-transfer resistance in the anion redox couple $[Fe(CN)_6]^{3-/4-}$. Taking AMP as a model, fast and sensitive detection is realized, with a response time of less than 5 minutes and a detection limit of 2×10^{-6} M. By such a TREAS mode, EIS signals are not transduced through the analytes themselves but through the changes in TREAS after binding with the analytes. It is clear that smart concepts will ultimately extend the EIS technique to wider applications.

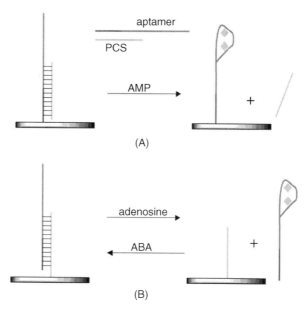

Figure 12.5 Principles of the two types of impedimetric aptasensors in TREAS mode. (A) An aptamer-containing strand is immobilized onto the electrode (B) A partlially complementary strand (PCS) is immobilized onto the electrode ABA represents the adenosine-binding aptamer. [(A) From (Zayats et al. 2006); (B) from (Li et al. 2007b).]

Dong's group developed a different but also interesting strategy (Figure 12.5B) (Li et al., 2007b). As an opposite route to the design described above (Figure 12.5A), it is the partially complementary strand that is immobilized on a gold electrode (through thiol modified on the strand), not the apatmer-containing strand (for adenosine). Correspondingly, when analytes are present, the aptamer-containing strand would fall off the electrode, which leads to decreased interfacial electron-transfer resistance in the anion redox couple $[Fe(CN)_6]^{3-/4-}$ (Figure 12.6A). The detection limit for adenosine in this work is as low as 5×10^{-7} M. Such a higher sensitivity compared to that of the sensor described above (Zayats et al., 2006) may be attributed to loss of the relatively longer aptamer-containing strand. This is just one of the advantages of this strategy (Li et al., 2007b). Meanwhile, the sensing interface is also endowed with regenerative ability by direct rehybridizing with the aptamer-containing strand. Through testing, the sensing interface can be recovered to more than 90% after one detection (Figure 12.6B). It is notable that the method proposed does not rely on molecule size or the aptamer's conformational change, so it may possess a potential for the wider application of more targets, even proteins.

Recently, such a strategy has been extended to protein detection (Du et al., 2008). In this technique a unimolecular DNA strand containing two types of aptamers for different analytes is designed. As an example, two aptamers are

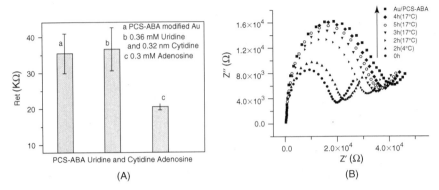

Figure 12.6 Comparison of Ret between 0.3 mM adenosine (c) and a mixed solution of control molecules (0.36 mM uridine and 0.32 mM cytidine) (b). In the presence of adenosine, obvious Ret decrease is observed; in the presence control molecules, no decrease is observed. The error bars represent the standard deviation of two measurements. [From Li et al. (2007), with permission. Copyright © 2007 Royal Society of Chemistry.]

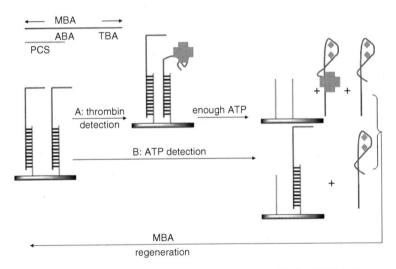

Figure 12.7 Principle of the multifunctional aptasensor. ABA, ATP-binding aptamer; TBA, α-thrombin-binding aptamer; MBA, strand containing the aptamers; PCS, strand partly complementary to MBA. [From Du et al. (2008).]

chosen for ATP and α-thrombin and combined in a unimolecular DNA strand called a mixed binding aptamer (MBA) (Figure 12.7). In this MBA, the aptamer for ATP is partially hybridized with the partially complementary strand, while the aptamer for α-thrombin is free beside the part duplex. So in the presence of the protein α-thrombin, the interfacial electron-transfer resistance (in the anion redox couple $[Fe(CN)_6]^{3-/4-}$) is increased as explained earlier, which could be used to recognize the analyte. Furthermore, a sensitive detection limit of 1×10^{-11} M is achieved. However, the specialty of this sensor rests with its regeneration method and potential for multianalysis. After the protein has been detected, the sensing surface is treated with a large amount of ATP. Following the principle introduced above, enough ATP could draw all the MBA (containing that bound with α-thrombin) away from the sensing surface and prepare it for regeneration. This process may also be used to detect ATP when treating protein-covered electrode with the required concentration of ATP. Correspondingly, after treatment with ATP, the electrode could be recovered again using MBA (Figure 12.7). However, such a fabricated sensing system could not be used for samples with multiple analytes in one solution. So improvements are still needed.

12.2.4   Electrochemical Aptasensors with Nonlabeled Redox Probes

An electrochemical sensor with a nonlabeled redox probe shakes off probes labeled on the apatmers but still requires that the probes indicate changes on the sensing interface. One commonly used probe, methylene blue (MB), belongs to the phenothiazine family and is an aromatic cationic dye showing optically and electrochemically active properties (Tuite and Norden, 1994). Usually, MB can bind with dsDNA or tRNA via intercalation, electrostatic absorption, or G-base binding, which has been used widely in DNA sensors and recently, in aptasensors (Tuite and Norden, 1994; Bang et al., 2005).

Kim's group uses MB as a probe in a molecule beacon such as the detection of α-thrombin (Figure 12.8A) (Bang et al., 2005). At first, an amino-functionalized ssDNA containing 15-mer α-thrombin aptamer is synthesized in a hairpinlike conformation with five base pairs on the arm. Then the ssDNA is linked to a carboxyl-modified gold electrode, followed by blocking the electrode surface with BSA. Such a modified electrode is treated further with an MB solution to produce an MB intercalated DNA-functionalized sensing surface. Through differential pulse voltammetry (DPV), a reduction peak of MB incorporated into the arms of the molecule beacon is observed at −0.2 V. In the presence of α-thrombin, more of the hairpinlike aptamer beacons are changed into a G-quadruplex conformation state due to aptamer/α-thrombin interaction. The hairpin structure is thus opened and the MB is released, which leads to a decreased MB current at −0.2 V. Using this sensor, a linear range between 0 and 50.8 nM ($r = 0.999$) is reached, with an estimated detection limit of 11 nM. The results demonstrate that the electrochemical method using an aptamer probe is indeed convenient and allows quantitative detection of target protein.

Using another route, Dong's group realized MB-based small-molecule detection using TREAS. Like the EIS sensor described earlier (Figure 12.5B), a part duplex consists of a partially complementary strand immobilized onto a gold electrode with an aptamer-containing strand for recognition (Figure 12.8B). Taking adenosine, for example, before treatment with the analyte, MB molecules will

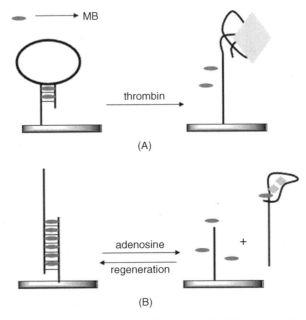

Figure 12.8 Two types of sensors with MB as a probe: (A) aptamer-beacon approach; (B) TREAS approach. [(A) From Bang et al. (2005); (B) from Li et al. (2007b).]

intercalate into the part duplex and produce an obvious current at -0.2 V. Once in the presence of adenosine, the aptamer–analyte interaction will destroy the duplex and reduce the amount of MB intercalating the DNA, which eventually results in decreased current (Figure 12.9). The sensor is sensitive, selective, and could be regenerated for later use.

Another commonly used redox probe, $[Ru(NH_2)]_6{}^{3+}$, could interact with DNA molecules through electrostatic adsorption and thus be used as an indicator for the amount or density of DNA on an electrode (Yu et al., 2003; Su et al., 2004). Also, protein is detected using $[Ru(NH_2)]_6{}^{3+}$ as a probe (Figure 12.10A) (Cheng et al., 2007). Antilysozyme DNA aptamers are immobilized on gold surfaces by means of self-assembly, for which the surface density of aptamers is determined by cyclic voltammetry (CV) for redox behavior of cations $[Ru(NH_2)]_6{}^{3+}$ bound to the surface via electrostatic interaction with the DNA phosphate backbone. Upon incubation of the electrode with a lysozyme solution, the CV response of surface-bound $[Ru(NH_2)]_6{}^{3+}$ decreases substantially. This results from the fact that the binding of lysozyme to an aptamer-modified gold electrode (with its inherent ability to bind in a $1:1$ ratio) should reduce the negative surface charges (contributed by the DNA backbone) since lysozyme has a net charge of $+8$. Then the relative decrease in the integrated charge of the reduction peak can be tabulated as a quantitative measure of protein concentration, with a detection limit of 0.5 μg mL^{-1}.

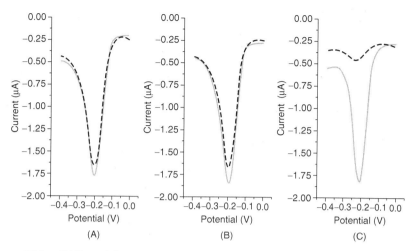

Figure 12.9   Differential pulse voltammetries of the sensing system before (solid line) and after (dashed-dotted line) being reacted with three nucleosides at 1 mM in 25 mM TRIS–HCl solution containing 20 mM NaCl: (A) uridine; (B) cytidine; (C) adenosine. [From Wang et al. (2008), with permission.]

TREAS is also used for small-molecule detection (Figure 12.10B) (Shen et al., 2007). In this case, the detectable signal is produced from discriminate electrostatic interaction between $[Ru(NH_2)]_6^{3+}$ and the part-duplex or aptamer-containing strand. A chronocoulometric technique is used to indicate the result. So when an analyte (AMP) exists, the signal is reduced by the amount by which $[Ru(NH_2)]_6^{3+}$ confined on the electrode surface decreased. Recently, a ferrocene-functionalized cationic polyelectrolyte poly(3-alkoxy-4-methylthiophene) was used as a novel redox probe (Le Floch et al., (2005)). Thiolated α-thrombin-binding aptamer is linked to an Au electrode. The electrostatic interaction of the polyelectrolyte with the aptamer yields a voltammetric response in the ferrocene. After binding with α-thrombin, the cationic polymer is blocked for proteins that are nearly neutral under binding conditions, so the electrochemical response is reduced. Although it is convenient to operate, the detection limit of this sensor is not satisfactory, being higher than 10^{-6} M.

12.3   FLUORESCENT MOLECULAR SWITCHES

Fluorescence spectroscopy is a traditional technique for fabricating biosensing systems and has been used widely in aptasensors for small molecules (Stojanovic et al., 2000a, 2001; Nutiu and Li, 2003), proteins (Hamaguchi et al., 2001; Yang et al., 2005; Li et al., 2007a), metal ions (Li and Lu, 2000; Lu et al., 2003; Nagatoishi et al., 2005; Shen et al., 2007), and even cells (Herr et al., 2006; Shangguan et al., 2006).

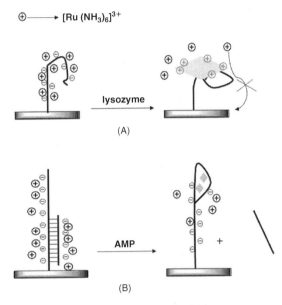

Figure 12.10 Two types of sensors with $[Ru(NH_2)]_6^{3+}$ as a probe: (A) Aptamer beacon approach; (B) TREAS approach. [(A) From Cheng et al., (2007); (B) from Shen et al. (2007).]

Until now, several strategies have been reported for converting an existing aptamer into a fluorescent probe, among which the methods used most frequently are the molecular beacon (aptamer beacon) approach (Figure 12.11A) and the duplex-to-complex switching approach (Figure 12.11B) (Navani and Li, 2006). Both approaches profit from the inherent properties of aptamers as oligonucleotide molecules. Thus, most aptasensors with fluorescence detection can be prepared smartly, which means that in SFALF modes, general steps such as separation and immobilization can easily be left out.

12.3.1 POSFALF Mode

Although in the SFALF mode, a majority of fluorescent aptasensors are still "probe-on" (POSFALF), which mode is the most popular for use in fluorescent assays. The aptamer beacon approach can often be divided into two commonly adopted aspects. One, like a molecular beacon for DNA sensing, places an aptamer sequence in a molecular beaconlike hairpin structure end-labeled with a fluorophore (F) and a quencher (Q) (Figure 12.11A) (Hamaguchi et al., 2001; Yamamoto and Kumar, 2000; Lu et al., 2003; Nagatoishi et al., 2005; Li et al., 2007; Shen et al., 2007) the binding of the target disrupts the hairpin stem and separates F from Q, leading to a detectable increased fluorescence. The other aspect often labels the aptamer with F1 and Q/F2 at each end, respectively

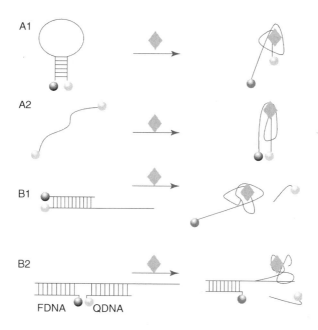

A1

A2

B1

B2

FDNA   QDNA

Figure 12.11 Two approaches to designing POSFALF fluorescent aptasensors. The aptamer beacon approach contains two modes of signal-on (A1) (Yamamoto and Kumar, 2000; Hamaguchi et al., 2001) and signal-off/FRET (A2) (Li et al., 2002; Ueyama et al., 2002). The duplex-to-complex switching approach contains two signal-on modes: B1 (Rupcich et al., 2005; Elowe et al., 2006) and B2 (Nutiu and Li, 2003).

(Figure 12.11A) (Li et al., 2002; Ueyama et al., 2002). It requires only conformational change of aptamers once they bind to targets to get a quenched or fluorescence resonance energy transfer (FRET) signal.

The duplex-to-complex switching approach (Figure 12.11B) is usually divided into two modes. One often places an F-labeled (Q-labeled) aptamer in a duplex structure with a (partial) complementary sequence labeled with a Q (F) (Figure 12.11B1) (Rupcich et al., 2005; Elowe et al., 2006). Addition of the target forces the departure of the partial complementary strand from the aptamer, accompanied by an increase in fluorescence. In the other mode, one often needs to design a duplex consisting of one aptamer-containing sequence and two partially complementary strands that are labeled with F (FDNA) and Q (QDNA), respectively, on the ends near each other (Figure 12.11B2) (Nutiu and Li, 2003). When target in present, the QDNA will be completed, due to the formation of an aptamer–target complex, and thus an increased fluorescence is observed. A representative aptasensor fabricated in this way was developed by Li's group, in which both small molecules of ATP and proteins of α-thrombin were detected successfully (Figure 12.11B2) (Nutiu and Li, 2003). Obviously, in comparison to the aptamer beacon approach, the duplex-to-complex switching approach is a little more complex but appears to be more applicable, for in this way the dependence on the design of a hairpin structure or conformational change of aptamers is reduced, which exposes it to wider targets. However, the sequences of duplexes in this approach should be more carefully optimized to guarantee a satisfactory sensing efficiency and high signal/noise ratio. Now, through continuous improvement, the duplex-to-complex switching approach has been used successfully in enzyme-related assays (Rupcich et al., 2006) and DNAzyme-related assays (Stojanovic et al., 2001b).

Recently, with the development of synthesis techniques, new fluorescent probes are employed continually as an alternative to commonly used F or Q to decrease the background or improve the practicality (Nagatoishi et al., 2005; Yang et al., 2005). For example, Tan's group has labeled PDGF-binding DNA aptamer with one pyrene molecule on each end to produce a time-resolved aptamer beacon (Yang et al., 2005). Upon PDGF binding, the aptamer switches its fluorescence emission from 400 nm (pyrene monomer, with a fluorescence lifetime of about 5 ns) to 485 nm (pyrene excimer, with a fluorescence lifetime of about 40 ns). Such wavelength shifting, especially the time-resolved capabilities, would benefit practical samples by decreasing the background signals coming from the native fluorescence of the biological environment (Yang et al., 2005).

Novel nanoparticles such as QDs are also used (Dwarakanath et al., 2004; Levy et al., 2005; Choi et al., 2006; Ikanovic et al., 2007; Liu et al., 2007). Ellington's group fabricated a duplex-to-complex switching aptasensor by labeling α-thrombin-binding aptamer with QDs and the complementary strand with Q, which could quench the fluorescence of QDs before sensing (Levy et al., 2005). The addition of α-thrombin will withdraw the QDs from Q, leading to a recovered fluorescence.

12.3.2 PFSFALF Mode

Although labeled probes might produce a higher signal-noise ratio and relatively wider applications in fluorescent aptasensors, smarter routes with no label steps are also promising. Their advantages include convenience, low cost, ease of preparation, and a sensitivity similar to that of labeled methods.

"We report the first examples of modular aptametric sensors which transduce recognition events into fluorescence changes through allosteric regulation of noncovalent interactions with a fluorophore" (Stojanovic and Kolpashchikov, 2004). These sensors are designed to access the PFSFALF mode in fluorescent aptasensing, which makes them readily applicable for intracellular applications. As shown in Figure 12.12, the main sensing systems contain three parts: (1) aptamer of analyte (ATP, etc.) as a recognition domain, which binds the analyte; (2) RNA aptamer of malachite green (MG, a dye) as a reporting domain, which increases the quantum yield of this dye up to 2000-fold upon binding and is thus used to signal the binding event of an analyte through binding to a fluorophore; and (3) a communication module, which is designed as a conduit

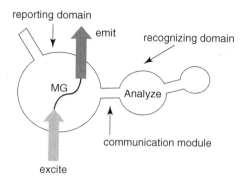

Figure 12.12 Parts of the sensing system: a recognizing domain to bind analyte, a reporting domain to bind MG molecule, and a communication module to combine the recognizing domain and the reporting domain. Induced by the interaction between analyte and its aptamer, the fluorescence intensity ($\lambda_{ex} = 610\,nm$) of MG will be enhanced. [From Stojanovic and Kolpashchikov (2004).]

between the recognition and signal domains. Once in the presence of analyte, the recognition process (at the recognition domain) will be transduced through a communication module to the reporting domain, leading to increased recognition of MG and concomitant fluorescence enhancement. By detecting increased fluorescence, ATP and flavine mononucleotide (FMN) are detected successfully and theophylline (TH) is distinguished from its closely related molecule, caffeine. Meanwhile, for better sensing, the communication module is carefully optimized for its key role in signal transduction. It is also proven that all the sensing systems are feasible in mimicking intracellular milieu, but only the MG-FMN sensor shows an impressively stronger signal than that of any of the aptameric sensors reported previously. Even so, the smart method shows the potential worth of greater efforts to use this sensor.

In fact, aptamers selected capable of reporting domains are limited. Another series of fluorescent aptasensors take different routes and serve as complementary ways to make the PFSFALF more easily prepared and available for wider applications. This type of sensor is more similar to the POSFALF mode and can also be divided into two parts: the aptamer beacon approach and the duplex-to-complex switching approach.

Using the aptamer beacon approach depends mainly on conformational changes in aptamers once binding to targets. In an example reported by Bai's group, they take [Ru(phen)$_2$(dppz)]$^{2+}$ (its structure is shown in Figure 12.13A) as a light switch complex and detect such proteins as IgE, α-thrombin, and PDGF-BB (Jiang et al., 2004). [Ru(phen)$_2$(dppz)]$^{2+}$ has a high binding affinity to duplex nucleic acid (K_d, ~10^6 M^{-1}). It has no luminescence (it is excited at 450 nm) in aqueous solution, as the triplet MLCT (metal-to-ligand charge transfer) excited state is effectively quenched by the hydrogen in the ligand. When it binds to dsDNA, the interaction between the ligand and the base pairs

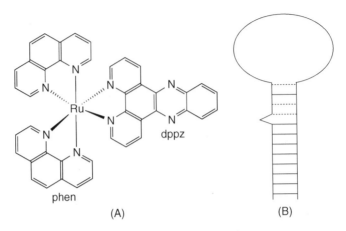

Figure 12.13 (A) Structure of $[Ru(phen)_2(dppz)]^{2+}$; (B) secondary structure of IgE aptamer. Dashed lines represent the non-Waston–Crick base pairs. [From Jiang et al. (2004).]

of duplex nucleic acid protects the phenazine nitrogen from water, leading to an emission at 610 nm up to 10^4-fold higher. Based on this principle, IgE and its 37-mer aptamer are at first chosen as a model system. As reported, IgE aptamer here has a predicted stem–loop structure (Figure 12.13B), including nine Waston–Crick base pairs and three non-Waston–Crick base pairs, which makes it suitable to bind with the dye and thereby leads to a 20-fold emission increase (the dye/DNA ratio is $8:1$). When the protein is added to the dye–aptamer solution, a significant luminescence intensity decrease is observed within 3 minutes. The reason for the decrease is hypothesized as being that upon protein binding the aptamer conformational change induced, as well as the blocking of $[Ru(phen)_2(dppz)]^{2+}$ intercalation by protein, would result in reduced dye molecules intercalating with the aptamer and thus in a significant protein-dependent fluorescent change. Through the fluorescent change, sensitive and simple detection is realized, with an IgE detection limit of 500 pM in the physiological buffer and almost no disturbance from other proteins.

"The method . . . importantly was applicable to RNA aptamers which have a larger population than DNA aptamers" according to Bai and co-workers (Jiang et al., 2004). Indeed, that may be the most notable advantage for this route and most other nonlabeled sensors for "most of the signalling aptamers developed so far are labeled DNA aptamers, but the fluorescent labelling of RNA aptamers, which have a larger population than DNA aptamers, is difficult because of the instability of RNA molecules" (Jiang et al., 2004). In the same work, detection in human serum is also undertaken. It is found that the detection efficiency in the 1% serum solution remains similar to that in PBS solution, but is much decreased in more concentrated serum solution because of the higher background caused by tailing of broad serum peak around 520 nm. That is just one of the challenges that most PFSFALF routes should confront and try best to resolve when related to practical detections.

Other systems of α-thrombin/RNA aptamer, PDGF-BB/DNA aptamer, are also proven viable in this work (Jiang et al., 2004) and the strategy is then used for small molecules such as ATP (J. Wang et al., 2005). As a further step, other dyes, such as TOTO, are also employed to improve the method (Zhou et al., 2006). A simple route has been fabricated in a duplex-to-complex switching approach by Dong's group (Li et al., 2007a): "...In principle, if the aptamers did not contain a secondary structure into which dyes intercalated, this method could be applicable." That is one of the aspects that makes a duplex-to-complex switching approach promising.

In this strategy, an α-thrombin/15-mer DNA aptamer system is taken as a model and ethidium bromide (EB) is used as a fluorescing molecular switch. The mechanism of the interaction between the dye and DNA has been delineated clearly. Following the neighbor exclusion principle, where every second site along the helix is unoccupied, EB could readily intercalate between the base pairs of dsDNA. When it is dissociative in aqueous solution EB, $[Ru(phen)_2(dppz)]^{2+}$ shows a low fluorescence intensity attributed to efficient quenching of the excited state by transferring an amino proton to a solvent (water) molecule. Whereas when EB intercalates to dsDNA it is shielded to some extent and shows obvious enhancement in fluorescence intensity (up to nearly 11 times that in its free state), a similar property is not evident when it is mixed with ssDNA or quadruplex DNA, especially with a 15-mer quadruplex, which is simply a more stable structure (than an unfolded conformation) formed when an anti-α-thrombin aptamer binds to α-thrombin. The design is based on this property. As shown in Figure 12.14A, a 15-mer anti-α-thrombin aptamer is hybridized with its complementary strand to form a duplex-to-complex switch. Obviously, in the absence of α-thrombin, an EB molecule intercalates to the DNA duplex with high efficiency, and correspondingly, high fluorescence intensity ($\lambda_{em} = 600$) can be detected. Accompanied by the addition of increasing amounts of α-thrombin, competition between the α-thrombin and the complementary strand for aptamer will lead to partial unwinding of the DNA double-helix structure and a more quadruplex/α-thrombin structure with ssDNA (complementary strands). This, to some extent, releases to the solvent (water) EB being shielded by dsDNA and ultimately makes the fluorescence intensity decrease. As shown in Figure 12.14B, after α-thrombin was added, decreased fluorescent intensity was observed and it took 30 minutes for the α-thrombin/dsDNA/EB system to attain equilibrium. However, if α-thrombin was replaced by a BSA molecule, no fluorescent decrease was observed. Thus, α-thrombin here is detected selectively and sensitively, with a detection limit of 2.28 nM. This method successfully inherits its design concept from labeled duplex-to-complex switching and keeps the majority of advantages that nonlabeled aptamer beacon methods possess. At the same time, it extends targets to systems in which targets have the ability to compete with the complementary strands of their aptamers, or the aptamers do not contain a secondary structure into which EB intercalates.

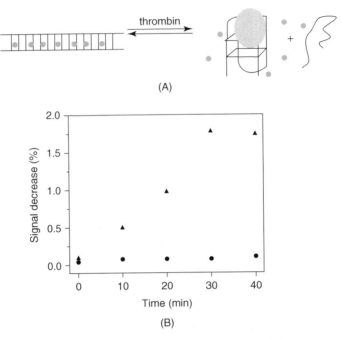

Figure 12.14   (A) Duplex-to-complex switching approach. (B) Time dependence of flu-orescence response after a 40-minute incubation. EB/dsDNA/α-thrombin (solid triangles) and EB/dsDNA/BSA (solid dots). [From Li et al. (2007a), with permission. Copyright © 2007 Royal Society of Chemistry.]

The key to the entire response efficiency in this type of sensor is the ability of aptamers to compete with the complementary strand. Although α-thrombin can bind its aptamer with K_d values as low as in the nanometer range, it is still difficult for it to compete with a completely complementary strand to an evident degree, which results in several defaults that limit the detection efficiency: for example, the relatively longer response time (30 minutes) and limited detectable range (to ca. 22.8 nM) (Li et al., 2007a). However, these problems can be resolved by redesigning the duplex-to-complex system and replacing the fluorescing molecular switch EB with other, more effective dyes.

12.4   COLORIMETRY

Due to their simple performance, and especially the potential ability to eliminate the use of analytical instruments, more and more colorimetric sensors are being developed, including DNA sensors (Elghanian et al., 1997), pH sensors (Gazda et al., 2004), metal ion sensors (Ghosh et al., 2006), and now, aptasensors (Huang et al., 2005; Lu and Liu, 2006, 2007; Wang et al., 2006; Zhao et al., 2007).

12.4.1 POSFALF Mode

A large part of colorimetric aptasensors are in this mode. Most of them employ AuNPs as sensing elements. For example, Chang's group has developed a highly specific colorimetric sensing system for PDGF and their reporters (PDGFRs) using AuNPs (Huang et al., 2005). The red dispersed AuNPs are modified using a PDGF-binding aptamer (apt-AuNP) that has two binding sites on one PDGF molecule. Taking PDGF-AA as an example (Figure 12.15A), its addition could draw AuNPs together through one (target) to two (aptamer) binding reactions. That leads to aggregations of AuNPs and a corresponding purple color. At an optimized condition of 8.4 nM apt-AuNPs in 200 mM NaCl, the color changes from red to purple in the presence of 10 nM PDGF-AA, and the linear range calculated by the extinction ratio A_{650}/A_{530} is 25 to 75 nM, with a response time of 60 minutes. Other proteins, such as PDGF-AB, PDGF-BB, and PDGFR-β, are detected sensitively in the same way.

Another series of AuNP-based colorimetric aptasensnors have been fabricated by Lu's group in a very smart design in which small molecules make use of a duplex-to-complex mode (Lu and Liu, 2006, 2007). In the example shown in Figure 12.15B, AuNPs are reduced to purple aggregates by a duplex consisting of an aptamer-contained chain and two complementary chains in which AuNPs are modified on the 3' end (3'-Adap$_{Au}$) or the 5' end (5'-Adap$_{Au}$), respectively

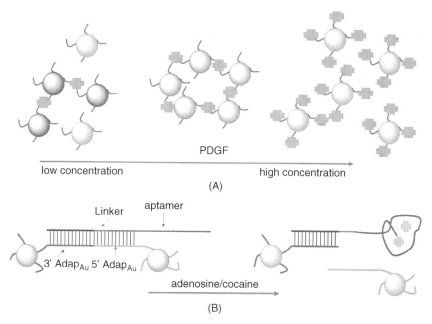

Figure 12.15   (A) Aggregation of aptamer-AuNPs in the presence of PDGFs at low, medium, and high concentrations; (B) colorimetric detection of small molecules. [(A) From Huang et al. (2005); (B) from Liu and Lu (2006).] (See insert for color representation.)

(Liu and Lu, 2006). In the presence of enough target molecules, the aggregates are destroyed by the competition between the target and the chain (5′-AdapAu), leading to a color change from purple to red. By this means, adenosine and cocaine are fast (less than 10 seconds) and easily detected with only the naked eye at some concentration range. "These aptamer-based tools could become a very useful addition to the nanostructure biodiagnostics work box. Imagine that simple litmus tests for every analyte for which aptamers exist were available!" (Famulok and Mayer, 2006).

In fact, such a duplex-to-complex concept is derived from a series of colorimetry processes developed for metal ions by the same group. Using the inherent properties of DNAzymes and a similar purple-to-red principle, metal ions such as Pb^{2+} (Liu and Lu, 2003), Cu^{2+} (Liu and Lu, 2007), and UO_2^{2+} (Liu et al., 2007) have been detected efficiently. This duplex-to-complex concept is now being transferred to the "dipstick" method as a further step in approaching practical applications (Liu et al., 2006).

12.4.2 PFSFALF Mode

In fact, POSFALF colorimetric aptasensors described above are very simply fabricated; however, the PFSFALF aptasensors are designed more smartly and provide very interesting strategies. One of the most popular nonlabel routes in colorimetric aptasensnors still employs AuNPs as sensing elements. This is due primarily to the fact that the dispersed degree of the nanoparticles can be influenced readily by the surrounding environment, which may lead to a corresponding color change between red and purple.

In Rothberg's germinal work, they note the selective adsorption of ssDNA on AuNPs and show that ssDNA can stablize AuNPs against aggregation at a salt concentration that would ordinarily screen the repulsive interactions of citrate ions (Li and Rothberg, 2004). Because of that, the color of AuNPs is determined principally by surface plasma resonance, and because it is dramatically affected by nanoparticle aggregation, a simple colorimetric hybridization assay can be realized by using the difference between ssDNA and dsDNA electrostatic properties.

Inspired by Rothberg's work, nonlabeled colorimetric aptasensors have been fabricated. By making use of aptamers as nucleic acid molecules, this concept has been extended to additional targets, such as metal ions (Wang et al., 2006) and proteins (Wei et al., 2007)

Dong's group reports a simple, sensitive, and label-free 17E DNAzyme-based sensor for Pb^{2+} detection. The catalytic activity of some DNAzymes is divalent metal ion–specific, just as the catalytic activity of some protein enzymes is metal ion cofactor–dependent (Santoro and Joyce, 1997; Carmi et al., 1998). In this work, the authors focus on 17E DNAzyme, which is a divalent Pb^{2+}-specific enzyme employed widely in Pb^{2+} sensors (Santoro and Joyce, 1997). As shown in Figure 12.16A, in the presence of Pb^{2+}, 17E DNAzyme could cleave the substrate 17DS, which could release ssDNA (including 17E and fragments from 17DS)

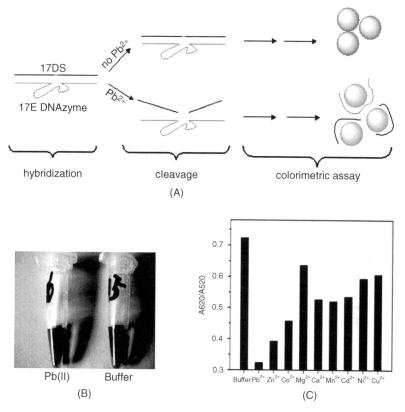

Figure 12.16 (A) AuNP-based colorimetric detection for Pb^{2+}. (B) 200 μL AuNPs/17E-17DS duplex solutions in the absence (right) and in the presence (left) of 50 μM Pb^{2+} after addition of 20 μL of 0.5 M NaCl. (C) Corresponding columns of the absorption ratio (A$_{620}$/A$_{520}$) of 200 μL of GNPs/17E-17DS duplex solutions in the presence of various 50 μM metal ions after addition of 20 μL of 0.5 M NaCl. [From Wei et al. (2008), with permission.] (See insert for color representation.)

from the hybridized 17E–17DS duplex. These ssDNA could be adsorbed on the red dispersed AuNPs, thus protecting the nanoparticles from aggregation in the presence of a given high concentration of salt (ca. 50 mM here). The characteristic SPR absorption band of AuNPs at about 530 nm is observed. However, the 17E–17DS duplex is retained in the absence of Pb^{2+}, which could not adsorb onto the AuNPs and thus could not stabilize the AuNPs under the same salt concentration. Accordingly, a shoulder band occurs at about 620 nm beside the SPR band at about 530 nm. When the Pb^{2+} is 50 μM, the color of AuNPs is changed from red to purple, which indicates that the Pb^{2+} can be monitored directly by the naked eye at this concentration (Figure 12.16B). Meanwhile, the absorption ratio (A$_{620}$/A$_{520}$) decreases gradually with an increase in Pb^{2+}

concentrations, presenting a positive correlation between decrease and Pb^{2+}. The selectivity of the sensors has already been proven through controlled experiments (Figure 12.16C). At a concentration of 50 μM, none of the other six metal ions, containing Mg^{2+}, Ca^{2+} Mn^{2+}, Cd^{2+}, Ni^{2+}, and Cu^{2+}, could elicit a response, which keeps the solution red. While in the absorption ratio, Zn^{2+} and Co^{2+} can respond slightly, which is in agreement with previous work (Brown et al., 2003).

Similar work does not depend on AuNP-based discriminability between ssDNA and dsDNA but on the significant structural variations of the aptamer upon binding with a metal ion (Wang et al., 2006). A DNA aptamer for K^+ that has the sequence 5′–GGG TTA GGG TTA GGG TTA GGG–3′ is employed as the model system in this work. The K^+ aptamer is a G-rich ssDNA and is random-coil-like in solution. Upon binding to its target (K^+), the aptamer folds to a four-stranded tetraplex structure (G-quartet) via intramolecular hydrogen bonds between guanines. It is found that like dsDNA, a G-quartet structure so formed still loses the ability to protect AuNPs from a high salt concentration (74 mM here) and results in a salt-induced aggregation of nanoparticles. The reason proposed is that on the one hand, DNA bases possess a high affinity to gold via coordination between the gold and nitrogen atoms (favoring DNA adsorption); on the other hand, negatively charged surfaces of AuNPs repel DNA phosphate backbones (disfavoring DNA adsorption) electrostatically. Such a G-quartet structure increases the surface charge density compared to unstructured ssDNA, and prevents the exposure of DNA bases to AuNPs, thus disfavoring adsorption of G-quartets on AuNPs in both facets. Unstructured ssDNA is soft and random-coil-like, which is in sharp contrast to the rigid structure of G-quartets. Therefore they possess greater freedom than G-quartets to wrap AuNPs. It may also contribute to the differentiation ability of AuNPs. So when the aptamer-AuNP solution is treated with increased K^+, an increased absorption ratio (A_{620}/A_{520}) is measured, due to the formation of increased G-quartets. At a concentration of 1.67 mM, an obvious color change from red to purple can be observed by the naked eye. These colormetric aptasensors were shown to be selective, for control sequences for both aptamers and other ions for K^+ do not respond.

Rationally, this method is applicable to most aptamers that undergo similar structural variations, from random coil to G-quartets, upon binding with targets. In another work, α-thrombin and its 29-mer aptamer are chosen as a model. The advantage of using a 29-mer prior to a 15-mer is evident (Wei et al., 2007). This 29-mer aptamer folds into a structure of G-quadruplex/duplex, which does not depend on the existence of K^+, an ion that stabilizes the G-quadruplex in a 15-mer aptamer. Therefore, the disturbance of K^+ (as described by Wei et al.) in the binding buffer can be neglected. Meanwhile, a 4-mer duplex formed together with a G-quadruplex (as shown in Figure 12.17A) makes the structure more stable while playing an important role in further repelling AuNPs as it does other ssDNAs.

In fact, as described by Wei et al., there is always equilibrium (for a 29-mer aptamer) between a random conformation and a G-quadruplex/duplex, whether or

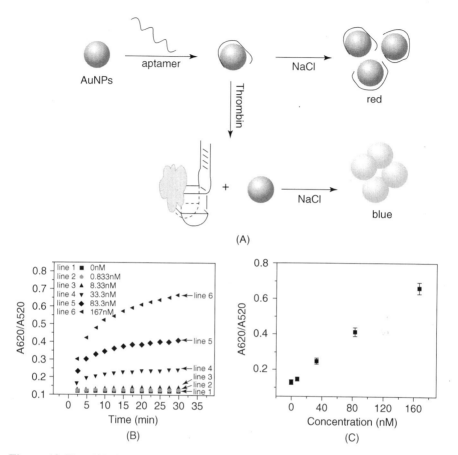

Figure 12.17 (A) AuNP-based colorimetric detection for α-thrombin. (B) 200 μL of AuNPs/17E-17DS duplex solutions in the absence (right) and in the presence (light) of 50 μM Pb^{2+} after addition of 20 μL of 0.5 M NaCl. (C) Corresponding columns of the absorption ratio (A_{620}/A_{520}) of 200 μL of GNPs/17E-17DS duplex solutions in the presence of 50 μM different metal ions after addition of 20 μL of 0.5 M NaCl. [From Wei et al. (2007), with permission. Copyright © 2007 Royal Society of Chemistry.] (See insert for color representation.)

not α-thrombin exists. But α-thrombin is favorable for this G-quadruplex/duplex conformation, so even through unfolded aptamer exists, the addition of α-thrombin would induce more aptamers to fold into a G-quadruplex/duplex conformation, which releases the AuNPs from being protected by ssDNA in a high salt concentration (166 mM NaCl). In this way, α-thrombin could be seen by the naked eye to detect selectively and sensitively in a concentration as low as 83 nM, and a linear range of 0 to 167 nM (as shown in Figure 12.17B and C) is obtained with a detection limit of 0.83 nM. Until this work, such smartly fabricated AuNPs-based colorimetric sensors had not been available for wider applications. The method can also be used in systems in which an aptamer changes in structures without AuNP stability after binding.

However, "refinement" brings "default." Because a lot of analytes could maintain or destroy the stability of AuNPs themselves, the AuNP-based methods depend heavily on the nature of targets. In addition, if the samples themselves are colored, direct observation by the naked eye may be invalid. Therefore, this method may now be applicable only for simply composed samples. Much improvement is required before more complicated systems, especially biological samples, can be used.

In addition to AuNPs, other complexes are used in simple colorimetric aptasensors. In 2002, Stojanovic and Landry developed another nonlabel method, following their earlier work on fluorescent aptasensors. After a smarter route was designed using a colorimetric strategy. The authors tried to choose an aptamer that would bind both a chromophore and an analyte of interest. Therefore, the binding of analyte may alter the microenviroment of the chromophore and produce a visible signal of that event. Cocaine and its aptamer were chosen as a model, with cocaine binding the aptamer via a hydrophobic pocket formed by a noncanonical three-way junction with one of the stems structured through non-Watson–Crick interactions. The authors screened a collection of 35 dyes for changes in visible spectra upon addition of a stock solution of cocaine to a mixture of a given dye and the aptamer. One of the cyanine dyes, diethythiotricarbocyanine iodide, was finally chosen, for it could bind the aptamer, and even through the concentration of cocaine is as low as in the micromolar range, it could still be replaced by the analyte and display both a significant attenuation of absorbance and a change in the ratio of two relative maxima (750 nm for the monomer and 670 nm for the dimer) that dominated the visible spectrum. The sensing process follows the path shown in Figure 12.18A. An optimized aptamer/dye ratio is mixed in the binding buffer, and after equilibration for a given time, cocaine is added little by little, followed by acquisition of the absorption spectra within 1 minute. It is found that at a dye/aptamer ratio of 7 : 4 from 2 to 600 μM cocaine could be detected through decreased absorption at 760 nm, whereas at a higher dye/aptamer ratio, 0.5 to 67 μM cocaine could be detected through both decreased absorptions at 760 nm and increased absorptions at 670 nm. The detection is selective, for control molecules such as cocaine metabolites show no response. The sensing principle seems complex but is well explained in the paper. It is proved that there are two maxima for the dye, due to binding to DNA molecules as monomer (760 nm) and as dimer (670 nm). While in this system, the aptamer inclines more to bind the dimer. In the presence of cocaine, the analyte first interacts rapidly with the monometric dye–aptamer complex to release the dye, which results in the reduction of absorbance at 760 nm within seconds. The increase in absorption at 670 nm at the higher concentrations of the dye and aptamers is due to competition with cocaine of the dye released for binding to the remaining monometric dye–aptamer complex. In fact, a solution of the dye–aptamer complex displays a blue color. When cocaine is added, the solution would be decolorized after 12 hours, due to the appearance of a blue precipitate from hydrolysis of the dye

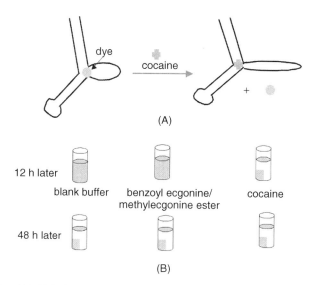

(A)

(B)

Figure 12.18   (A) Colorimetric method. (B) Mimic color of a dye–aptamer complex in the presence of: nothing, benzoyl ecgonine/ecgonine methyl ester, and cocaine, from left to right. [From Stojanovic and Landry (2002).]

in a slightly basic buffer, whereas for the solution itself or for added cocaine metabolites, this decolorizing reaction would happen after 48 hours. The difference between these times makes the visible colorimetric recognition applicable (Figure 12.18B). According to the authors, although the low-micromolar dissociation constant and selectivity for cocaine of this aptamer does not allow the determination of picomolar concentrations of cocaine metabolites in urine, the method is sufficient for handheld colorimetric field tests used in drug interdiction (Stojanovic and Landry, 2002).

New materials are still taken as recognition elements. A cationic, water-soluble, electroactive, photoactive polymer [one of the poly(3-alkoxy-4-methylthiophene)s] has been synthesized and used in protein detection (Ho and Leclerc, 2004). This polymer can exhibit chromic properties due to the conformational changes in the flexible conjugated backbone. When it is complexed to ssDNA or dsDNA, it displays important optical changes that can be used to transduce binding of an aptamer to a given target. Although an aqueous solution of polymer is yellow (with maximum absorption at 402 nm), a red color (with maximun absorption at 527 nm) is observed in the presence of ssDNA (X1–5′–GGTTGGTGGTTGG–3′ is used). This red shift is supposed to relate to a stoichiometric complexation between unfolded anionic ssDNA and the cationic polymer, and such polyelectrolyte complexes tend to be insoluble in the medium in which they are formed. The optical properties are different when K^+ (KCl) is present. This cation could facilitate the formation of a quadruplex state of X1 and stabilize it. In this case, polymer is allowed to wrap this folded structure through electrostatic interactions, leading to an orange color (obviously, a blue shift in maximal absorption). Similar results are obtained when the chloride counterion is replaced by a bromide or iodide anion, indicating the selectivity of detection toward K^+.

X1 is also the 15-mer aptamer for α-thrombin, so this route is applicable for α-thrombin detection too. In the presence of K^+, the quadruplex state of X1 is also promoted by α-thrombin binding. Accordingly, for the $1:1:1$ complex between polymer, X1, and α-thrombin, an orange color is observed, with the same absorption spectrum as that induced by K^+. In this way, α-thrombin can be recognized directly by observing the color of the sample solution. However, such detection has a relatively poor detection limit of 1×10^{-7} M, which is much higher than most reported α-thrombin sensors (most have a detection limit no higher than about the nanomolar range. Interestingly, when taking the fluorescent properties of polymer instead, very small quantities of α-thrombin can be detected in aqueous solutions. The fluorescence of the polymer is quenched in the presence of random-coil ssDNA (X1), whereas when the DNA is in a quadruplex state by binding α-thrombin, the yellow form of the polymer is fluorescent (the maximum emission at 525 nm). Detection is sensitive, under these conditions, as 1×10^{-11} M α-thrombin can be detected, which is 10,000-fold more sensitive than can be achieved using the UV–visible spectrum. This work is shown schematically in Figure 12.19.

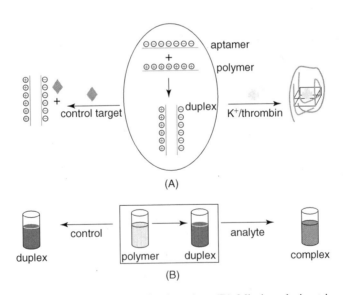

(A)

(B)

Figure 12.19   (A) Principle of specific detection. (B) Mimic colorimetric recognition of analytes such as K^+ and α-thrombin. It is observed that after interacting with DNA, the color of the polymer is changed from yellow to red. If analyte is added, the color is blue-shifted. However, if control molecules such as LiCl/BSA are added, no color change can be observed. [From Ho and Leclerc (2004).]

Another type of widely used colorimetric aptasensor employs hemin and its aptamer as a DNAzyme complex. Since the DNAzyme complex catalyzes chemical reactions to produce colored products, which is different from the colorimetric detection described above, we will discuss this type of aptasensor in a separate section.

12.5 HEMIN–APTAMER DNAzyme-BASED APTASENSOR

The DNAzyme itself is used as a biocatalyst. Another interesting example of a catalytic DNA is that of certain peroxidase-like activities, including a supramolecular complex between hemin and a single-stranded guanine-rich nucleic acid aptamer (Travascio et al., 1998). It is suggested that the supramolecular docking of the guanine–quadruplex layers facilitates the intercalation of hemin into the complex and significantly enhances the biocatalytical activity of the hemin center. Using these peroxidase-like activities, another type of colorimetric aptasensor in PFSFALF can be developed. With nothing labeled, the sensing processes proceed smartly.

Willner's group has been engaged in applying this method to DNA sensing and has made many improvements (Xiao et al., 2004; Yi et al., 2004). Usually, they have employed a peroxidase-like catalytic reaction in which hemin–aptamer complex catalyzes the reaction oxidation of 2,2'-azino-bis(3-ethylbenzthiazoline)-6-sulfonic acid (ABTS) by H_2O_2 (Xiao et al., 2004). The product (ABTS$^+$) is a colored molecule that can transfer signals directly by its absorbance at 414 nm.

Figure 12.20 depicts one of the designs of Willner's group. A molecular beacon ssDNA is synthesized in a hairpin structure consisting of two parts. One is a template segment that is complementary to the target ssDNA. The other is an antihemin aptamer, part of which is complementary to the template part in forming a duplex in the hairpin. At this conformation, even in the presence of hemin, the formation of catalytic DNAzyme is prohibited. Once a fully complementary target DNA exists, it will hybridize with the template part and release the hemin aptamer to a completely free configuration. Then a catalytic DNAzyme can be formed and oxidation of the ABTS by H_2O_2 can be realized. The result can be read out through ABTS$^+$ accumulation at 414 nm. Although hemin itself also possesses catalytic activity, it is proven that at the pH value used, 8.1, this activity is too poor to lead to an effective oxidation reaction. It indicates that a detectable signal results primarily from the catalytic activity of DNAzyme, with no disturbance from hemin itself. In this way, a detection limit of 8 μM is reached. Meanwhile, it is found that the response of a completely complementary target (for the template part) is about eightfold higher than even that of a single-mismatched target, which exhibits the ability of the sensor to discriminate among gene mutants. The method was later improved by combining it with PCR (Xiao et al., 2004). Although the experiments are more complex, much higher sensitivity is achieved. Forty molecules could be detected in a 50-μL sample, which displayed an ultrasensitive detection.

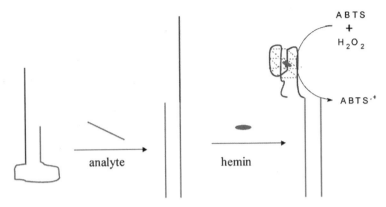

Figure 12.20   Analysis of DNA by opening of a beacon nucleic acid and the generation of DNAzyme. [From Xiao et al. (2004).]

Dong's group has adopted a different DNAzyme-based strategy for DNA sensing (Li et al., 2007d). As shown in Figure 12.21A, the antihemin aptamer is divided into two separate parts (S1 and S2), with a template ssDNA tail on each end. In the presence of hemin, the two parts self-assemble on the hemin site to form a layered G-quadruple supramolecular complex (DNAzyme, $K_d \sim 130$ nm). At a pH of 8.0, this DNAzyme can catalyze the oxidation of ABTS by H_2O_2, and the disturbance from hemin itself can still be neglected. The authors adopted two separate sensing processes, adding hemin either following or preceding the target (S3) and then comparing different results observed. In the former process, similar to earlier work (Yi et al., 2004), DNAzyme is not formed when the template part has hybridized with the target. Therefore, it is a signal-off process in which the more target that is added, the lower the signal that is produced. However, the opposite results appear when the hemin is added before the target. The authors hypothesize that the supramolecular complex here is not as stable as the hemin/G-quartet structure formed directly by hemin and its 18-mer DNA aptamer due to its ternary structure and two free tails. However, in the presence of target DNA, the two parts (of the supramolecular complex) are attached further through part hybridization (Figure 12.21A) and become more stable. Thus, the catalytic activity is enhanced, leading to a signal-on process. It should be noted that in this condition, the target DNA is not entirely hybridized with the template DNA, due to the block coming from the supramolecular complex formed. Therefore, this signal-on phenomenon is applicable only for targets at low concentration, ranging from 0.005 to 0.3 μM ($r = 0.997$) (Figure 12.21B). When the concentration is increased further, the signal decreases. This may be attributed to the fact that more and more targets begin to bind with the template parts in an entirely hybridized conformation, which blocks the formation of a supramolecular complex, just as in the process in which hemin is added after the target. So as shown in Figure 12.21B, when the concentration ranges from 0.3 to 1.0 μM, the signal

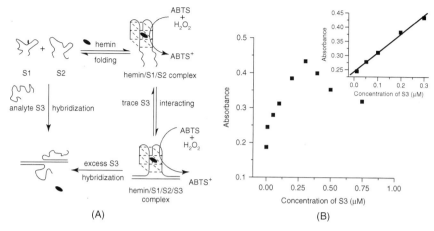

(A)

(B)

Figure 12.21 (A) Schematic of label-free colorimetric approaches for sensing DNA by using a supramolecular hemin/G-quartetcomplex with two free nucleic acid parts as the sensing element (template DNA). The DNAzyme is formed by incubating hemin with two guanine-rich single-strand DNAs (S1, S2). The complementary DNA (S3) is used as the analyte. (B) Investigation of the effect of S3 on the catalytic activity of the hemin/S1/S2 complex by using ABTS/H_2O_2 colorimetric detection. The figure represents the dependence of the absorbance of ABTS$^+$ on the S3 concentration. The insert reveals a linear relation between the absorbance and S3 concentration. [From Li et al. (2007d), with permission. Copyright © 2007 Royal Society of Chemistry.]

is gradually decreased. Obviously, the signal-on process is more sensitive, but the detection range is limited.

Recently, such hemin–aptamer DNAzyme-based sensors have been used for with more application binding with aptamers of other analytes. Taking lysozyme detection, for example, a part duplex is first hybridized from two elaborately designed DNA strands (Di et al., 2007). One is a sensing sequence containing a region of lysozyme aptamer and another of hemin aptamer. The second is a blocking sequence, used to include two separate nine-base complementary domains in a sensing sequence, resulting in cooperative binding. With hemin but no analyte (lysozyme here), formation of the DNAzyme is seriously inhibited by the blocking sequence, and low catalytic activity can be tested. If lysozyme is added, the affinity coming from both lysozyme–aptamer and hemin–aptamer will obviously reduce the melt temperature of the duplex, which finally leads to dehybrization of the duplex and the formation of two aptamer–substrate complexes. Correspondingly, the DNAzyme will display its catalytic activity. In this work, an ABTS–H_2O_2 detection system is adopted, so the signal can be read out directly from the increased absorbance at 414 nm, and a detection limit of 1×10^{-13} M is achieved, which has been one of the most sensitive detections for this analyte until now. Meanwhile, small-molecule AMP is detected in the same way (detection limit: 4×10^{-6} M), indicating the universality of this strategy.

12.6 LIQUID CHROMATOGRAPHY, ELECTROCHROMATOGRAPHY, AND CAPILLARY ELECTROPHORESIS APPLICATIONS

Aptamers have also been employed in various chromatography techniques, such as liquid chromatography (Romig et al., 1999), electrochromatograpy (Connor and McGown, 2006), and capillary electrophoresis (Berezovski and Krylov, 2005). The advantages of aptamers applied in this field are obvious: Wide targets enlarge the range of analytes that can be used; easy synthesis reduces the cost; selected as functional oligonucleotides decreases the adsorption (compared with antibodies); and high affinity guarantees a high selectivity and separation effect. Based on these advantages, processes such as molecular recognition (Cho et al., 2004; Li et al., 2007a, 2007c), quantitative detection (German et al., 1998; Pavski and Le, 2001), analogy separation (Deng et al., 2001), and even chiral selection (Michaud et al., 2003; Brumbt et al., 2005) have been undertaken using aptamer-based chromatography analysis. Chromatography techniques are also used to assist the SELEX procedure and exhibit increasing potential in this field.

Some analytical applications in chromatography using apamers are summarized in Table 12.3. One can see the rapid development of the use of aptamers from the table. However, most aptamer-based chromatography techniques are not PFSFALF. First, for example, if the aptamers are used to enhance the separation effect, they usually need to be modified onto the inner walls of the column (Connor and McGown, 2006). Second, due to relatively limited detection methods coupled with chromatography (e.g., fluorescence), even when aptamers are used as recognition and seperation materiels simply by being added to running buffers or sample solutions, the aptamers or targets themselves still need to be derived or labeled to produce detectable signals (German et al., 1998). Even so, there is still a chance to make PFSFALF available in aptamer-based chromatography techniques. That is what we focus on in this section. In the following descriptions, we give several examples of the use of the CE technique. Most chromatography techniques possess different operational principles but similar functions, such as separation.

Recently, Dong's group used analytes as indicators of themselves and has developed a series of selective aptamer-based PFSFALF coupling CE separations using different detection methods (T. Li et al., 2007a). A small molecule of cocaine is first taken as a model molecule to be recognized and separated from its analogs (T. Li et al., 2007a). Both cocaine and its analog, ecgonine, contain a tertiary amino group and can generate strong electrochemiluminesence (ECL) emission on a platinum electrode in the presence of tris(2,2'-bipyridyl)ruthenium(II), $[Ru(bpy)_3]^{2+}$. So when the two analogs are mixed and separated through CE, two peaks of ECL appear (Figure 12.22A). Once cocaine-binding aptamers are added to the sample, the net amount of cocaine will be reduced, due to the formation of an aptamer–cocaine complex. Following CE, such a complex can be separated from the surplus cocaine as well, which ultimately leads to a reduced cocaine peak. By contrast, the ecgonine is not affected by the aptamer and

TABLE 12.3 Aptamers in Chromatography Techniques

Target	Oligo-nucleotide	Separation System[a]	Detection System[a]	Model	Ref.
L-Selectin	DNA	LC	UV	PFSOALF	Romig et al., 1999
α-Thrombin	DNA	CEC	FL	PFSOALF	Connor et al., 2006
HCVRNA polymerase	RNA	LC/chip	MALDI-TOF	PFSOALF	Cho et al., 2004
HCVRNA replicase	RNA	LC/chip	FL	PFSOALF	Chung et al., 2005
Adenosine and analogs	DNA	Nano-LC	UV	PFSOALF	Deng et al., 2001, 2003
Cocaine and analogs	DNA	CE	ECL	PFSFALF	Li, T. et al., 2007a
Amino acid amides	DNA	CE	EC	PFSFALF	Li, T. et al., 2007c
HRP and hemin	DNA	CE	CL	PFSFALF	Li, T. et al., 2007b
FMN and FAD	RNA	CEC	UV	PFSOALF	Clark et al., 2003a
FMN and thiourea	RNA	CEC	UV	PFSOALF	Clark et al., 2003b
IgE and α-thrombin	DNA	CEC	LIF	POSFALF	German et al., 1998; Buchanan et al., 2003
HIV-1 RTase	DNA	CEC	LIF	POSFALF	Pavski et al., 2001; Wang, H. L. et al., 2005; Fu et al., 2006
α-Thrombin and anti-α-thrombin III	DNA	ACE	LIF	POSFALF	Huang et al., 2004
α-Thrombin	DNA	ACE	LIF	POSFALF	Berezovski et al., 2003
Taq DNA polymerase	DNA	ACE	LIF	POSFALF	Berezovski et al., 2005
Nontarget proteins					
β-Lactoglobulin A and B	G-quartet DNA	CEC	UV	PFSOALF	Rehder et al., 2001
Bovine milk proteins	G-quartet DNA	CEC	UV	PFSOALF	Rehder-Silinski et al., 2003
Albumins	G-quartet DNA	CEC	UV	PFSOALF	Dick et al., 2004

(continued)

285

TABLE 12.3 (*Continued*)

Target	Oligo-nucleotide	Separation System[a]	Detection System[a]	Model	Ref.
Nontarget species					
Binary mixtures of amino acids(D-trp and D-tyr), enantiomers (D-trp and L-trp), and polycyclic aromatic hydrocarbons	G-quartet DNA	CEC	UV	PFSOALF	Kotia et al., 2000
Isomeric dipeptides Trp-Arg and Arg-Trp	G-quartet DNA	CEC	UV	PFSOALF	Charles et al., 2002
Homodipeptides and alanyl dipeptides	G-quartet DNA	CEC	UV	PFSOALF	Vo et al., 2004
Fibrinogen peptides	G-quartet DNA	CEC	UV	PFSOALF	Vo et al., 2006
Enantiomers					
Arginine-vasopressin	DNA	LC	UV	PFSOALF	Michaud et al., 2003
Adenosine	DNA	Micro-LC	UV	PFSOALF	Michaud et al., 2004
Tyrosinamide	DNA	Micro-LC	UV	PFSOALF	Michaud et al., 2004
Arginine	RNA	Micro-LC	UV	PFSOALF	Brumbt et al., 2005
Amino acids and derivatives	RNA	Micro-LC	UV	PFSOALF	Ravelet et al., 2005

[a] **ACE**, affinity capillary electrophoresis; **CEC**, capillary electrochromatography; **CL**, chemiluminescence; **EC**, electrochemistry; **ECL**, electrochemiluminescence; **FAD**, flavin adenine dinucleotide; **FMN**, flavin mononucleotide; **HCV**, hepatitis C virus; **HIV-1 RTase**, reverse transcriptase of human immunodeficiency virus type 1; **LC**, liquid chromatography; **LIF**, laser-induced fluorescence; **MALDI-TOF**, matrix-assisted laser desorption ionization–time-of-flight mass spectrometry; **UV**, UV–visible spectrum.

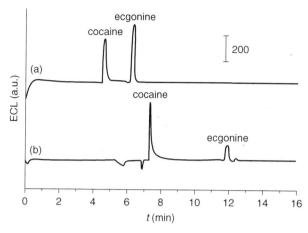

Figure 12.22 CE-ECL analyses for a mixture of 600 mM cocaine and 400 mM ecgonine using 20 mM phosphate (pH 7.0) in the (a) absence and (b) presence of 1% BMIMBF$_4$ running buffer. The upright bar refers to an ECL intensity of 200 (a.u.). [From T. Li et al. (2007a), with permission. Copyright © 2007 Wiley.]

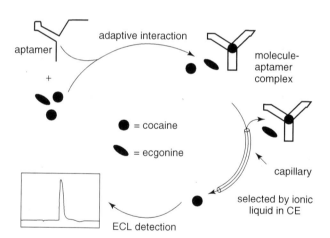

Figure 12.23 Label-free method for aptamer-based recognition of cocaine from its hydrolysate (ecgonine) using CE-ECL analysis assisted by an ionic liquid selector. An ionic liquid was used as the selector for the sample injection in the CE process, enriching cocaine and excluding the cocaine–aptamer complex and ecgonine from the capillary. [From T. Li et al. (2007a), with permission. Copyright © 2007 Wiley.]

keeps its peak without change (Figure 12.23). What should be noted is that no peak of aptamer–cocaine complex is observed. The explanation for this phenomenon is as follows: First, interaction between the aptamer and redox-active sites on the cocaine molecules could be taken up through aptamer binding, which results in disappearance of the ECL emission. This phenomenon may be applicable for many other molecules as well. Second, in the running buffer, 1% room-temperature ionic liquid [1-butyl-3-methylimidazolium tetrafluoroborate (BMIMBF$_4$)] is added to serve as a selector during the sample injection process. As in Dong's group's earlier report, when a selector is used, positive molecules such as cocaine (pH 7.0) can be enriched during sample injection; while a negative molecule of ecgonine (pH 7.0), aptamer, and cocaine–aptamer complex will be repelled out of the capillary to a large degree (as in Figure 12.24). So in this system no cocaine–aptamer complex peak is observed and the adsorption of DNA onto the capillary inner wall is further reduced. At the same time, the ecgonine peak is greatly reduced (Figure 12.22B). Through the peak decrease, cocaine was sensitively recognized using itself as a signal probe in the presence of its analog and a detection limit as low as 2 μM. This method has also been proven feasible in a biological system of 25% fetal bovine serum.

The advantages of this method are obvious. First, taking analytes themselves as indicators binding with a CE separation technique, all possible label processes were avoided, which makes recognition very simple and convenient. Second, aptamers as recognition elements could be a very useful alternative to the traditional spike method, especially when peaks of analytes are overlapped to a large degree with other coexisting materials. Here just depending on whether the peak is lowered in the presence of aptamers, we could judge the existence of the target and even its exact amount. Finally, but no less important, a lot of molecules may produce detectable signal according to their intrinsic properties. The method has a promising potential for wider targets using various detection techniques, such as electrochemistry, chemiluminescence, UV, and LIF.

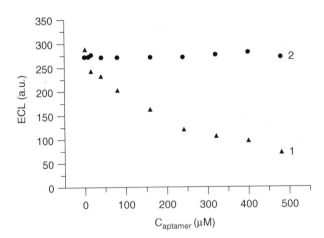

Figure 12.24   Dependence of ECL intensities of cocaine and its analog on the concentrations of aptamer A1 after incubation for 1 hour at room temperature in 20 mM phosphate of pH 7.0. (1) 800 mM cocaine, (2) 1.3 mM ecgonine. [From T. Li et al. (2007a), with permission. Copyright © 2007 Wiley.]

In recent work using a similar principle, the amino acid amides argininamide (Arm) and tyrosinamide (Tym) were recognized successfully from their analogs as well, using CE electrochemical detection (T. Li et al., 2007c). Then the some hypothesis is made regarding conformation of the Tym-binding aptamer. Many guanines are spaced in a regular pattern as a G-doublet in the consensus sequence of the Tym aptamer used, so it is naturally hypothesized that it may fold into a G-quartet structure for binding Tym. However, no signal differences are found when the binding reaction is carried out in a buffer with 20 mM K^+. It is well known that the G-quartet structure depends strongly on the existence of K^+, so no effect by K^+ on binding affinity may lead to the hypothesis that no G-quartet formed in this process.

Another work focuses on a small molecule of hemin with CE chemiluminescence detection and has been extended to aptamer-based DNA assays (T. Li et al., 2007b). Both hemin and a hemin-18-mer/aptamer complex (DNAzyme) can catalyze the chemiluminescent generation of luminol in the presence of H_2O_2. Coupled with CE separation, hemin is first sensitively distinguished from the control molecule HRP at pH 11. Under these conditions, the catalytic activity of hemin is optimized but that of synthetic DNAzyme is seriously inhibited, which reduces signal interference for the entire system. This property is then applied to the detection of target ssDNAs that could form G-quartets with the templet ssDNA (Figure 12.25). In this case, both hemin and templet are in excess. In the presence of target DNA, the templet DNA composes to form a G-quartet that could bind with hemin. Thus, the free hemin is reduced. Following CE, free hemin is well separated from the target–templet–hemin

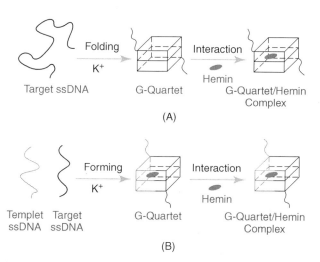

Figure 12.25 Label-free approach to DNA detection using CE-CL analysis: (A) Detection of ssDNAs that can fold into the G-quartet structure; (B) Detection of ssDNAs that can form the G-quartets with the templet ssDNA. [From T. Li et al. (2007b), with permission.]

complex and the a decreased CL signal is observed for the free hemin. In this way, ssDNA is quantitively detected even at conditions under which both hemin and target–templet–hemin complex will produce a CL signal. The detection limit of ssDNA is 0.1 μM. Although it is not as sensitive as many other DNA sensors, it does provide a smart way to achieve label-free DNA detection using the CE technique and shows promising potential for future study.

12.7 OTHER APTASENSORS

In addition to the methods described above, other techniques, such as ion-selective field-effect transistors (ISFETs) (Zayats et al., 2006), quartz crystal microbalance (QCM) (Tombelli et al., 2005b), love waves (Schlensog et al., 2004), FM (Basnar et al., 2006), and SPR (Balamurugan et al., 2006), are also used for aptasensor fabrication. Convenient and smart aptamer-based detections are easily realized with these techniques because that all of them depend directly on changes happening on the substrate surfaces to produce detectable signal: for example, charge change (and thereby potential change) in ISFETs and the mass change associated with crystals according to the Sauerbrey equation.

Using these techniques, both protein and small-molecule are successfully detected.

12.8 CONCLUSIONS

In this chapter we discuss mainly the development of easily fabricated aptasensors: label-free aptasensors. As novel functional nucleic acids, aptamers have displayed many advantages over traditional recognition elements, especially in simplifying the entire detection process. Thus, usually, such simple routes are fast, sensitive, and selective, and defaults must be considered. It is found that many of the easy routes are not applicable to practical detection, or rather, detection in complex biological conditions. Sensors used when no labeling is required, especially, sometimes seem to be too smartly designed to overcome disturbances in practical samples that contain various types of complexes. Therefore, more effort will be needed before these sensors can be used successfully in practical samples.

REFERENCES

Baker, B. R., Lai, R. Y., Wood, M. S., Doctor, E. H., Heeger, A. J., Plaxco, K. W. (2006). An electronic, aptamer-based small-molecule sensor for the rapid, label-free detection of cocaine in adulterated samples and biological fluids. *J Am Chem Soc* 128, 3138–3139.

Balamurugan, S., Obubuafo, A., Soper, S. A., McCarley, R. L., Spivak, D. A. (2006). Designing highly specific biosensing surfaces using aptamer monolayers on gold. *Langmuir* 22, 6446–6453.

Bang, G. S., Cho, S., Kim, B. G. (2005). A novel electrochemical detection method for aptamer biosensors. *Biosens Bioelectron* 21, 863–870.

Basnar, B., Elnathan, R., Willner, I. (2006). Following aptamer–thrombin binding by force measurements. *Anal Chem* 78, 3638–3642.

Berezovski, M., Krylov, S. N. (2005). Thermochemistry of protein-DNA interaction studied with temperature-controlled nonequilibrium capillary electrophoresis of equilibrium mixtures. *Anal Chem* 77, 1526–1529.

Berezovski, M., Nutiu, R., Li, Y. F., Krylov, S. N. (2003). Affinity analysis of a protein–aptamer complex using nonequilibrium capillary electrophoresis of equilibrium mixtures. *Anal Chem.* 75, 1382–1386.

Breaker, R. R. (1997). DNA enzymes. *Nat Biotechnol* 15, 427–431.

Breaker, R. R., Joyce, G. F. (1994). DNA enzyme that cleaves RNA. *Chem Biol* 1, 223–229.

Breaker, R. R., Joyce, G. F. (1995). A DNA enzyme with Mg^{2+}-dependent RNA phosphoesterase activity. *Chem Biol* 2, 655–660.

Brown, A. K., Li, J., Pavot, C. M. B., Lu, Y. (2003). A lead-dependent DNAzyme with a two-step mechanism. *Biochemistry* 42, 7152–7161.

Brumbt, A., Ravelet, C., Grosset, C., Ravel, A., Villet, A., Peyrin, E. (2005). Chiral stationary phase based on a biostable L-RNA aptamer. *Anal Chem* 77, 1993–1998.

Buchanan, D. D., Jameson, E. E., Perlette, J., Malik, A., Kennedy, R. T. (2003). Effect of buffer, electric field, and separation time on detection of aptamer-ligand complexes for affinity probe capillary electrophoresis. *Electrophoresis* 24, 1375–1382.

Carmi, N., Balkhi, S. R., Breaker, R. R. (1998). Cleaving DNA with DNA. *Proc Natl Acad Sci USA.* 95, 2233–2237.

Centi, S., Tombelli, S., Minunni, M., Mascini, M. (2007). Aptamer-based detection of plasma proteins by an electrochemical assay coupled to magnetic beads. *Anal Chem* 79, 1466–1473.

Charles, J. A. M., McGown, L. B. (2002). Separation of Trp-Arg and Arg-Trp using G-quartet-forming DNA oligonucleotides in open-tubular capillary electrochromatography. *Electrophoresis* 23, 1599–1604.

Cheng, A. K. H., Ge, B., Yu, H. Z. (2007). Aptamer-based biosensors for label-free voltammetric detection of lysozyme. *Anal Chem* 79, 5158–5164.

Cho, S., Lee, S. H., Chung, W. J., Kim, Y. K., Lee, Y. S., Kim, B. G. (2004). Microbead-based affinity chromatography chip using RNA aptamer modified with photocleavable linker. *Electrophoresis* 25, 3730–3739.

Choi, J. H., Chen, K. H., Strano, M. S. (2006). Aptamer-capped nanocrystal quantum dots: a new method for label-free protein detection. *J Am Chem Soc* 128, 15584–15585.

Chung, W. J., Kim, M. S., Cho, S., Park, S. S., Kim, J. H., Kim, Y. K., Kim, B. G., Lee, Y. S. (2005). Microaffinity purification of proteins based on photolytic elution: toward an efficient microbead affinity chromatography on a chip. *Electrophoresis* 26, 694–702.

Clark, S. L., Remcho, V. T. (2003a). Electrochromatographic retention studies on a flavin-binding RNA aptamer sorbent. *Anal Chem* 75, 5692–5696.

Clark, S. L., Remcho, V. T. (2003b). Open tubular liquid chromatographic separations using an aptamer stationary phase. *J Sep Sci* 26, 1451–1454.

Connor, A. C., McGown, L. B. (2006). Aptamer stationary phase for protein capture in affinity capillary chromatography. *J Chromatogr A* 1111, 115–119.

Cuenoud, B., Szostak, J. W. (1995). A DNA metalloenzyme with DNA–ligase activity. *Nature* 375, 611–614.

Daniels, J. S., Pourmand, N. (2007). Label-free impedance biosensors: opportunities and challenges. *Electroanalysis* 19, 1239–1257.

Deng, Q., German, I., Buchanan, D., Kennedy, R. T. (2001). Retention and separation of adenosine and analogues by affinity chromatography with an aptamer stationary phase. *Anal Chem* 73, 5415–5421.

Deng, Q., Watson, C. J., Kennedy, R. T. (2003). Aptamer affinity chromatography for rapid assay of adenosine in microdialysis samples collected in vivo. *J Chromatogr A* 1005, 123–130.

Di, L., Shlyahovsky, B., Elbaz, J., Willner, I. (2007). Amplified analysis of low-molecular-weight substrates or proteins by the self-assembly of DNAzyme–aptamer conjugates. *J Am Chem Soc* 129, 5804–5805.

Dick, L. W., Swinteck, B. J., McGown, L. B. (2004). Albumins as a model system for investigating separations of closely related proteins on DNA stationary phases in capillary electrochromatography. *Anal Chim Acta* 519, 197–205.

Dwarakanath, S., Bruno, J. G., Shastry, A., Phillips, T., John, A., Kumar, A., Stephenson, L. D. (2004). Quantum dot-antibody and aptamer conjugates shift fluorescence upon binding bacteria. *Biochem Biophys Res Commun* 325, 739–743.

Du, Y., Li, B. L., Wei, H., Wang, Y. L., Wang, E. K. (2008). Multifunctional label-free electrochemical biosensor based on an intergrated aptamer. *Anal. Chem.* 80, 5110–5117.

Elghanian, R., Storhoff, J. J., Mucic, R. C., Letsinger, R. L., Mirkin, C. A. (1997). Selective colorimetric detection of polynucleotides based on the distance-dependent optical properties of gold nanoparticles. *Science* 277, 1078–1081.

Ellington, A. D., Szostak, J. W. (1990). In Vitro selection of RNA molecules that bind specific ligands. *Nature* 346, 818–822.

Elowe, N. H., Nutiu, R., Allah-Hassani, A., Cechetto, J. D., Hughes, D. W., Li, Y. F., Brown, E. A. (2006). Small-molecule screening made simple for a difficult target with a signaling nucleic acid aptamer that reports on deaminase activity. *Angew Chem Int Ed Engl* 45, 5648–5652.

Famulok, M., Mayer, G. (2006). Chemical biology: aptamers in nanoland. *Nature* 439, 666–669.

Famulok, M., Mayer, G., Blind, M. (2000). Nucleic acid aptamers: from selection in vitro to applications in vivo. *Acc Chem Res* 33, 591–599.

Fu, H., Guthrie, J. W., Le, X. C. (2006). Study of binding stoichiometries of the human immunodeficiency virus type 1 reverse transcriptase by capillary electrophoresis and laser-induced fluorescence polarization using aptamers as probes. *Electrophoresis* 27, 433–441.

Gazda, D. B., Fritz, J. S., Porter, M. D. (2004). Multiplexed colorimetric solid-phase extraction: determination of silver(I), nickel(II), and sample pH. *Anal Chem* 76, 4881–4887.

German, I., Buchanan, D. D., Kennedy, R. T. (1998). Aptamers as ligands in affinity probe capillary electrophoresis. *Anal Chem* 70, 4540–4545.

Ghosh, T., Maiya, B. G., Samanta, A. (2006). A colorimetric chemosensor for both fluoride and transition metal ions based on dipyrrolyl derivative. *Dalton Trans* 795–801.

Guo, W. W., Yuon, J. P., Li, B. L., Du, Y., Yim, E. B., Wong, E. K. (2008). Nanoscale enhanced Ru(bpy)$_3^{2+}$ electrochemiluminescence labels and related aptamer-based biosensing system. *Analyst* 133, 1209–1213.

Hamaguchi, N., Ellington, A., Stanton, M. (2001). Aptamer beacons for the direct detection of proteins. *Anal Biochem* 294, 126–131.

Hansen, J. A., Wang, J., Kawde, A. N., Xiang, Y., Gothelf, K. V., Collins, G. (2006). Quantum-dot/aptamer-based ultrasensitive multi-analyte electrochemical biosensor. *J Am Chem Soc* 128, 2228–2229.

Herr, J. K., Smith, J. E., Medley, C. D., Shangguan, D. H., Tan, W. H. (2006). Aptamer-conjugated nanoparticles for selective collection and detection of cancer cells. *Anal Chem* 78, 2918–2924.

Ho, H. A., Leclerc, M. (2004). Optical sensors based on hybrid aptamer/conjugated polymer complexes. *J Am Chem Soc* 126, 1384–1387.

Huang, C. C., Cao, Z. H., Chang, H. T., Tan, W. H. (2004). Protein–protein interaction studies based on molecular aptamers by affinity capillary electrophoresis. *Anal Chem* 76, 6973–6981.

Huang, C. C., Huang, Y. F., Cao, Z. H., Tan, W. H., Chang, H. T. (2005). Aptamer-modified gold nanoparticles for colorimetric determination of platelet-derived growth factors and their receptors. *Anal Chem* 77, 5735–5741.

Ikanovic, M., Rudzinski, W. E., Bruno, J. G., Allman, A., Carrillo, M. P., Dwarakanath, S., Bhahdigadi, S., Rao, P., Kiel, J. L., Andrews, C. J. (2007). Fluorescence assay based on aptamer-quantum dot binding to *Bacillus thuringiensis* spores. *J Fluoresc* 17, 193–199.

Ikebukuro, K., Kiyohara, C., Sode, K. (2005). Novel electrochemical sensor system for protein using the aptamers in sandwich manner. *Biosens Bioelectron* 20, 2168–2172.

Jayasena, S. D. (1999). Aptamers: an emerging class of molecules that rival antibodies in diagnostics. *Clin Chem* 45, 1628–1650.

Jiang, Y. X., Fang, X. H., Bai, C. L. (2004). Signaling aptamer/protein binding by a molecular light switch complex. *Anal Chem* 76, 5230–5235.

Katz, E., Willner, I. (2003). Probing biomolecular interactions at conductive and semiconductive surfaces by impedance spectroscopy: routes to impedimetric immunosensors, DNA-sensors, and enzyme biosensors. *Electroanalysis* 15, 913–947.

Kotia, R. B., Li, L. J., McGown, L. B. (2000). Separation of nontarget compounds by DNA aptamers. *Anal Chem* 72, 827–831.

K'Owino, I. O., Sadik, O. A. (2005). Impedance spectroscopy: a powerful tool for rapid biomolecular screening and cell culture monitoring. *Electroanalysis* 17, 2101–2113.

Lai, R. Y., Plaxco, K. W., Heeger, A. J. (2007). Aptamer-based electrochemical detection of picomolar platelet-derived growth factor directly in blood serum. *Anal Chem* 79, 229–233.

Le Floch, F., Ho, H. A., Leclerc, M. (2006). Label-free electrochemical detection of protein based on a ferrocene-bearing cationic polythiophene and aptamer. *Anal Chem* 78, 4727–4731.

Lee, M., Walt, D. R. (2000). A fiber-optic microarray biosensor using aptamers as receptors. *Anal Biochem* 282, 142–146.

Levy, M., Cater, S. F., Ellington, A. D. (2005). Quantum-dot aptamer beacons for the detection of proteins. *ChemBioChem* 6, 2163–2166.

Li, B. L., Dong, S. J. (2007). Amplified electrochemical aptasensor taking AuNPs based sandwich sensing platform as a model *Biosens Bioelectron* 23, 965–970.

Li, B. L., Wei, H., Dong, S. J. (2007a). Sensitive detection of protein by an aptamer-based label-free fluorescing molecular switch. *Chem Commun* 73–75.

Li, B. L., Du, Y., Wei, H., Dong, S. J. (2007b). Reusable, label-free electrochemical aptasensor for sensitive detection of small molecules. *Chem Commun* 3780–3782.

Li, H. X., Rothberg, L. (2004). Colorimetric detection of DNA sequences based on electrostatic interactions with unmodified gold nanoparticles. *Proc Natl Acad Sci U S A.* 101, 14036–14039.

Li, J., Lu, Y. (2000). A highly sensitive and selective catalytic DNA biosensor for lead ions. *J Am Chem Soc* 122, 10466–10467.

Li, J. W. J., Fang, X. H., Tan, W. H. (2002). Molecular aptamer beacons for real-time protein recognition. *Biochem Biophys Res Commun* 292, 31–40.

Li, T., Li, B. L., Dong, S. J. (2007a). Adaptive recognition of small molecules by nucleic acid aptamers through a label-free approach. *Chem Euro J* 13, 6718–6723.

Li, T., Li, B. L., Dong, S. J. (2007b). Aptamer-based label-free method for hemin recognition and DNA assay by capillary electrophoresis with chemiluminescence detection. *Anal Bioanal Chem* 389, 887–893.

Li, T., Du, Y., Li, B. L., Dong, S. J. (2007c). CE with electrochemical detection for investigation of label-free recognition of amino acid amides by guanine-rich DNA aptamers. *Electrophoresis* 28, 3122–3128.

Li, T., Dong, S. J., Wang, E. K. (2007d). Enhanced catalytic DNAzyme for label-free colorimetric detection of DNA. *Chem Commun* 4209–4211.

Li, Y. F., Breaker, R. R. (1999). Phosphorylating DNA with DNA. *Proc Natl Acad Sci U S A* 96, 2746–2751.

Li, Y. F., Sen, D. (1996). A catalytic DNA for porphyrin metallation. *Nat Struct Biol* 3, 743–747.

Liss, M., Petersen, B., Wolf, H., Prohaska, E. (2002). An aptamer-based quartz crystal protein biosensor. *Anal Chem* 74, 4488–4495.

Liu, J. W., Lu, Y. (2003). A colorimetric lead biosensor using DNAzyme-directed assembly of gold nanoparticles. *J Am Chem Soc* 125, 6642–6643.

Liu, J. W., Lu, Y. (2006). Fast colorimetric sensing of adenosine and cocaine based on a general sensor design involving aptamers and nanoparticles. *Angew Chem Int Ed Engl* 45, 90–94.

Liu, J. W., Lu, Y. (2007). A DNAzyme catalytic beacon sensor for paramagnetic Cu^{2+} ions in aqueous solution with high sensitivity and selectivity. *J Am Chem Soc* 129, 9838–9839.

Liu, J. W., Mazumdar, D., Lu, Y. (2006). A simple and sensitive "dipstick" test in serum based on lateral flow separation of aptamer-linked nanostructures. *Angew Chem Int Ed Engl* 45, 7955–7959.

Liu, J. W., Lee, J. H., Lu, Y. (2007a). Quantum dot encoding of aptamer-linked nanostructures for one-pot simultaneous detection of multiple analytes. *Anal Chem* 79, 4120–4125.

Liu, J. W., Brown, A. K., Meng, X. L., Cropek, D. M., Istok, J. D., Watson, D. B., Lu, Y. (2007b). A catalytic beacon sensor for uranium with parts-per-trillion sensitivity and millionfold selectivity. *Proc Natl Acad Sci U S A* 104, 2056–2061.

Lu, Y., Liu, J. W. (2006). Functional DNA nanotechnology: emerging applications of DNAzymes and aptamers. *Curr Opin Biotechnol* 17, 580–588.

Lu, Y., Liu, J. W. (2007). Smart nanomaterials inspired by biology: dynamic assembly of error-free manomaterials in response to multiple chemical and biological stimuli. *Acc Chem Res* 40, 315–323.

Lu, Y., Liu, J. W., Li, J., Bruesehoff, P. J., Pavot, C. M. B., Brown, A. K. (2003). New highly sensitive and selective catalytic DNA biosensors for metal ions. *Biosens Bioelectron* 18, 529–540.

Michaud, M., Jourdan, E., Villet, A., Ravel, A., Grosset, C., Peyrin, E. (2003). A DNA aptamer as a new target-specific chiral selector for HPLC. *J Am Chem Soc* 125, 8672–8679.

Michaud, M., Jourdan, E., Ravelet, C., Villet, A., Ravel, A., Grosset, C., Peyrin, E. (2004). Immobilized DNA aptamers as target-specific chiral stationary phases for resolution of nucleoside and amino acid derivative enantiomers. *Anal Chem* 76, 1015–1020.

Mir, M., Vreeke, M., Katakis, L. (2006). Different strategies to develop an electrochemical thrombin aptasensor. *Electrochem Commun* 8, 505–511.

Nagatoishi, S., Nojima, T., Juskowiak, B., Takenaka, S. (2005). A pyrene-labeled G-quadruplex oligonucleotide as a fluorescent probe for potassium ion detection in biological applications. *Angew Chem Int Ed Engl* 44, 5067–5070.

Navani, N. K., Li, Y. F. (2006). Nucleic acid aptamers and enzymes as sensors. *Curr Opin Chem Biol* 10, 272–281.

Nutiu, R., Li, Y. F. (2003). Structure-switching signaling aptamers. *J Am Chem Soc* 125, 4771–4778.

Nutiu, R., Mei, S., Liu, Z. J., Li, Y. F. (2004). Engineering DNA aptamers and DNA enzymes with fluorescence-signaling properties. *Pure Appl Chem* 76, 1547–1561.

Osborne, S. E., Ellington, A. D. (1997). Nucleic acid selection and the challenge of combinatorial chemistry. *Chem Rev* 97, 349–370.

Pavlov, V., Shlyahovsky, B., Willner, I. (2005). Fluorescence detection of DNA by the catalytic activation of an aptamer/thrombin complex. *J Am Chem Soc* 127, 6522–6523.

Pavski, V., Le, X. C. (2001). Detection of human immunodeficiency virus type 1 reverse transcriptase using aptamers as probes in affinity capillary electrophoresis. *Anal Chem* 73, 6070–6076.

Pejcic, B., De Marco, R. (2006). Impedance spectroscopy: over 35 years of electrochemical sensor optimization. *Electrochim Acta* 51, 6217–6229.

Polsky, R., Gill, R., Kaganovsky, L., Willner, I. (2006) Nucleic acid–functionalized Pt nanoparticles: catalytic labels for the amplified electrochemical detection of biomolecules. *Anal Chem* 78, 2268–2271.

Radi, A. E., Sanchez, J. L. A., Baldrich, E., O'Sullivan, C. K. (2005). Reusable impedimetric aptasensor. *Anal Chem* 77, 6320–6323.

Radi, A. E., Sanchez, J. L. A., Baldrich, E., O'Sullivan, C. K. (2006). Reagentless, reusable, ultrasensitive electrochemical molecular beacon aptasensor. *J Am Chem Soc* 128, 117–124.

Ravelet, C., Boulkedid, R., Ravel, A., Grosset, C., Villet, A., Fize, J., Peyrin, E. (2005). A L-RNA aptamer chiral stationary phase for the resolution of target and related compounds. *J Chromatogr A* 1076, 62–70.

Ravelet, C., Grosset, C., Peyrin, E. (2006). Liquid chromatography, electrochromatography and capillary electrophoresis applications of DNA and RNA aptamers. *J Chromatogr A* 1117, 1–10.

Rehder, M. A., McGown, L. B. (2001). Open-tubular capillary electrochromatography of bovine beta-lactoglobulin variants A and B using an aptamer stationary phase. *Electrophoresis* 22, 3759–3764.

Rehder-Silinski, M. A., McGown, L. B. (2003). Capillary electrochromatographic separation of bovine milk proteins using a G-quartet DNA stationary phase. *J Chromatogr A* 1008, 233–245.

Robertson, D. L., Joyce, G. F. (1990). Selection in vitro of an RNA enzyme that specifically cleaves single-stranded DNA. *Nature* 344, 467–468.

Rodriguez, M. C., Kawde, A. N., Wang, J. (2005). Aptamer biosensor for label-free impedance spectroscopy detection of proteins based on recognition-induced switching of the surface charge. *Chem Commun* 4267–4269.

Romig, T. S., Bell, C., Drolet, D. W. (1999). Aptamer affinity chromatography: combinatorial chemistry applied to protein purification. *J Chromatogr B* 731, 275–284.

Rupcich, N., Nutiu, R., Li, Y. F., Brennan, J. D. (2005). Entrapment of fluorescent signaling DNA aptamers in sol-gel-derived silica. *Anal Chem* 77, 4300–4307.

Rupcich, N., Nutiu, R., Li, Y. F., Brennan, J. D. (2006). Solid-phase enzyme activity assay utilizing an entrapped fluorescence-signaling DNA aptamer. *Angew Chem Int Ed Engl* 45, 3295–3299.

Santoro, S. W., Joyce, G. F. (1997). A general purpose RNA-cleaving DNA enzyme. *Proc Natl Acad Sci U S A* 94, 4262–4266.

Schlensog, M. D., Gronewold, T. M. A., Tewes, M., Famulok, M., Quandt, E. (2004). A Love-wave biosensor using nucleic acids as ligands. *Sens Actuat B* 101, 308–315.

Sen, D., Geyer, C. R. (1998). DNA enzymes. *Curr Opin Chem Biol* 2, 680–687.

Shangguan, D., Li, Y., Tang, Z. W., Cao, Z. H. C., Chen, H. W., Mallikaratchy, P., Sefah, K., Yang, C. Y. J., Tan, W. H. (2006). Aptamers evolved from live cells as effective molecular probes for cancer study. *Proc Natl Acad Sci U S A* 103, 11838–11843.

Shen, L., Chen, Z., Li, Y. H., Jing, P., Xie, S. B., He, S. L., He, P. L., Shao, Y. H. (2007). A chronocoulometric aptamer sensor for adenosine monophosphate. *Chem Commun* 2169–2171.

Shen, Y. T., Mackey, G., Rupcich, N., Gloster, D., Chiuman, W., Li, Y. F., Brennan, J. D. (2007). Entrapment of fluorescence signaling DNA enzymes in sol-gel-derived materials for metal ion sensing. *Anal Chem* 79, 3494–3503.

Stojanovic, M. N., Kolpashchikov, D. M. (2004). Modular aptameric sensors. *J Am Chem Soc* 126, 9266–9270.

Stojanovic, M. N., Landry, D. W. (2002). Aptamer-based colorimetric probe for cocaine. *J Am Chem Soc* 124, 9678–9679.

Stojanovic, M. N., de Prada, P., Landry, D. W. (2000). Fluorescent sensors based on aptamer self-assembly. *J Am Chem Soc* 122, 11547–11548.

Stojanovic, M. N., de Prada, P., Landry, D. W. (2001a). Aptamer-based folding fluorescent sensor for cocaine. *J Am Chem Soc* 123, 4928–4931.

Stojanovic, M. N., de Prada, P., Landry, D. W. (2001b). Catalytic molecular beacons. *ChemBioChem* 2, 411–415.

Su, L., Sankar, C. G., Sen, D., Yu, H. Z. (2004). Kinetics of ion-exchange binding of redox metal cations to thiolate-DNA monolayers on gold. *Anal Chem* 76, 5953–5959.

Tombelli, S., Minunni, A., Mascini, A. (2005a). Analytical applications of aptamers. *Biosens Bioelectron* 20, 2424–2434.

Tombelli, S., Minunni, A., Luzi, E., Mascini, M. (2005b). Aptamer-based biosensors for the detection of HIV-1 Tat protein. *Bioelectrochemistry* 67, 135–141.

Travascio, P., Li, Y. F., Sen, D. (1998). DNA-enhanced peroxidase activity of a DNA aptamer–hemin complex. *Chem Biol* 5, 505–517.

Tuerk, C., Gold, L. (1990). Systematic evolution of ligands by exponential enrichment: RNA ligands to bacteriophage T4 DNA polymerase. *Science* 344, 505–510.

Tuite, E., Norden, B. (1994). Sequence-specific interactions of methylene-blue with polynucleotides and DNA: a spectroscopic study. *J Am Chem Soc* 116, 7548–7556.

Ueyama, H., Takagi, M., Takenaka, S. (2002). A novel potassium sensing in aqueous media with a synthetic oligonucleotide derivative: fluorescence resonance energy transfer associated with guanine quartet-potassium ion complex formation. *J Am Chem Soc* 124, 14286–14287.

Vo, T. U., McGown, L. B. (2004). Selectivity of quadruplex DNA stationary phases toward amino acids in homodipeptides and alanyl dipeptides. *Electrophoresis* 25, 1230–1236.

Vo, T. U., McGown, L. B. (2006). Effects of G-quartet DNA stationary phase destabilization on fibrinogen peptide resolution in capillary electrochromatography. *Electrophoresis* 27, 749–756.

Wang, H. L., Lu, M. L., Le, X. C. (2005). DNA-driven focusing for protein-DNA binding assays using capillary electrophoresis. *Anal Chem* 77, 4985–4990.

Wang, J., Jiang, Y. X., Zhou, C. S., Fang, X. H. (2005). Aptamer-based ATP assay using a luminescent light switching complex. *Anal Chem* 77, 3542–3546.

Wang, L. H., Liu, X. F., Hu, X. F., Song, S. P., Fan, C. H. (2006). Unmodified gold nanoparticles as a colorimetric probe for potassium DNA aptamers. *Chem Commun* 3780–3782.

Wang, Y. L., Wei, H., Li, B. L., Ren, W., Guo, S. J., Dong, S. J., Wang, E. K. (2007). SERS opens a new way in aptasensor for protein recognition with high sensitivity and selectivity. *Chem Commun*. 5220–5222.

Wang, J. L., Wang, F. A., Dong, S. J. (2008). Methylene blue as an indicator for sensitive electrochemical detection of adenosine based on aptamer switch. *J. Electronal. Chem*. in press.

Wei, H., Li, B. L., Li, J., Wang, E. K., Dong, S. J. (2007). Simple and sensitive aptamer-based colorimetric sensing of protein using unmodified gold nanoparticle probes. *Chem Commun* 3735–3737.

Wei, H., Li, B. L., Li, J., Dong, S. J., Wang, E. K. (2008). DNAzyme-based colorimetric sensing of lead (Pb^{2+}) using unmodified gold nanoparticle probes. *Nanotechnology* 19, 095501.

Willner, I., Zayats, M. (2007). Electronic aptamer-based sensors. *Angew Chem Int Ed Engl* 46, 2–13.

Xiao, Y., Pavlov, V., Niazov, T., Dishon, A., Kotler, M., Willner, I. (2004). Catalytic beacons for the detection of DNA and telomerase activity. *J Am Chem Soc* 126, 7430–7431.

Xiao, Y., Lubin, A. A., Heeger, A. J., Plaxco, K. W. (2005a). Label-free electronic detection of thrombin in blood serum by using an aptamer-based sensor. *Angew Chem Int Ed Engl* 44, 5456–5459.

Xiao, Y., Piorek, B. D., Plaxco, K. W., Heeger, A. J. (2005b). A reagentless signal-on architecture for electronic, aptamer-based sensors via target-induced strand displacement. *J Am Chem Soc* 127, 17990–17991.

Xiao, Y., Rowe, A. A., Plaxco, K. W. (2007). Electrochemical detection of parts-per-billion lead via an electrode-bound DNAzyme assembly. *J Am Chem Soc* 129, 262–263.

Xu, D. K., Xu, D. W., Yu, X. B., Liu, Z. H., He, W., Ma, Z. Q. (2005). Label-free electrochemical detection for aptamer-based array electrodes. *Anal Chem* 77, 5107–5113.

Xu, Y., Yang, L., Ye, X. Y., He, P. A., Fang, Y. Z. (2006). An aptamer-based protein biosensor by detecting the amplified impedance signal. *Electroanalysis* 18, 1449–1456.

Yamamoto, R., Kumar, P. K. R. (2000). Molecular beacon aptamer fluoresces in the presence of Tat protein of HIV-1. *Genes Cells* 5, 389–396.

Yang, C. J., Jockusch, S., Vicens, M., Turro, N. J., Tan, W. H. (2005). Light-switching excimer probes for rapid protein monitoring in complex biological fluids. *Proc Natl Acad Sci U S A* 102, 17278–17283.

Yi, X., Pavlov, V., Gill, R., Bourenko, T., Willner, I. (2004). Lighting up biochemiluminescence by the surface self-assembly of DNA–hemin complexes. *Chembiochem* 5, 374–379.

Yu, H. Z., Luo, C. Y., Sankar, C. G., Sen, D. (2003). Voltammetric procedure for examining DNA-modified surfaces: quantitation, cationic binding activity, and electron-transfer kinetics. *Anal Chem* 75, 3902–3907.

Zayats, M., Huang, Y., Gill, R., Ma, C. A., Willner, I. (2006). Label-free and reagentless aptamer-based sensors for small molecules. *J Am Chem Soc* 128, 13666–13667.

Zhang, H. Q., Wang, Z. W., Li, X. F., Le, X. C. (2006). Ultrasensitive detection of proteins by amplification of affinity aptamers. *Angew Chem Int Ed Engl* 45, 1576–1580.

Zhao, W. A., Chiuman, W., Brook, M. A., Li, Y. F. (2007). Simple and rapid colorimetric biosensors based on DNA aptamer and noncrosslinking gold nanoparticle aggregation. *Chembiochem* 8, 727–731.

Zheng, J., Lin, L., Cheng, G. F., Wang, A. B., Tan, X. L., He, P. G., Fang, Y. Z. (2007). Study on an electrochemical biosensor for thrombin recognition based on aptamers and nano particles. *Sci China Ser B Chem* 50, 351–357.

Zhou, C. S., Jiang, Y. X., Hou, S., Ma, B. C., Fang, X. H., Li, M. L. (2006). Detection of oncoprotein platelet-derived growth factor using a fluorescent signaling complex of an aptamer and TOTO. *Anal Bioanal Chem* 384, 1175–1180.

Zhou, L., Ou, L-J., Chu, X., Shen, G-L., Yu, R-Q. (2007). Aptamer-based rolling circle amplification: a platform for electrochemical detection of protein. *Anal Chem* 79, 7492–7500.

Zuo, X. L., Song, S. P., Zhang, J., Pan, D., Wang, L. H., Fan, C. H. (2007). A target-responsive electrochemical aptamer switch (TREAS) for reagentless detection of nanomolar ATP. *J Am Chem Soc* 129, 1042–1043.

INDEX

Aptamers in Bioanalysis, Edited by Marco Mascini
Copyright © 2009 John Wiley & Sons, Inc.